Public Lands History

GENERAL EDITORS

Ruth Alexander

Mark Fiege

Adrian Howkins

Janet Ore

Jared Orsi

Sarah Payne

National Parks beyond the Nation

Global Perspectives on
"America's Best Idea"

Edited by

ADRIAN HOWKINS, JARED ORSI,
AND MARK FIEGE

UNIVERSITY OF OKLAHOMA PRESS : NORMAN

Library of Congress Cataloging-in-Publication Data

Names: Howkins, Adrian. | Orsi, Jared. | Fiege, Mark.
Title: National parks beyond the nation : global perspectives on "America's
  best idea" / edited by Adrian Howkins, Jared Orsi, and Mark Fiege.
Description: Norman : University of Oklahoma Press, 2016. | Series: Public
  lands history series ; volume 1 | Includes bibliographical references and index.
Identifiers: LCCN 2015037638 | ISBN 978-0-8061-5225-7 (hardcover : alkaline paper)
Subjects: LCSH: National parks and reserves. | National parks and reserves—
  History. | National parks and reserves—Philosophy. | National parks and reserves—
  United States.
Classification: LCC SB481 .N324 2016 | DDC 363.6/8—dc23
LC record available at http://lccn.loc.gov/2015037638

National Parks beyond the Nation: Global Perspectives on "America's Best Idea" is Volume 1
in the Public Lands History series.

The paper in this book meets the guidelines for permanence and durability of the
Committee on Production Guidelines for Book Longevity of the Council on Library
Resources, Inc. ∞

1 2 3 4 5 6 7 8 9 10

For Alison, Becky, and Janet
and for the friends and benefactors of the
Public Lands History Center

CONTENTS

ILLUSTRATIONS

ACKNOWLEDGMENTS

National Parks beyond the Nation is the product of a collaboration involving many people and much generous support. From the beginning, we envisioned this project as a means not only to produce innovative scholarship but also to expand the intellectual community that we and our colleagues have fostered at Colorado State University's Public Lands History Center. As much as possible, we wanted to encourage scholarly inquiry through cooperation and critical, civil dialogue. We wanted the process to be as important as the product; we wanted to get people to talk to one another and help each other; and we wanted to foster enduring relationships that mattered as much as, if not more than, the book we intended to publish. Undertaking the combined goals of community building and scholarship is challenging but rewarding, and we are grateful to students, university administrators, benefactors, agency professionals, colleagues, spouses and families, and friends who joined us in the endeavor.

Our work would have been much more demanding, perhaps even impossible, without the assistance of the students, staff, and colleagues with whom we created and sustain the Public Lands History Center. Much PLHC activity revolves around our students. For indispensable help with planning, publicity, research, ideas, typing and formatting, and expert driving, we thank Avana Andrade, Ashley Baranyk, Jim Bertolini, Nichelle Frank, Stephan Greenway, Brandon Luedtke, Kayla Steele, and Clarissa Trapp. Our colleague Ruth Alexander, now PLHC Council Chair, expertly led one of the symposium seminars and contributed ideas, advice, and encouragement. In organizing a scholarly event, no one is more talented than Maren Bzdek, the PLHC Program Manager, and we deeply appreciate the phenomenal range of administrative, intellectual, and interpersonal skills that she devoted to making the symposium a success and bringing this book to fruition.

At Colorado State University, we benefited mightily from the support of energetic, thoughtful, and far-sighted administrators. Dean Ann Gill and Associate

Dean Stephan Weiler of the College of Liberal Arts have championed the PLHC and its projects from the beginning. We thank them for their advice and help, and for the high standards that they set for themselves and for the entire college. We are grateful to Provost Rick Miranda and President Tony Frank for their support; we thank Dean Joyce Berry of Warner College of Natural Resources for encouragement; and we would like to acknowledge Director Diana Wall and Operations Manager Jarvis Choury of the School of Global Environmental Sustainability for reaching out to historians and providing crucial support at a critical moment.

We offer special thanks to the benefactors whose support of the William E. Morgan Chair of Liberal Arts enabled the programs that led to *National Parks beyond the Nation.* The generosity of the Monfort Family Foundation, the Bohemian Foundation, Tom and Jean Sutherland, Robert and Joyce Everitt, and Bryan and Axson Morgan allowed us to host a first-rate team of scholars and channel their efforts into the book. It is especially humbling for us to acknowledge the generosity and humanity of Tom and Jean Sutherland. We hope that *National Parks beyond the Nation* and the activities of the PLHC will, in a small way, augment the efforts of people everywhere who seek a more open, just, and peaceful world.

Many colleagues and friends assisted us in ways large and small. In the CSU Department of History, we thank chairs Diane Margolf and Doug Yarrington, and we offer heartfelt gratitude to Nancy Rehe, whose devotion to administrative details and help with travel arrangements and other logistics were crucial. One of the great outcomes of our work has been a deeper relationship with colleagues in the National Park Service, and for inspiration, encouragement, advice, ideas, and help, we thank Ben Bobowski, Kathy Brown, Don Irwin, Dave Louter, Don Stephens, Stephanie Toothman, and Judy Visty. For providing venues for public events in which we and our authors aired some of our ideas, we thank CSU's Lory Student Center, the Fort Collins Public Library, Rocky Mountain National Park, and New Belgium Brewery. Many colleagues from CSU and elsewhere provided conversation, ideas, editorial help, and friendship, and in particular we thank David Cooper, Patty Limerick, Bruce MacBryde, Linda Meyer, Coll Thrush, Louis Warren, and Kim Workman. We thank Jennifer Fish-Kashay, Ann Little, Sarah Payne, and our colleagues in the CSU Department of History, and in particular we are grateful to Eli Alberts, Nate Citino, Prakash Kumar, and Thaddeus Sunseri for helping us focus our thinking on levels beyond the nation-state. Jay Dew and Chuck Rankin at the University of Oklahoma Press expressed interest in this project from the start, and we appreciate their and Steven Baker's efforts to shepherd the book to publication. Two anonymous reviewers immeasurably improved the manuscript with their suggestions and criticisms, and we thank them for their help.

Perhaps the greatest reward of *National Parks beyond the Nation* was getting to know our authors. They are a lively, interesting, deeply intelligent, and provocative bunch, and all share a trait that too often is missing in academic work: a sense of humor. Jane, Mark, Ted, Chris, José, Patrick, Alan, Ann, Karen, Steve, Paul, and Emily—thanks so much for working with us from your homes and research sites around the world and for accompanying us on a great intellectual adventure for one week when we gathered in Fort Collins in 2011.

Historical scholarship takes place not just in libraries, archives, seminar rooms, and national parks, but also in households, and we thank our families for their love and support. In particular, we wish to acknowledge our respective partnerships with Alison Hicks, Becky Orsi, and Janet Ore. Alison, Becky, and Janet are colleagues and muses as well as spouses, and they help us sustain the personal protected areas in which the life of the mind can flourish.

National Parks beyond the Nation

Introduction

ADRIAN HOWKINS, MARK FIEGE, AND JARED ORSI

MANY PEOPLE IN THE UNITED STATES—in the official mind and in the popular imagination—believe that their nation led the world in the creation of national parks and other protected areas. "The idea of a national park was an American invention of historic consequences marking the beginning of a worldwide movement that has subsequently spread to more than one hundred countries," states the 2006 edition of the National Park Service's *Management Policies.*[1] Rather than immediately dismissing or merely debunking the exceptionalist claim that national parks are "America's Best Idea," *National Parks beyond the Nation* uses it as a starting point for thinking about a truly international history of national parks. Bringing together the work of fifteen scholars, this book reveals the tremendous diversity of the global national park experience. In some cases the United States influenced, and was influenced by, the histories of national parks in other parts of the world. In other cases, the U.S. experience was not at the center of international national park history or was not directly involved at all.

A fundamental premise of the essays in this collection is that an understanding of national park history benefits from thinking "beyond the nation."[2] Historical interactions and influences within and between national park systems have been intellectual, political, and material: people have thought differently about national parks at different times and in different places; the activities permitted within national parks have varied from one case study to the next; and neat physical boundaries have been disrupted by wandering animals, human movements, the spread of diseases, and climate change. The interaction of these three categories—intellectual, political, material—at numerous scales and across national frontiers provides a fascinatingly rich subject for historical scholarship. By taking a multifaceted approach to the international history of national parks, we join other scholars who seek unconventional perspectives on a topic of global significance and who already are moving national park historiography beyond familiar confines.[3]

This collection of essays comes out of a symposium held at Colorado State University (CSU) in Fort Collins in September 2011. While the collection reflects trends in the field of history toward an increasingly international approach, the location of this workshop was no coincidence. CSU has a long history of involvement with the National Park Service. As early as the 1920s, the institution gained a reputation as the "Ranger Factory" because of the comparatively large number of its graduates who had careers with the U.S. National Park Service (NPS) and other federal government land agencies.[4] Much of this close connection can be explained by CSU's strong reputation in natural resource management, its location in the U.S. West with its abundance of public lands administered by the federal government, and the proximity of Rocky Mountain National Park, one of the most visited units in the NPS system. The CSU Department of History has helped build the university's national park emphasis by hiring international faculty, by maintaining faculty positions and offering courses in fields such as environmental history and world history, by sustaining a graduate program in public history, and by founding the Public Lands History Center in 2007.[5]

The recent trend toward a global perspective within international environmental history raises new questions about national parks. During informal discussions in the CSU History Department, we—the editors of this volume—found ourselves making comparisons between U.S. parks and other protected areas and similar areas in the rest of the world. This led us to think about possible influences and connections. Are similarities between parks around the world the result of shared environmental characteristics, or are they more closely tied to a common culture? What role has the NPS played in U.S. diplomatic history, and how has this role changed from before the Cold War to the present? Rocky Mountain National Park in Colorado, for example, has a sister park relationship with Polish and Slovakian national parks on both sides of the Tatra Mountains, a branch of the Carpathians (figure 1).[6] The Tatra and Rocky Mountain parks have alpine ecologies and were shaped by pastoralism, livestock husbandry, and conservation practices, but the similarities and differences between the two places came to light only after the Cold War, when scientists and park personnel could exchange visits.[7]

With funding from the William E. Morgan Chair of Liberal Arts, we thought it would be an interesting and useful exercise to host a symposium at Colorado State University to examine in more detail and from different perspectives the questions we were asking. Our aim was to bring together a variety of scholars from the United States and around the world working on themes connected to the international history of national parks. Some of those invited to participate identify themselves specifically as park historians, while others do not. As might be expected, national park scholars emphasize environmental history, but not everyone with an essay in this collection is an environmental historian. The choice of participants was determined by the desire to include a diversity of approaches within the time and means available, and within the constraints inherent in bringing together a cosmopolitan group of researchers, thinkers, and writers. We aimed to have as broad a geographical scope as possible, including scholars working from the United States and in other parts of the world, and we are pleased that all continents are represented in this volume. We also sought a combination of established and emerging scholars. No less important, we sought to put all of these participants in dialogue—reading and critiquing each other's contributions in order to allow the resulting anthology to cohere around a common conversation.

Despite the book's geographical breadth, our intention in bringing a group of scholars together was not to achieve encyclopedic global coverage, but rather to explore similarities, differences, and above all relationships in national park practices. Much as we eschewed a nationalistic exploration of *primacy* ("Who established the first national park?"), we avoided the impulse—so characteristic of area studies during the Cold War—to achieve comprehensive *coverage.* While it would certainly have been productive to have included park historians working on Russia, China, India, Central America, and other nations and regions, any effort at coverage instantly bogs down in irresolvable dead ends: If India, then why not Pakistan? If Central America, then why not the Balkans or the Middle East? If China, then why not Mongolia, the Koreas, or Japan? Each inclusion inevitably invites questions about exclusions. While remaining rooted in the nation-state framework, we desire to move beyond its confines to explore fresh comparisons and underexamined or missed relationships on the global level. Our goal is to take the history of national parks beyond the nation, and we hope that other historians will join us in this ambitious project.[8] To this end, we invite readers to consider the ways in which the approaches modeled here also might apply to other places.

It has long been recognized that national parks offer excellent locations for "doing environmental history" since they provide relatively contained case studies of the interaction of human ideas, human actions, and the material environment over time.[9] What has worked well within a national framework promises to work equally well in a global context. One of the claims often made about environmental history is that it transcends national boundaries: the nation is seldom the central actor in histories of acid rain, global warming, or the loss of biodiversity.[10] This "global turn" in historical scholarship, however, is not confined to environmental history. In the past decade or so, American historians have considered ways in which the United States has fit into global trends. Collections such as Thomas Bender's *Rethinking American History in a Global Age* (2002) and Leon Fink's *Workers across the Americas* (2011) led the way in bringing a global perspective to the study of American history.[11] More than simply challenging the concept of American exceptionalism, works such as

National Parks and Protected Areas

CANADA
1. Auyuittuq
2. Jasper
3. Banff
4. Yoho
5. Glacier (British Columbia)
6. Waterton Lakes

UNITED STATES
7. Adirondack
8. Badlands
9. Devils Tower
10. Glacier (Montana)
11. Glacier Bay
12. Grand Canyon
13. Grand Teton
14. Great Smoky Mountains
15. Hot Springs
16. Isle Royale
17. Mackinac
18. Mesa Verde
19. Mount Rainier
20. Mount Rushmore
21. Organ Pipe Cactus
22. Rocky Mountain
23. Washington Monument
24. Yellowstone
25. Yosemite

MEXICO
26. El Tepozteco

PERU
27. Huascarán

ARGENTINA
28. Iguazú
29. Nahuel Huapi

BRAZIL
30. Aparados da Serra
31. Araguaia
32. Brasília
33. Caparaó
34. Chapada dos Veadeiros
35. Emas
36. Iguaçu
37. Itatiaia
38. Paulo Afonso
39. Pico da Neblina
40. São Joaquim
41. Serra da Canastra
42. Serra dos Órgãos
43. Sete Cidades
44. Sete Quedas
45. Tapajós
46. Tijuca
47. Ubajara
48. Xingu

POLAND
49. Tatra
SWITZERLAND
50. Swiss
ITALY
51. Gran Paradiso
SLOVAKIA
52. Tatra

Mentioned in the Book

JAPAN
70 Fuji-Hakone-Izu

NEPAL
71 Chitwan
72 Sagarmatha

INDONESIA
73 Komodo
74 Kutai
75 Lorentz
76 Bromo Tengger Semeru
77 Ujung Kulon

CONGO
53 Virunga (Albert)
54 Garamba

KENYA
55 Mount Longonot
56 Maasai Mara
57 Mau Highlands

TANZANIA
58 Gombe Stream
59 Kilimanjaro
60 Ngorongoro
61 Serengeti

MOZAMBIQUE
62 Gorongosa

SOUTH AFRICA
63 Gough Island
64 Kalahari Gemsbok
65 Kruger
66 Marion Island
67 Mont-Aux-Sources
68 Royal Natal
69 Richtersveld

AUSTRALIA
78 Royal
79 Kakadu
80 Sydney Opera House
81 Uluru-Kata Tjuta
82 Willandra Lakes

NEW ZEALAND
83 Abel Tasman
84 Aoraki/Mount Cook
85 Arthur's Pass
86 Fiordland
87 Paparoa
88 Te Urewera
89 Taranaki/Egmont
90 Tongariro
91 Westland Tai Poutini

these seek to add depth to our understanding of U.S. history by incorporating broader perspectives. The aim of such scholarship is generally not to replace a national paradigm for writing history with a global one—which uncritically validates today's supposedly globalized world order in place of the nation-state—but rather to offer more nuanced interpretations of a past that combine local, national, transnational, and international scales.[12]

Despite these trends toward global perspectives, both in environmental history and in recent historiography more generally, much of the historical scholarship on national parks remains rooted in a national framework. This is often appropriate. If not quite an oxymoron, the concept of "international national park history" at the very least contains a number of tensions and contradictions. The United States offers a good example. The nation provides the necessary context for individual park histories, because policy emanating from Washington, D.C., interacts with local political, cultural, and environmental realities. The national context also provides a useful way of thinking about general trends in national park history as demonstrated by Richard West Sellars's *Preserving Nature in the National Parks,* which looks at the history of ecological concepts in U.S. national parks.[13] And the nation-state provides an appropriate context for critical histories of national parks, which examine the consequences—both intended and unintended—of the interaction of national policy with local conditions. Mark David Spence's *Dispossessing the Wilderness,* for example, examines the history of American Indian removal in Yellowstone, Yosemite, and Glacier National Parks, putting these episodes in the context of wider federal government policy.[14] But while the framework of the nation has yielded much insight into national park history, the lack of an international perspective leaves many illuminating questions unasked and important connections unstudied. And all too often, the lack of a broader approach allows an unwarranted exceptionalism to endure.

Some historians writing about parks and park systems in various parts of the world have supplemented their national foci with references to scales larger than the nation.[15] In recent years, a few scholars have made an explicitly international approach the focus of national park history. Terence Young and Lary M. Dilsaver have pioneered the study of the international work of the U.S. National Park Service over the course of the twentieth century.[16] Their research is intentionally one-directional, seeking to understand the processes through which the U.S. National Park Service administration undertook a number of projects in different parts of the world.

Other studies have attempted to be truly international and comparative rather than being centered on the United States or any other single national framework.[17] These works have shown that national parks have become important institutions around the world, revealing the sheer diversity of the national park experience. Although such a genuinely international approach to national park history can be useful, it also can be disorienting, since it is not always clear how international case studies relate to each other. Ironically, there is a tendency for comparative scholarship to reinforce distinctions between nations and thus fall back into national categories.[18]

We asked every author in this collection to consider what role, if any, the United States has played in the history of parks in the particular parts of the world that they study. In this way, *National Parks beyond the Nation* offers a thorough examination of the international history of national parks while retaining some focus on the history of U.S. national parks. Our choice to emphasize the United States in this collection arises from its origins in a symposium organized by historians working on national park history at a U.S. institution with strong connections to the U.S. National Park Service. A similar approach could take as its starting point virtually any other nation's experience with the international history of national parks. The results likely would be different, but equally revealing. Taken together, the chapters in this collection challenge the uncritical assumption of U.S. centrality (or any nation's centrality, for that matter) in the global history of national parks, while allowing for a thorough examination of historical relationships across different scales.[19] By taking an approach to the history of U.S. national parks that is both international and fundamentally relational, all of the contributors to this volume raise fresh questions and open up new fields of study. In this sense, *National Parks beyond the Nation* makes a methodological statement with potential for wider application.

This volume adopts a roughly chronological organization. It is divided into three parts, each loosely united around a single theme. The first part investigates the alleged U.S. birth of the idea of national parks and asks whether origins even matter. The second part explores how the national park idea developed an international life of its own and intersected with a number of important world historical themes. Finally, the essays in part 3 examine the ways in which parks both construct national and cultural boundaries and provide opportunities to break them down. Many of the chapters, if not all, could just as easily be fit into one or both of the other parts. The book contains a diversity of writing styles—ranging from the personal essay employed by Chris Conte to the social science approach of José Drummond—to reflect the diversity of national park scholarship around the world. Not every author is in agreement, and in some instances we disagree with an author. Rather than suggesting confusion, however, we hope that these differences come across as authentic expressions of individual perspectives that convey the richness of international national park history.

The collection begins with an essay by the three editors that highlights the approach taken in the book as a whole. Mark Fiege's sketch of Mount Rainier National Park in Washington State shows that it is difficult if not impossible to frame a park's history entirely within a national context. For example, the geological location of the park on the Pacific Ocean's "Ring of Fire" creates a material connection with landscapes and parks in Asia, the Pacific Islands, and

South America. Adrian Howkins then traces U.S. National Park Service involvement in Antarctica. Shortly after the signing of the Antarctic Treaty in 1959, the National Park Service created a fleeting possibility of an alternative future for the southern continent, and also revealed a diplomatic history of national parks. Finally, Jared Orsi discusses Organ Pipe Cactus National Monument in Arizona, infamous for being America's most dangerous park. Its location on the U.S.–Mexico border highlights the ways in which parks and national boundaries mutually construct—but also destabilize—one another.

Alan MacEachern's surprising chapter addresses the exchange of national park ideas across the U.S.–Canadian border in the 1910s. Animated by a wry sense of humor, it argues that both cooperation and rivalry shaped the park services in the two countries. Theodore Catton's essay focuses on the role of the state in national park histories in an examination of the historical relationship between national parks in the United States and New Zealand. The chapter argues that New Zealand took a more minimal approach to park management than did the United States, despite increased concerns about the problem of invasive species. Building on Theodore Roosevelt's visit to South America, Emily Wakild examines how North American ideas of tropical wilderness have helped shape the development of national parks in South America. It is often claimed that "science knows no boundaries," and Patrick Kupper's chapter examines the transnational history of national park science. He argues that the United States was relatively slow to adopt a "scientific" approach to national park management, but that when it did so, it had widespread influence around the world.

Jane Carruthers's chapter examines historical interactions between the U.S. National Park Service and South African national parks. She contrasts the perceived popularity of U.S. national parks with the widespread association of South African national parks with the former apartheid regime. In a moving description of his personal experiences in Gombe Stream National Park, Tanzania, and Gorongosa National Park, Mozambique, Chris Conte explores how violence has shaped national parks. Among other questions, he asks how violence can become part of a landscape's history, focusing on the civil war that took place between 1976 and 1992 within the wider context of the Cold War. Steve Rodriguez examines the themes of imperialism and nationalism in his study of the development of national parks during the Suharto regime in Indonesia between 1967 and 1998. This history takes place against the backdrop of the Cold War, and Rodriguez suggests a strong connection between the development of national parks and Indonesian nationalism.

Karen Routledge analyzes Glacier-Waterton International Peace Park between 1932 and 1970, and the problems that grizzly bears posed when they crossed the very borders that divided the U.S. and Canadian jurisdictions. The chapter highlights the tension between a transnational ecosystem and national systems of regulation. José Drummond examines the historical development of "landscape units" in Brazil, making comparisons to the United States and to international standards more generally. Ann McGrath assesses similarly problematic histories in her comparison of climbing controversies at Uluru, Northern Territory, Australia, and Devils Tower, Wyoming. The comparative approach reminds us that U.S. national parks have not been exempt from questions of exclusion and potentially abusive access that have been much debated in other parts of the world. The final chapter is Mark Carey's probing inquiry into the effects of global warming on national parks in Peru and the United States. Perhaps more than any other phenomenon, climate change exemplifies the need for a global approach to historical scholarship. But Carey also demonstrates that our understanding of climate change remains rooted in specific places and that our responses to the problem of climate change have often taken place within national frameworks.

Moving beyond the exceptionalism of "America's Best Idea," Paul Sutter's conclusion points to a richness of national park experiences in which primacy—who came up with the idea first?—becomes almost meaningless. Rather than the "best idea" phrase, popularly and erroneously attributed to Wallace Stegner alone, Sutter suggests that another Stegnerian expression, "geographies of hope," more aptly describes national parks around the planet.

The essays in this collection are just the beginning of a process in which scholars move the history of national parks beyond the nation. But both individually and collectively, they reveal the importance of a global perspective on national park history while connecting the authors to a long internationalist tradition in which people from many lands work together to realize a shared vision of a better world.

Notes

1. U.S. Department of the Interior, National Park Service, *Management Policies 2006* (Washington, D.C.: Government Printing Office, 2006), 7. See also Ken Burns's documentary extravaganza *The National Parks: America's Best Idea* (2009) at www.pbs.org/nationalparks/ (accessed July 17, 2014), and its companion volume by Dayton Duncan and Ken Burns, *The National Parks: America's Best Idea* (New York: Knopf, 2009).

2. Thinking about parks beyond the nation inevitably takes us into uncharted terminological taxonomies. The aim of this volume is not to wade into the debates over the exact meanings of the terms historians apply to describe phenomena not contained within a single nation-state. Nor is it to advance new terminology. Because the practice of looking at history in frames larger than nation-states is still young and fertile, it lacks an agreed-upon set of definitions to distinguish its terms, and scholars have therefore been left to use language largely as they choose. The editors have opted for a few simple, basic definitions. For the purposes of the introduction and chapter 1, when the editors say "international" we mean exchange among two or more nations or their peoples. The exchange might be

economic, intellectual, diplomatic, material, or other. When something crosses a national border, say, an animal or a migrant, we use the term "transnational." "Global" and "world" we reserve for things such as climate change, whose scope truly takes in the entire planet, or at least significant multinational chunks of it. Individual authors in this book sometimes use these and related terms somewhat differently. For a critique of scholars' use of terminology to describe phenomena beyond the nation, see Joseph E. Taylor III, "Boundary Terminology," *Environmental History* 13 (July 2008): 454–81.

3. See, for example, the recent exchange between Ian Tyrell, Paul Sutter, Thomas Dunlap, and Astrid Swenson in the *Journal of American Studies* 46 (February 2012): 1–49; Terence Young and Lary M. Dilsaver, "Collecting and Diffusing 'the World's Best Thought': International Cooperation by the National Park Service," *George Wright Forum* 28, no. 3 (2011): 269–78; and Patrick Kupper, "Science and the National Parks: A Trans-Atlantic Perspective on the Interwar Years," *Environmental History* 14 (January 2009): 58–81.

4. James E. Hansen II, *Democracy's University: A History of Colorado State University, 1970–2003* (Fort Collins: Colorado State University, 2007), 14.

5. Originally the Center for Public History and Archaeology, the Public Lands History Center assumed its current name in 2010.

6. http://www.nps.gov/oia/topics/sisterparks/maps/sister_map.html (accessed July 22, 2015).

7. Ben Bobowski and Vaughn Baker, "Rocky Mountain National Park's Sister Park Relationship," *George Wright Forum* 28, no. 3 (2011): 296–98.

8. Particularly suggestive in this regard is Young and Dilsaver, "Collecting and Diffusing 'the World's Best Thought.'"

9. For a theoretical discussion of the field of environmental history, see Donald Worster, "Doing Environmental History," in *The Ends of the Earth: Perspectives on Modern Environmental History*, ed. Donald Worster (Cambridge: Cambridge University Press, 1989). See also J. Donald Hughes, *What is Environmental History?* (Cambridge: Polity, 2006).

10. See, for example, John Robert McNeill and Erin Stewart Mauldin, eds., *A Companion to Global Environmental History* (Hoboken, N.J.: Wiley, 2012); and Erika Marie Bsumek, David Kinkela, and Mark Atwood Lawrence, eds., *Nation-States and the Global Environment: New Approaches to International Environmental History* (New York: Oxford University Press, 2013).

11. Thomas Bender, ed., *Rethinking American History in a Global Age* (Berkeley: University of California Press, 2002); Leon Fink, ed., *Workers across the Americas: The Transnational Turn in Labor History* (New York: Oxford University Press, 2011). See also, for example, Thomas Bender, *A Nation among Nations: America's Place in World History* (New York: Hill & Wang, 2006); Edward J. Davies II, *The United States in World History* (New York: Routledge, 2006); "Rethinking History and the Nation-State: Mexico and the United States as a Case Study; A Special Issue," *Journal of American History* 86 (September 1999): 438–697; "The Nation and Beyond: Transnational Perspectives on United States History; A Special Issue," *Journal of American History* 86 (December 1999): 965–1307.

12. See, for example, Richard White, "The Nationalization of Nature," *Journal of American History* 86 (December 1999): 976–86; and Taylor, "Boundary Terminology."

13. Richard West Sellars, *Preserving Nature in the National Parks: A History* (New Haven, Conn.: Yale University Press, 1997).

14. Mark David Spence, *Dispossessing the Wilderness: Indian Removal and the Making of the National Parks* (New York: Oxford University Press, 1999).

15. See, for example, Theodore Catton, *Inhabited Wilderness: Indians, Eskimos, and National Parks in Alaska* (Albuquerque: University of New Mexico Press, 1997); Emily Wakild, *Revolutionary Parks: Conservation, Social Justice, and Mexico's National Parks, 1910–1940* (Tucson: University of Arizona Press, 2011).

16. Young and Dilsaver, "Collecting and Diffusing 'the World's Best Thought.'"

17. See, for example, Bernhard Gissibl, Sabine Höhler, and Patrick Kupper, eds., *Civilizing Nature: National Parks in Global Historical Perspective* (New York: Berghahn, 2012); Ethan Carr, Shaun Eyring, and Richard Guy Wilson, eds., *Public Nature: Scenery, History, and Park Design* (Charlottesville: University of Virginia Press, 2013).

18. The problem of nationalism in national comparisons is addressed by Ian Tyrrell's chapter in Bender, ed. *Rethinking American History in a Global Age*. See also Ian Tyrrell, "America's National Parks: The Transnational Creation of National Space in the Progressive Era," *Journal of American Studies* 46, no. 1 (2012): 1–21.

19. White, "The Nationalization of Nature." See also Richard White, "The Problem with Purity," Tanner Lectures on Human Values, University of California, Davis, May 10, 1999, http://tannerlectures.utah.edu/lectures/documents/white00.pdf (accessed October 26, 2012).

ONE | # Beyond the Best Idea
A Look at Mount Rainier, Antarctica,
and the Sonoran Desert

MARK FIEGE, ADRIAN HOWKINS, AND JARED ORSI

IT MIGHT BE SAID that the volcanic cone and the national park that bears its name are symbols of American greatness. Swathed in glaciers, Mount Rainier soars to 14,410 feet over some 50 miles, the sharpest rise from saltwater to summit in the lower forty-eight states. On rare occasions when the dawn light and the clouds are just right, the mountain casts an enormous shadow—an inverted conical silhouette—across the sky. From a distance of one hundred miles and more, a vista possible on a clear day, Rainier seems to float above the Earth—a peak among peaks, a god above gods, solemn, transcendent, and white. No wonder people flock to it on July mornings, ascend its broad shoulders on winding roads to rustic lodges at Paradise and Sunrise, and from there look up and out and glimpse eternity as it unfolds in all directions. It is a classic scene in which landscape and emotion fuse in a timeless vision of nature and nation, a reminder that on the list of America's achievements, the national park idea might be at the top.

It might also be said that the peak and park are something else besides—not just icons, but unstable icons whose national symbolism threatens to crumble and dissolve at every turn. From the west entrance, the road passes through meadow and forest, along ridges and a rock-strewn river, and on to Paradise. About fifteen miles in, a stream spills over a ledge and roars into mist that drips from eyeglasses and pastes shirt to skin. A crowd of tourists revels in the fury and freshness, and their delight rises above the din in a symphony of languages—English, but also German, French, Spanish, Russian, Japanese, and others. On the trail below, where the trees and moss soften the thunder, a middle-aged couple and their adult children stop for conversation. Their appearance—the wife is wearing a sari—and British accents suggest roots in South Asia. Is their presence here a coincidence? Perhaps not—perhaps they are here because the falls goes by the name of Narada, a Hindu word meaning "pure" or "uncontaminated," and, perhaps, a reference to the Hindu musician, sage, seer, wanderer, and giver of knowledge who leads humankind on the true path.[1]

Nothing along the highway becomes more stable the higher the pavement goes—not the nation, not the park, not even the mountain itself. Short of Paradise is a campground, fragrant with the smell of subalpine fir and deceptively tranquil. Despite the mountain's apparent solidity, it is weak. Water trickling from the glaciers saturates the volcanic soil such that the slightest seismic shudder can trigger a muddy cataclysm. The evidence reposes in valley bottoms miles away—ash, rock, and ancient logs, a stratified record of ruin. The greatest event, the Osceola Mudflow about 5,600 years ago, removed half a cubic mile of earth from the mountain, reached as far as Puget Sound, and left deposits as deep as 70 feet.[2] The danger is impossible to ignore. In the tent that night, the mind conjures a nightmare: sirens and screams and flashlights as campers run, half-dressed, for the higher ground that is their only hope. Daylight brings relief, but also thoughts of people on the other side of the planet who know all too well how the quaking earth, with little or no warning, can summon monsters to crush bodies, drown children, and bury dreams. Rainier is in the United States, but geologists and park rangers refer to the mudflow using the Indonesian word *lahar*. The lahar is a global common denominator, a grim reaper that cuts, without remorse, across time and space, cultures and nations.

Beyond Paradise, the road leads down the mountain to a confluence of streams that is less ominous, but still unsettling. This is Ohanapecosh, "standing at the edge," a site once occupied by Taidnapam (upper Cowlitz) people and now a ranger station, campground, and picnic area. The facilities at Ohanapecosh stand in a glade of ancient fir and cedar where hot springs and glacial melt converge—and they stand at the terminus of a human movement that brought colonialism to a Native land. Traces of the story appear in the uniforms of the rangers who distribute brochures, maps, smiles, and advice to the public. The gray and green clothing, epaulettes, badges, and campaign hats are characteristic of constabulary forces that, for a century and more, have policed the borders of nation-states and nature preserves around the planet. Although the Department of the Interior once acknowledged the right of Natives to hunt and fish on and around Rainier for subsistence purposes, park rangers prohibited them from doing so.[3] A place of staggering beauty and deeply felt patriotism, Mount Rainier also is Indian country, a landscape of overlapping jurisdictions and political fault lines where cultures coexist in complex, sometimes uneasy relationships. "Ohanapecosh," the visitor says, testing a word that sounds at once strange and familiar. *Ohanapecosh*.

Mount Rainier exemplifies the ways that the nation is not the only lens through which to examine and interpret national park history. It is true that national parks are patriotic symbols. The national park is uniquely American, citizens of the United States assert; it is the country's "best idea" and a model for protected areas around the world. Yet the popular story obscures an equally compelling reality.[4] As a brief visit to Rainier shows, national parks have histories that are not merely products of nation building and nationalism, but also of transnational movements and international relationships, including colonialism, cross-cultural conflict and exchange, the migrations of plants and animals and people, and climate change. Even the place names and scientific terminology of Mount Rainier National Park—even the name of the mountain itself—reveal that beyond the best idea is a global story of unappreciated significance and unimagined drama.

A focus on the supranational history of Mount Rainier highlights a little-known but striking feature of many national parks, which is their tendency to appear in places where the modern nation-state is not strongest, but weakest. The characteristic loci of national parks are frontiers and borderlands between nations; landscapes of rural cultures and homelands of tribal societies; and remote, often extreme environments poorly suited to easy survey, settlement, and economic development. The creation of Mount Rainier National Park in 1899 can be interpreted as an important moment in a larger process through which the United States consolidated its hold on far-flung territories and began the buildup of its modern governmental apparatus.[5] Yet that process also can be interpreted as contingent, conflicted, and tentative. The mountain itself might be one source of that fragility. Rainier is a high, windy, frigid, active volcano buried for much of the year under heavy accumulations of snow. The hand of the U.S. government is evident here, but, of necessity, much less than in other places. And when Rainier finally erupts, as it surely will, it will tax—like an earthquake or hurricane, fire or flood, tornado or terrorist attack—the power of the nation to rescue the thousands of citizens who stand in harm's way.

The premise that informs an alternative version of Mount Rainier's history can inform an understanding of the entire U.S. national park experience. Antarctica—an extreme environment far from the environmental imaginations of most Americans—provides a case in point. Under the 1959 Antarctic Treaty, the U.S. State Department invited the National Park Service (NPS) to prepare conservation plans for the continent. Rather than simply applying conventional American assumptions about nature to Antarctica, however, NPS personnel had to confront questions regarding imperialism, nationalism, and decolonization in the context of the Cold War. Did other nations signatory to the treaty want a U.S.-style park, or did they prefer alternative arrangements? In the case of Antarctica, "America's Best Idea" was not easily exported. But perhaps nowhere was the national park concept more problematic—perhaps nowhere did a national park better illustrate a history beyond the nation—than at the border between the United States and Mexico. From its inception in 1937, Organ Pipe Cactus National Monument was never simply a national space, because wildlife, livestock, and people moving through the Sonoran Desert regularly crossed the international boundary and passed into its confines. The ensuing controversies spoke volumes about a national park history rooted in, but not completely contained by, the modern nation-state. Surveys of Rainier, Antarctica, and the Sonoran Desert thus bring national parks into view, but ultimately lead to destinations beyond national bounds.

We begin in the United States, at Mount Rainier National Park, and from there proceed to Antarctica and the U.S.-Mexico border before heading to parks in many other places around the world. In Rainier's name and in its geology, in the people who have gathered on it and used it as a base camp for imagining distant places, if not the entire planet, Rainier is strikingly global. Although it first erupted well before *Homo sapiens* arrived in the Western Hemisphere, an important moment in the mountain's global development took place some 220 years ago, in late May 1792. From the deck of HMS *Discovery* as it plied the waters of a cold, deep, saltwater passage, Captain George Vancouver looked south and spied a majestic white cone looming on the horizon.

Where the Mountains Have No Names

Vancouver and his men were on an expedition to claim distant lands for the British Empire. From the Pacific, the *Discovery* and its companion, the *Chatham*, had sailed east through a strait rimmed by forested shorelines below snowy peaks. At the far end of the strait, the vessels had turned south into a narrower channel or inlet. Landing at a sandy point below a cliff, Vancouver and his men surveyed the water and terrain that lay before them, including the mountain range that extended north and south along the eastern horizon. The captain "was the friend and acquaintance of many men who had taken part for their country in the disputes and the war with the American colonies," wrote the historian Edmond Meany, and "it was perfectly natural that he should compliment those men as he discovered or rediscovered the places that needed naming."[6] Vancouver already had named a prominent volcano to the north after Joseph Baker, his third lieutenant. Gazing on "the round snowy mountain" at the "southern extremity" of the range, he thought of a friend and fellow Royal Navy officer. From that moment on, the peak would be "distinguished by the name of MOUNT RAINIER" in honor of Rear Admiral Peter Rainier. A descendant of a French Huguenot refugee family named Regnier, Admiral Rainier had been wounded in battle against the Patriot naval forces of the United States of America.[7]

Vancouver continued his imperial naming as he proceeded southward. Whidbey, Penn, Townshend, Hood, Burrard, Vashon, Gardner, Puget, Bellingham, and more joined Baker and Rainier on the geographic honor roll. Vancouver perhaps best expressed his purposes in the name he attached to a sound that separated "Whidbey's Island" from the mainland to the east: Possession. The Royal Navy had arrived to possess the place for King George III, and thus all that the captain could see from his deck—land, water, and all that they contained—would pass into the light of civilization under the name of New Georgia. Vancouver was a principled name-giver, in Meany's judgment. "He was scrupulously magnanimous and generous and always recognized the geographic names bestowed by his predecessors of any nation whatsoever," the historian wrote.[8] Vancouver, for example, recognized "Juan de Fuca" as the Spanish name for the strait into which the *Discovery* and the *Chatham* had sailed eastward from the open Pacific Ocean. But to the Royal Navy captain, mostly this was an unmapped and unclaimed place, a land where the mountains and virtually all else needed naming.

The people who actually lived there and who received Vancouver and his men as guests likely would have disagreed had they known the expedition's political and geographical intent. Everywhere the explorers went, they made contact with Natives who lived in villages on the beaches, bluffs, and forest edges that lined the sound to which Vancouver affixed the name of Third Lieutenant Peter Puget. "These good people conducted themselves in the most friendly manner," wrote Vancouver of an encounter in which the explorers exchanged gifts with, and received food from, approximately sixty Indians, "including . . . women and children."[9] Although this was a densely inhabited area thick with indigenous meanings, Vancouver recorded none of the names the Indians attached to the many places where they came into the world and at which they lived, labored, fought, worshipped, traded, conducted politics, and died. "It is undoubtedly true that he would have preserved many Indian names of places if he could have learned them," Meany asserted. "Besides the difficulty of the language, subsequent investigators have found that Indians rarely have fixed or permanent names for places."[10]

Nothing could have been further from the truth. If Vancouver exchanged gifts with Indians and dined with them, then he could have attempted phonetic translations of the names that they attached to the world, however strange to his ear. He might have noted the identities of the people, the Suquamish, Duwamish, Shilshoolabsh, Whulshootseed, and Nisqually, among the dozens of groups in whose midst he had appeared. He might have recorded a few of their sites—qWátub, Qéélbeed, XaXaboos, CHuXáydoos—that numbered in the thousands along the shore. He might have written down the name of the saltwater body—XwulcH (Whulge)—on which he and his men sailed. And he might have asked the people about the towering white volcano to the south, which they knew by numerous toponyms such as Puskehouse, Tuahku, and Tiswauk. But learning about local cultures was not Vancouver's intent, and he began a process in which English names (or, in the case of Rainier, a French name translated into English) replaced those of the first inhabitants. A few indigenous names, such as Seattle, Snohomish, Whatcom, Tacoma, and Puyallup, appeared on official maps, but these were Anglicized, often romanticized derivations of the originals. A colonial landscape obscured if not erased Native geographies.[11]

In the coming decades, European Americans time and again imposed names on the region and its landmarks, including its gleaming volcanoes. In some instances, the power to name was determined by political contests. Even as Professor Meany—ensconced in his office at the University of Washington in Seattle—researched and wrote about Vancouver, a struggle was under way over the name of the majestic peak to the south. That struggle revealed much about the forces beyond the nation that shaped the meaning and identity of national parks.

Although the Americans who explored and settled the country around Puget Sound during the nineteenth century generally referred to the mountain as Rainier, the name was not so strongly attached as to prevent movements for an alternative. Several times between 1883 and 1924, boosters from Tacoma proposed the name of their city for the prominent volcano. The word and its numerous Native linguistic variants—Tahoma, Tacoma, and Tachoma were but three—generally meant "white mountain" or "snowy mountain" and applied to any number of volcanic peaks in the region. Even though nearby tribes had specific names for the landmark, the proponents of the name change insisted that Tacoma was the true Native designation. As Tacoma, the mountain would bring greater attention and, presumably, population and business to the namesake city and the Northern Pacific Railroad that terminated there.[12]

The agitation to rename Mount Rainier was related to a larger cultural and political struggle in which citizens of the United States resisted the imposition of British names on the landscape of the American West. The creation and naming of Washington Territory in 1853 and Washington State in 1889 was a means by which the far northwest corner of the contiguous United States—an area acquired from Britain in 1846—claimed a revolutionary heritage that stretched back some 2,000 miles and a hundred years. A similar sensibility appropriated Native names throughout the American West. British botanists proposed *Wellingtonia gigantea,* after the Duke of Wellington, for the immense trees that grew on the west slope of the Sierra Nevada and that became known to outsiders in the aftermath of the 1848–49 California gold rush. American scientists countered with *Washingtonia gigantea* before compromising on a Native-derived name. They did not take inspiration from Indian leaders such as Tecumseh and Tenskwatawa, who resisted Americans, but from a man who accommodated them. With *Sequoia gigantea,* after the chief who devised the Cherokee syllabary, the scientists named the tree that eventually became a national park icon.[13]

Both Washington and almost, but not quite, Tacoma became the name of the national park that Congress created in 1899; and almost, but not quite, Tacoma became the name of the peak that defined the park. That these efforts failed spoke legions about the manner in which Mount Rainier and its national park had become fixed in a global, not just national, geography. In 1924 the U.S. Geographic Board rejected a final plea to rename the mountain. The evidence the board cited to buttress its decision included scientific papers and natural history specimens containing "thousands of references to Mount Rainier" and stored in libraries and museums across the United States and Europe. Even more telling were the numerous English, German, French, Italian, Spanish, Russian, Dutch, Belgian, and Arabic gazetteers and atlases in the Library of Congress that referred to Mount Rainier. C. Hart Merriam, chairman of the Geographic Board and a scientist who often worked in the national parks, wrote,

> No geographic feature in any part of the world can claim a name more firmly fixed—fixed by right of discovery, by right of priority, by right of international usage, and by the conspicuous place it holds in the literature, atlases, and official charts of the civilized nations of the earth. . . . To change it would be a blow to the stability of geographic and historical nomenclature, and a reflection on the intelligence of the American people. The name has become the property of the world and is no more a local matter than the name of the Andes of South America or the Himalaya of India. Think of the chaos in geography, history, and science that would result if new names were given to the world's most prominent landmarks.[14]

Thus did Rainier become a global peak in a global park and fixed in a global geographical imagination.

Ring of Fire

Geological conditions supported the idea of the global Mount Rainier. Beneath place names and national boundaries, absent from the atlases and gazetteers, is a deep earthen structure that gives rise to Rainier and hundreds of other volcanoes. The Ring of Fire, as scientists call it, is a band of volcanic activity that curves around the Pacific Ocean from South America to North America, the Aleutian chain to Japan, the Philippines to Malaysia and Indonesia, New Guinea to Polynesia and New Zealand, and south to Antarctica. Along the ring, crustal pressures force oceanic plates beneath lighter continental plates, creating trenches, shaking earth, and building mountains. As the water-saturated crust and upper mantle of the plates plunge downward, they heat and turn to magma, which erupts in fiery explosions that yield a spectacular array of volcanoes, some 70 percent of the world's total.[15] Amid cultural differences and national rivalries, the Ring of Fire is a volcanic unifier of a common global humanity.

The United States and Japan are among the disparate nations that the Ring of Fire unites. Along the edge of North America, the Explorer, Juan de Fuca, and Gorda plates push up the Cascade Mountains, including Rainier and other volcanoes. Although the Cascade Range is some 36 million years old, its violent, unstable nature—a nature of simultaneous destruction and creation—makes its volcanoes relatively young. Rainier, for example, has existed for only about a half million years. Far across the ocean, the Pacific and Philippine plates slide under the Asian continent, opening the Ryukyu and Japan trenches, forming the Japan archipelago, and building volcanoes such as Mount Fuji. Unlike Rainier, Fuji is not carved by glaciers and so is not heavily eroded. Like Rainier, it is young, its earliest form created about 200,000 years ago. And like Rainier, it presents a picturesque white cone that towers above the salt water. Only twenty miles from the Pacific, it soars to 12,389 feet.

No wonder the Japanese who came to the United States felt an attraction to Rainier. From the moment the first immigrants entered Puget Sound on steamboats, they looked south in awe and wonder at the white cone that formed a dramatic backdrop to Seattle, Tacoma, and other cities and towns. Filled with nostalgia, they called the mountain Tacoma-Fuji, thereby "Fujifying" it and bringing it, as Captain Vancouver had done more than a century before, into their cultural horizon.[16] Soon enough, they made their way to its slopes, where they camped, hiked, skied, picnicked, sat around campfires, marveled at the wildflowers, aimed their cameras at the peak and just about everything else, hunted for mushrooms, and created the memories that are so important in attaching people to place. "Looking to the north, you can see Mt. Rainier, appearing majestically—like our king of mountains, Mt. Fuji," wrote Iwao Matsushita during a visit in the summer of 1931. "I sit on a patch of heather. As I eat my pack lunch, I look at the peaks of the Tatoosh range—from Pinnacle Peak to Castle and Unicorn to the east from behind me. I marvel at the speed with which clouds are changing their shape. I crouch by a stream fed by the remaining snow to enjoy a cold drink of water and open my sushi."[17]

Matsushita may have been exceptional in the frequency of his trips to Rainier, but many Japanese immigrants shared his fondness for the mountain. He and his wife, Hanaye, arrived in Seattle in 1919, ostensibly so that he could study English language and literature at the University of Washington before returning to Japan to resume his teaching career. Within two weeks of his arrival, a friend, Kyo Koike, led him on his first outing to Rainier. Koike was a physician who had come to Seattle two years earlier and later helped establish the Rainier *Ginsha,* a haiku society. Koike also was a talented photographer, and in 1924 he, Matsushita, and others founded the Seattle Camera Club, with Rainier as their favorite subject.[18] "You know Mt. Rainier is one of the best national parks," Koike told some neophyte photographers. "I will tell you how to observe the holy mountain which old Indians worshipped as a god."[19] Koike and Matsushita and their families and friends made dozens of trips to Rainier. The more Matsushita went, the more he lost interest in his original plan. A job with Mitsui and Company, a giant trading firm, provided him with an income that supported his camera work and his visits to his beloved mountain—and gave him an excuse not to enroll at the University of Washington.[20]

The volcanic ties between the United States and Japan tightened as the two nations descended into war. Part of the impetus came from the Japanese national parks movement; part came from park proponents in the United States. In response to American national parks, to encourage economic development and national health and fitness, and to express pride in its cultural distinctiveness, Japan gradually took steps to establish national parks. Although the Japanese government in principle committed to a national park system in 1911, not until the National Parks Law of 1931 did it lay out an official policy for creating such a system. Between March 1934 and February 1936, when war mobilization ended the program, Japan established twelve sites, including Fuji-Hakone National Park in 1936.[21] Japan's actions prompted a round of national park diplomacy. Mount Rainier National Park superintendent Owen Tomlinson presented a sample of Rainier pumice in a cedar box to Issaku Okamoto, the Japanese consul in Seattle. Mountaineers from the National Parks Association of Japan presented a Fuji stone, encased in cherry wood, to the American ambassador in Tokyo for transfer to Rainier.[22]

For Iwao and Hanaye Matsushita, Kyo Koike, and other Japanese Americans, there would be no going back to Japan. Matsushita's devotion to Rainier had intensified over the years. At night he fell asleep looking at maps of the mountain. When the weekend arrived, he felt restless and longed for the peak. In 1940, when Mitsui and Company proposed that he transfer to the home office in Tokyo, he resigned his position. "I enjoy my life in Seattle," he wrote to his superiors. "I have so many happy memories with nice people—both Japanese and Americans. Especially I enjoy photography and mountain climbing. I have visited Mt. Rainier, my lover, more than 190 times. I cannot leave Seattle when I think of the beautiful views of Mount Rainier."[23] Hours after the attack on Pearl Harbor, he was under arrest. Months later, authorities detained Hanaye and Kyo. While the three waited out the war in prison—Matsushita at Fort Missoula, Montana, before joining Hanaye and Koike in the desert at Minidoka, Idaho—they dreamed of the mountain that had enriched their lives and to which they had grown deeply attached.[24]

In March 1943 an old Rainier acquaintance of Koike's, an official with the American Friends Service Committee, showed up at Minidoka. Floyd Schmoe was a Kansas native and a Quaker who served as a stretcher bearer and relief worker during the First World War and, from 1922 to 1928, as a ranger and chief naturalist at Mount Rainier National Park. While at Rainier, Schmoe met Koike and perhaps the Matsushitas, maybe on one of the Seattle Camera Club outings.[25] Rainier was a special place to Schmoe and his wife, Ruth, for it provided them with a quiet refuge—"of space, and calm, and silent places," he later wrote—as they reckoned with their conviction "that war is a crime for which we are all responsible and for which we all bear a heavy burden of guilt."[26] After World War II, Schmoe would organize housing reconstruction projects in Japan and other war-torn countries. For his efforts, he would be nominated three times for the Nobel Peace Prize, the only National Park Service employee to be so honored. At Minidoka, he checked on the condition of the Japanese Americans and tried to lift their spirits. An avid cameraman himself, he showed Koike and others his motion picture films, of a trip to Alaska, underwater marine life, Hawaiian volcanoes—and a skiing adventure at Mount Rainier.[27]

The emotional bonds and common humanity built upon the Ring of Fire outlasted the fires of war. The Rainier stone was lost during the conflict, and another was presented in its place. The Fuji stone, which survived, was exhibited in the superintendent's office at Rainier. Japanese and Americans formally proclaimed Rainier and Fuji to be sister mountains, siblings on opposite sides of the Pacific. Iwao and Hanaye Matsushita and Kyo Koike returned to Seattle; after Kyo died in 1947 from a cerebral hemorrhage, Iwao buried his friend's ashes at the foot of a tree on the slopes of the mountain they revered.[28] Fifty-two years later, museum curators inventoried belongings that Japanese Americans, preparing for imprisonment, had stored and then forgotten in the basement of an old hotel in Seattle. Among the piles of suitcases and documents, kimonos and rice cookware, cans of soy sauce, bottles of Aqua Velva aftershave, and folded American flags was "a brocade pillowcase of Mount Rainier National Park."[29]

On the Edge of Outer Space

Thousands of miles away, on the morning of May 1, 1963, fierce winds buffeted a tent on the south face of the world's highest mountain. Two climbers huddled inside, waiting for dawn and the moment when they would emerge into the freezing air to ascend the 1,578 vertical feet that separated them from the summit, 29,028 feet above sea level. At 4:00 A.M., Jim Whittaker signaled to his partner, Nawang Gombu, that they should prepare to leave. Whittaker was far from the mountain on which he had learned to climb and on which he had made his reputation, yet his experience of ice and powerful winds on Rainier had prepared him and other fellow climbers for the conditions they encountered on Everest.[30] In a howling blizzard on the sharp slope of a nearly impossible peak, he, Gombu, and the other expedition members demonstrated in dramatic fashion the global nature of national parks.

Similar indigenous meanings and shared ties to British colonialism linked Rainier and Everest in world history. Just as American Indian tribes had names for Rainier, the peoples of Tibet and Nepal had names for the immense peaks that loomed above their native landscapes. The tallest of these the Tibetans called Chomolungma, "mother goddess of the universe"; to the Nepalese, it was Sagarmatha, "goddess of the sky." And just as the British attached a colonial name to Rainier, so they affixed a comparable colonial name to Everest. During the 1830s and 1840s, while Britain still claimed the territory that encompassed Rainier, explorers for the Great Trigonometrical Survey of India pushed deep into the Himalayas. In 1847 Andrew Waugh, superintendent of the survey, spied the mountain on the horizon, although its topographical significance was not yet known. Three years later, the survey designated the landmark with the abstract label of Peak XV. In 1857 Waugh officially announced that it was the highest in the world, and he named it in honor of George Everest, who had directed the survey from 1833 to 1843. For decades, Everest remained an unwinnable mountaineering prize. Finally, in 1953, the ninth British expedition made it to the top. Edmund Hillary was a New Zealander who later reached the South Pole as a member of the Commonwealth Trans-Antarctic Expedition. After that, he visited the North Pole, which made him the first person to summit Everest and reach both poles. Hillary's companion on the arduous final push to the top was a Sherpa guide named Tenzing Norgay, uncle of eighteen-year-old mountaineer Nawang Gombu.[31]

Cold War politics, less the striving for territorial conquests, framed the American Mount Everest Expedition (AMEE). Jim Whittaker and nineteen other American climbers joined for adventure, physical challenge, and professional standing, and to boost their personal fortunes. Norman Dyhrenfurth, the Swiss American who organized the expedition, also aligned it with President John F. Kennedy's concept of the New Frontier. Kennedy stressed the importance of youthful vigor, competitiveness, and self-sacrifice in meeting the challenge of the Soviet Union and its allies. Much of the political symbolism of the Cold War centered on the application of those virtues to the attainment of extreme environments—the deep ocean, outer space, the moon, the poles. Putting an American at the top of the world's highest peak, perhaps by an untried route, Dyhrenfurth thought, would be a noteworthy Cold War achievement and might attract governmental support. His hunch paid off, as Secretary of the Interior Stewart Udall helped him find sources of federal government assistance, including the U.S. Air Force Office of Scientific Research, the National Aeronautics and Space Administration, and the Office of Naval Research. Udall also arranged the permit from the Department of the Interior that allowed the AMEE to train at Mount Rainier National Park in the autumn of 1962.[32]

Rainier was a practical site from which to launch the Everest expedition.[33] Not only was it one of the tallest mountains in the contiguous forty-eight United States, but it also was an extreme environment characterized by high winds, heavy snow, glaciers, crevasses, icefalls, and freezing temperatures in high summer. "Such summers have prompted veteran Himalayan climbers to comment that Mount Rainier's upper weather exceeds the fury of anything they experienced on Mount Everest at nearly twice the elevation," reported Ruth Kirk, a prominent nature writer, in 1968. "Although Mount Rainier is only half as high as Mount Everest, altitude is all it lacks," wrote Whittaker at the end of the twentieth century. "Its glaciers and icefalls, its ferocious winds, and dangerous weather make it a near-perfect training ground for Himalaya expeditions, and it put us to the test."[34] Perhaps such statements exaggerated the similarities between Rainier and Everest. Yet the climbers' practical experience of the volcano and their competence in the face of its rigors probably influenced the success of the AMEE. All five Americans who made it to the top of Everest had guided or climbed on Rainier before 1962–63.[35]

The affinities between Rainier and Everest deepened after 1963. Jim Whittaker and other Rainier climbers returned multiple times to the Himalayas and Everest. The Sherpa mountaineers who accompanied the Americans and other visitors soon made their way to Rainier. In 1964 Whittaker, Lute Jerstad, and Tom Hornbein, all of whom reached the summit of Everest the previous year, invited Tenzing Norgay to scale the glacier-clad volcano. Norgay's wife, Dhaku, joined them, and she became the first woman to ascend via the Nisqually Icefall. In 1969 Phursumba Sherpa came to Rainier to serve as a guide for Rainier Mountaineering, Inc., operated by Lou Whittaker, Jim's brother.[36] More Sherpa guides soon followed, including Nawang Gombu, Phursumba's brother-in-law and Jim Whittaker's Everest partner. Gombu arrived at Rainier in 1971, and he worked there, off and on, until 2003. A gentle, kind man who had trained to become a Buddhist priest, he was a memorable character. At Camp Muir, situated at 10,080 feet on the mountain, climbers named one of the huts after him.[37] In such ways, Gombu and the other Sherpas contributed to the cosmopolitan mix of peoples who defined Mount Rainier National Park and whose ties extended far beyond it.

The connection between Rainier and Everest was not exceptional. Many other nations sent expeditions to Nepal and fostered relationships as important as the one that developed between the Sherpas and their American counterparts on Rainier. More than the United States, New Zealand, the native home of Edmund Hillary, influenced the creation of a national park at Mount Everest. In 1971 the United Nations Food and Agriculture Organization recommended the creation of a park, and in 1973 Prince Gyanendra of Nepal committed the nation to its establishment. The Nepalese government appealed to New Zealand for assistance, in part because of New Zealand's experience with mountain parks. Advisers from New Zealand arrived in 1975, and Sherpas went to New Zealand for training in park management. In 1976 Nepal officially established Sagarmatha National Park. Sagarmatha incorporated Sherpa communities much more than Mount Rainier included American Indian tribes. Not only did Sherpas help manage Sagarmatha, but the primary purpose of the park included the conservation of Sherpa culture and landscapes.[38] If Sagarmatha National Park was the expression of any nation's best idea, it must have been New Zealand's.

As Whittaker and Gombu neared the summit in May 1962, national distinctions broke down almost completely. Precedence—who was first—became meaningless in that extreme environment, and the two men decided to take the final steps side by side. At the top, they photographed each other and planted flags and a Buddhist friendship scarf. "The sky above us was that deep, dark blue you only see when you've climbed above most of the earth's atmosphere," Whittaker recalled. "We were on the edge of space." They felt no sense of conquest, but only their frailty. "The mountain is so huge and powerful, and the climber so puny, exhausted, and powerless. The mountain is forever; Gombu and I, meanwhile, were dying every second we lingered."[39] Cold, alone, gasping for breath in spite of their bottled oxygen, they looked down on the world from a place where, for all practical purposes, nations had no boundaries, and mountains, no names.

Parks to the End of the World

An extreme environment of a different sort awaited the sixty-four-year-old George Baggley, director of the Midwestern Region of the National Park Service, when, late in 1962, he left his office in Omaha, Nebraska, to travel to Antarctica with a South African resupply expedition. Under the terms of the 1959 Antarctic Treaty, the twelve signatory nations were encouraged to exchange personnel in an effort to foster transparency and promote Antarctica as a continent "dedicated to peace and science." Following the successful ratification of the treaty in 1961, the State Department sent out a letter of consultation to various government departments and agencies, including the NPS. Park Service officials replied that they were particularly interested in the preservation and conservation of living resources and the preservation of historic sites in Antarctica, two topics closely related to the NPS philosophy and objectives. They also noted that they were interested in what they called the "preservation of the natural scene." The Park Service response concluded, "It seems reasonable to suggest that we be equally concerned about preventing irreparable damage to natural environments in the only remaining area of considerable size on earth which has not yet experienced appreciable modification by the impact of modern civilization."[40]

NPS interest in the Antarctic continent in the early 1960s offers one of the most extreme examples of national parks beyond the nation. Not only was Antarctica geographically a long way from the continental United States, but also the Antarctic Treaty had created a system of government that did not follow the nation-state model. While NPS interest in Antarctica was clearly motivated by "traditional" national park themes such as environmental protection and historical preservation, involvement with this international region meant that Park Service officials had to deal directly with broader diplomatic and political questions related to imperialism, nationalism, decolonization, and the Cold War. The State Department clearly valued the opinions of Park Service officials, and NPS interest in Antarctica reveals that there was diplomatic history to the history of national parks that went far beyond the history of the nation-state.[41]

George Baggley was chosen to be an official U.S. observer in Antarctica on the strength of his exemplary Park Service career.[42] Following a course in forestry at Colorado A&M College in Fort Collins (now Colorado State University), Baggley took the Park Service exam and became a ranger at Yellowstone National Park in the summer of 1928. It was here that he met his wife, Herma Albertson Baggley, who was a pioneering female botanist in the National Park Service.[43] After nine years at Yellowstone, Baggley moved to Michigan, where he played a seminal role in the establishment of Isle Royale National Park in 1940. By the early 1960s, Baggley had risen to become director of the Midwestern Region of the NPS, and it was from this position that the opportunity to visit Antarctica arose.

Nothing in his experience of U.S. parks could quite have prepared Baggley for the environment he encountered in the earth's far southern latitudes. In the near constant light of the southern summer, Baggley endured numerous days sailing across the South Ocean, one of the roughest seas in the world. The 1962–63 South African resupply expedition visited Marion Island and Gough Island, South Africa's two established stations on the periphery of the Antarctic region. These are two of most isolated and inhospitable specks of land anywhere on the planet. Buffeted by constant winds and frequent snow, sleet, and rain, neither of these islands offered Baggley and his companions much respite from the elements, but at least they provided a break from the incessant motion of the sea. The death of the South African station commander at Gough Island during the season offered a tragic reminder of the treacherous nature of the Antarctic environment.[44] On the Antarctic continent itself the resupply expedition helped construct a new research station for South African scientists, known as SANAE I, a simple wood structure erected on a wooden raft. Built on a moving ice-shelf, SANAE I would have a lifespan of less than ten years, highlighting the fragility of human presence in contrast to the sublime grandeur of the Antarctic environment.

International Politics of Parks

Baggley's visit to Antarctica with the South Africans implicitly raised a number of contentious political issues. The inclusion of apartheid South Africa within the Antarctic Treaty had elicited criticism from postcolonial nations such as India. More generally, the relationship between the United States and the explicitly racist regime in South African created a problematic balancing act for U.S. officials as a result of not wanting to alienate a potentially valuable anti-Communist ally in the Cold War. None of these questions appeared to worry Baggley or his superiors unduly. Upon his return from Antarctica, Baggley submitted a report that suggested that he had a good time with the South Africans, and that he found his experience of being an Antarctic observer a positive one. The report also indicated that he had taken the opportunity to visit national parks in southern and central Africa during his travels.[45]

The report of George Baggley's visit to Antarctica provoked an internal debate within the Park Service about involvement in Antarctic affairs. Daniel B. Beard, the Assistant Director of Public Affairs, questioned the value of future National Park Service involvement in the Antarctic Observer Program, suggesting that priorities lay elsewhere and that ties to the Antarctic program were "rather remote."[46] "With our limited personnel," he argued, "it seems to me that we could make a much more substantial contribution to international affairs and more ably represent our Government if we confine ourselves to those fields in which we are especially competent."[47] A strong voice against Beard and in favor of continued NPS involvement in the Antarctic observer program came from Robert Rose, the acting chief of the Division of Natural History.[48] At the heart of Rose's argument was the idea that the NPS had much to offer and much to learn: "Abandonment of interest now is to accept a defeatist attitude and the surrender of the role of positive, authoritative leadership in a field in which the Service is supposed to excel. . . . The Service could gain much knowledge through close association with . . . [National Science Foundation] research programs which would be helpful in a great many ways in the high altitude and northern latitude areas of the National Park System."[49] Despite his apparently persuasive arguments, however, Robert Rose lost the internal debate, and NPS participation in the 1963–64 season of the Antarctic Observer Program was abandoned.[50] Despite withdrawing from this direct engagement with Antarctica, the NPS continued occasionally to contribute to Antarctic policy discussions.[51] But in general, future NPS involvement in Antarctic policy making was minimal, and certainly nothing approaching the scale envisioned by Robert Rose.

The unwillingness of the National Park Service to push for a greater role in Antarctica did not signify a retreat from international engagement more generally. As Beard had argued, there was a belief that NPS skills could be even more useful in other parts of the world, especially the recently independent former European colonies in Asia and Africa, where a Cold War conflict for hearts and minds was rapidly developing. George Baggley himself continued to pursue international opportunities as he argued in 1964 for increased connections between U.S. international aid and the diffusion of national park expertise:

> It seems to be, with United States' aid going to so many countries [about 100], that there should be a very definite tie-in to natural resource use and conservation. An effort should be made to sell parks and conservation of resources on the basis of their proven social and economic returns to a country and also that most of these backward countries *can* establish parks, forest reservations, or conservation areas with very little difficulty. They may have to be shown how to do so, but in every nation there are resources, which if properly designated and managed would bring pride, as well as social and economic benefit to the people. What I am getting at here is that we are furnishing aid to quite a number of countries that have resources suitable for parks, but no effective effort is being made by the United States, so far as I can determine, to bring parks into the picture as economic assets or to try to get some consideration of conservation measures by the aided country.[52]

Baggley's reference to "backward countries" highlights a certain arrogance in U.S. attitudes toward the rest of the world. But his memo nevertheless demonstrates that a diplomatic and developmental role was an important part of the National Park Service's Cold War history beyond the nation. Branded as "America's Best Idea," national parks offered a highly positive view of the United States that could be exported along with Hollywood films and popular music to foster perceptions of the "American dream."

Parks Strike Back

Alongside the perceived distortions and appropriations of U.S. park ideas as they were projected internationally, national park ideas also came back to affect U.S. government policy, raising the interesting question of how national parks around the world influenced U.S. policy during this period. As many of the essays in this volume demonstrate, despite American attempts to control the park idea, it took on a momentum of its own, often completely independently of the United States. Such a situation was highlighted by calls for the continent of Antarctica to be turned into a world park. In September 1972 the Second World Conference on National Parks was held at Yellowstone/Grand Teton National Parks in honor of the hundredth anniversary of Yellowstone's establishment. The very occurrence of this meeting might be seen as a U.S. attempt to continue to control and shape the international national parks movement. At the meeting, however, numerous discussions took place that were not in line with U.S. official thinking, including a fairly extensive debate about the continent of Antarctica.[53] Among the various resolutions passed by the conference at the end of the meeting was Recommendation Five: "that the nations party to the Antarctic Treaty should negotiate to establish the Antarctic Continent and the surrounding seas as the first world park . . . under the auspices of the United Nations."[54] While the United States was more flexible than some members of the Antarctic Treaty System, such a proposal was not fully in line with State Department thinking about the southern continent, which essentially sought to preserve the status quo created by the 1959 Antarctic Treaty.[55]

Calls for the conservation of Antarctica were nothing new. In the early nineteenth century, the Russian explorer Thaddeus von Bellingshausen had lamented the indiscriminate slaughter of Antarctica's seals and called for some sort of restraint or regulation of the industry. "As other sealers also were competing in the destruction of the [fur] seals," he wrote, "there could be no doubt that round the South Shetland Islands just as at South Georgia . . . the number of these sea animals will rapidly decrease."[56] In the early twentieth century, the British Empire had made the first formal sovereignty claim to Antarctica in a self-proclaimed (if ultimately failed) attempt to prevent the Antarctic whaling industry from following the same "tragedy of the commons" pattern as Antarctic sealing. Sovereignty claims to Antarctica provoked something of a backlash, and the middle decades of the twentieth century saw a number of suggestions for the internationalization of the southern continent, which included calls by

India for the United Nations to take control of the continent. It was partially in opposition to these calls for a genuine internationalization of the continent that the twelve countries with an active scientific program in Antarctica during the 1957–58 International Geophysical Year signed the 1959 Antarctic Treaty. But although the conduct of science was central to this treaty, it said nothing explicitly about conservation. Instead, a variety of conservation measures were passed by members of the treaty at their regular consultations.

By combining conservation with calls for internationalization, the notion of a world park raised at the Yellowstone/Grand Teton meeting was qualitatively different from other proposals for Antarctica's future. The growing momentum behind the idea of Antarctica as a world park was highlighted by plans for a "grand book" to commemorate the anniversary celebrations with the prospective title *The Great Parks: From Feudal Reserves to Antarctica.*[57] The idea of an Antarctic world park raised a number of interesting questions. Who would be the world park rangers, and what uniform would they wear? Who would get to visit Antarctica and for what purpose? Would there be a repetition of the seemingly contradictory goals of the U.S. National Park Service's 1916 Organic Act, which had called both for visitation and preservation? Would the dominant role of science be diminished in the new world park Antarctica? None of these questions were directly addressed by the 1972 resolution. Given previous objections of the Antarctic Treaty Consultative Parties to oppose U.N. involvement in the affairs of the southern continent, this was an ambitious and controversial recommendation that was out of line with the official policy of most member governments of the Antarctic Treaty System.

Rather than becoming enthusiasts for the idea of Antarctica as a world park, the consultative members of the Antarctic Treaty—including the United States—ignored the recommendation calling for a conservation-driven internationalization under the auspices of the United Nations. Instead, the members of the Antarctic Treaty System began to negotiate among themselves a minerals regime for the Antarctic continent in order to preemptively agree on the regulation and distribution of possible extractive activities. In thinking about the future of Antarctic minerals activities, scientists and policy makers were influenced by deep history. Developments in the field of geophysics gave support to the theories of plate tectonics and continental drift, which suggested that Antarctica had been at the center of the Gondwana supercontinent between 510 and 180 million years before present. Stretching from the Antarctic Peninsula to Ross Island, the West Antarctic Rift marks the southern limit of the Pacific Ring of Fire, and its volcanism attests to the dynamism of the world's tectonic plates. By putting together the pieces of the Gondwana puzzle, geophysicists could speculate about past geological connections and continuities. Given that several of the world's most productive mineral regions—including the Bushveld region of South Africa, which is among the most mineral-rich areas on the planet— appeared to have abutted continental Antarctica in the earth's geological past, such scientific developments led to mineral speculation by association.[58] During the 1970s a number of mining and petroleum companies began to pay attention to Antarctica, and the members of the Antarctic Treaty System wanted to make sure that they had in place a regulatory framework for the extraction of natural resources before they were found in what its members believed to be a far-sighted strategy to keep the peace.

The idea of establishing a legal regime for the peaceful and environmentally friendly extraction of mineral resources was very much in keeping with the conservationist ethos of the Antarctic Treaty System. But internationally, it was seen with a great deal more suspicion, and outsider nations led by Malaysia began to accuse the Antarctic Treaty countries of carving up the riches of Antarctica among themselves. These countries called the Antarctic Treaty System an exclusive club and asked for it to be handed over to U.N. authority.[59] While this demand had similarities to the recommendation made at the Second World Conference on National Parks, the Non-Aligned Movement had developmentalist goals at the forefront of their minds: they wanted to participate in the spoils of any Antarctic mineral bonanza through the "common heritage of mankind" principle.

Running alongside the postcolonial critique of the Antarctic Treaty System was an environmentalist critique led by organizations such as Greenpeace and the Antarctic and Southern Oceans Coalition (ASOC). In making their case against the Antarctic Treaty's minerals negotiations, these environmental organizations made extensive use of the world park idea. In January 1987 Greenpeace sailed from New Zealand to establish what was then the only nongovernmental research station in Antarctica close to the U.S. McMurdo Station and New Zealand's Scott Base. The Greenpeace station was named World Park Base in an effort to reinforce their case for a complete ban on any minerals activities in Antarctica. The research conducted by Greenpeace was focused on demonstrating the adverse environmental impacts of human activity in Antarctica. At McMurdo, for example, Greenpeace discovered "wheels, oil drums, and a pipe discharging brightly colored liquid straight into the water." The idea of Antarctica as a world park drew upon Antarctica as the earth's last—and perhaps only—terrestrial wilderness. "The public see the Antarctic as a symbol for salvation for the world," claimed a Greenpeace spokesperson, "because it is a zone of peace in a world of conflict; because it is mysterious in its isolation and inaccessibility; because we live our daily lives in an overcrowded and polluted world; Antarctica is gaining a special place in people's hearts."[60] Such rhetoric has clear parallels with earlier arguments made by John Muir and many others in advocating for national parks and wilderness preservation in the United States.[61] Many of the criticisms of wilderness that have been raised by the historiography of national parks and wilderness in the United States can also apply to Antarctica, although with important differences. Maybe as a continent without an indigenous population, Antarctica does have a claim to being a genuine "wilderness" region, if not quite "a symbol of salvation for the world."

Despite opposition from postcolonial states and environmental organizations, the members of the Antarctic Treaty went ahead and negotiated a minerals regime for Antarctica. The Convention on the Regulation of Antarctic Mineral Resource Activities (CRAMRA) was ready for signature in 1989. Following the successful negotiation of CRAMRA, however, Australia and France began to have second thoughts and announced that they would not sign. The reasons for this *volte face* involved a mixture of altruism and self-interest: some politicians in Australia, for example, worried that Antarctic minerals exports might compete with their own. But the environmentalist campaign against the minerals regime based on the idea of World Park Antarctica undoubtedly had some impact on the Australian and French decisions. Antarctic environmentalism offered opponents to the minerals regime important "green credentials" in exchange for their opposition, which really did not involve significant economic sacrifice. One by one the other members of the Antarctic Treaty System changed their position to oppose CRAMRA and begin the process of negotiating an environmental protocol in its place, which led to the Madrid Environmental Protocol after a series of meetings in 1991. Among its provisions was a fifty-year prohibition on any mineral activity in Antarctica, including prospecting. The administration of George H. W. Bush was the last government to accept the shift from a conservationist agenda to a preservationist agenda. Interestingly, President Bush's change of heart came during a visit to Mount Rushmore National Monument in South Dakota.[62] Without altering the fundamental structure of the Antarctic Treaty—and certainly without handing Antarctica over to the United Nations—many of the environmental protection measures associated with the campaign for a world park were put in place.

The period from the 1940s to the early 1990s reveals a history of national parks beyond the nation. On a continental scale the history of Antarctica reflects many of the themes in critical national park scholarship. What would Antarctica look like today if the National Park Service (and other park agencies from around the world) had gained a significant role in the administration of the southern continent in the 1960s or 1970s? Would a genuine "World Park Antarctica" really have worked? Of course, we really do not know what would have happened if Recommendation Five of the Second World Conference on National Parks in 1972 had been taken seriously, and if Antarctica had been handed over to the United Nations as a world park. But there is little need for "what if" speculation: the National Park Service's interest in Antarctica highlights an outward-looking agenda, which certainly deserves more study by environmental historians in other parts of the world. In terms of Antarctic history, the creation of what might be thought of as a "pseudo world park" by the terms of the Madrid Environmental Protocol certainly helped stabilize and strengthen the political structures of the Antarctic continent, and this was very much in the interests of U.S. Antarctic diplomacy. Interestingly, when the U.S. Palmer research station in the Antarctic Peninsula region found itself facing increasing numbers of tourists in the early 1990s, the U.S. government once again turned to the NPS and responded by stationing a park ranger at Palmer in an effort to control these visits.[63]

Lines in the Sand

No formal park materialized in Antarctica, but one did emerge in the Sonoran Desert in southern Arizona. Even on U.S. soil, however, sovereignty has been tested, and the nation has turned to a national park to control visits and other challenges that come from afar. National parks help bound the nation both physically and culturally. Physically, nation-states have used parks to define their territorial sovereignty. Culturally, parks supply powerful narratives that identify who is in and who is out with regard to imagined national communities. For these reasons people in many parts of the world have come to equate national parks with nations. Mount Rainier, Antarctica, and the other regions this book explores, however, reveal the myopia of framing national parks that way. Parks are at once much bigger and much smaller than the nation-states that house them. By refusing to stop at the borders, the essays productively and humanely draw our attention beyond the nation.

For example, at Arizona's Organ Pipe Cactus National Monument, which abuts the Mexican state of Sonora, the presence of an international border has transformed the management of a national park. Among other things, while most of the roughly 300,000 annual recreational visitors will encounter a cheery face under the NPS's trademark hat leading a campfire program or providing helpful information at an entrance kiosk, it is not impossible that they might also glimpse camouflage-clad rangers toting assault rifles through the desert, looking for some of the nonrecreational visitors who cross the border into the park seeking a job, a reunion with loved ones, or in some cases a payoff for smuggling drugs. How many of these visitors traverse the park is unknown, but for most of the last decade they have certainly exceeded the number of tourists, and possibly doubled or tripled it. At the very least, many of the recreational visitors will stop at the Kris Eggle Visitor Center and learn about the hot day in August 2002, when Ranger Eggle joined U.S. Border Patrol officials in pursuit of a drug smuggler who had rumbled over the border from Sonoyta, Mexico, in a stolen SUV.[64]

Just a few months out of the Federal Law Enforcement Training Center, where he finished first in his class, Eggle found himself on the front lines of America's wars on immigrants, terror, and drugs, without the field training or mentorship period that officers in other border agencies routinely get. By all accounts, the twenty-eight-year-old Eggle was extremely well liked by his colleagues at Organ Pipe, his generous, upbeat spirit buoying the morale of a crew assigned to the place that the National Park Rangers Lodge of the Fraternal Order of Police has labeled "America's most dangerous park." Perhaps the training honors or his friends' admiration or his widely acknowledged leadership skills and firm sense of duty gave him confidence beyond his true abilities that afternoon. Or perhaps it was the bulletproof Kevlar vest he wore. Whatever the reason, Eggle had plenty of cause for self-assurance behind the wheel of an NPS vehicle hurtling down the dirt road that parallels the international line. When the smuggler's SUV stalled, Eggle charged out into the 110-degree heat to chase him on foot through thorny tentacles of ocotillos. The smuggler crouched under a mesquite tree and waited for the ranger to catch up. He wore no shirt, and dirt caked his sweaty body. Eggle saw the fugitive and reached for his gun—too late. A bullet from the smuggler's rifle hit the radio or some other metallic piece on Eggle's belt and veered under the vest, through his skin to pierce an artery.[65]

Eggle died defending a park perimeter that federal policy and national ideology require be impermeable, but the challenges that Kris Eggle and his fellow rangers faced originated well beyond that perimeter. A bullet whizzing from a smuggler's rifle collided with the body of a spirited but ill-prepared ranger because of events that originated far from the park boundaries. In the 1990s, federal border policy shut down urban ports of entry while shifting the battle zone to remote and dangerous southwestern deserts. A globalizing economy displaced Latin America's poor and sent them over the U.S. border. The lucrative transnational narcotics market brought the trafficker to the border, and the international war on drugs impelled him and his weapon into the mesquite shrubbery of Arizona. The federal budget battles that limited the number of law enforcement rangers in the park to four and that stunted the length and scope of Eggle's training were fought and lost not in Arizona but in Washington, D.C. Even the endangered Sonoran Desert pronghorn and other wildlife, which Eggle had a mandate as a park ranger to conserve, routinely crossed between Mexico and the United States. Nothing but American imagination actually stops at the border. Eggle's death was partly a tragic result of viewing national parks as coterminous with the nation.

Building Fences

The tight link between national parks and nations, however, is an old one. It was embedded in the very founding of the park. President Franklin Roosevelt designated Organ Pipe Cactus National Monument in 1937. U.S. national parks are created by acts of Congress, but the 1906 Antiquities Act authorizes presidents to create national monuments by executive order. Most monuments such as Organ Pipe (and Devils Tower National Monument, which Ann McGrath discusses later in this volume) and national parks are both managed by the NPS. Historically Organ Pipe's lands had been part of a larger ecological and cultural area known as the Sonoran Desert.[66] Roosevelt's proclamation, however, severed this tie and integrated the parks' lands into a federal administrative system. To promote "the public interest" in preserving "historic landmarks" and "various objects of historic and scientific interest," his executive order "reserved" and "*set apart*" "certain lands in the state of Arizona" as Organ Pipe Cactus National Monument.[67] The history of the rest of the Sonoran Desert would unfold according to other logics, but the lands within the national monument would evolve under the "supervision, management, and control" of the NPS. From the beginning, then, Organ Pipe constituted a federal effort to separate land from its larger ecological and cultural contexts and to direct its history according to the mandates of the nation-state—a considerable exercise of federal power.

Since the founding of Organ Pipe, the United States has attempted to inscribe physical markers of its power right on the landscape. To demarcate the park physically and symbolically from the patchwork of federal, private, Mexican, and Indian lands that surround it and to carry out its conservation mission, the first superintendent William Supernaugh wanted to construct a fence around the entire park—on the assumption that the park started and stopped at its boundaries.[68] Although the plan quickly proved unfeasible, fencing the park's thirty-mile international border remained a reachable goal. By the middle of the 1940s, somewhere in the neighborhood of 1,000 head of Mexican stock—wild horses and burros along with privately owned stray cattle—had crossed the boundary and were foraging on the park's vegetation.[69] By comparison, the park's largest permitted rancher, the Gray family, grazed 1,050 head on the park's land. No matter who owned the stock, grazing was clearly impeding the park's conservation mandate. Livestock mowed down native vegetation, competed with wild animals for forage, and aggravated soil erosion.[70] They were also an embarrassment. The only road carrying tourists into the mountains coursed through a deathscape of dust, damaged vegetation, manure, and cattle carcasses, provoking complaints from visitors about the highly visible degradation. Worse still, the park's official master plan had slated the site for a public campground. In the estimation of one official, however, overgrazing had "ruined [it] as a public use spot."[71] For the moment, little could be done to eliminate the cattle belonging to the Grays, who had secured grazing rights at the park's inception, but officials targeted the reduction of Mexican stock as a means to mitigate the grazing problem. It was a matter, the Park Service's southwest region associate director wrote, of protecting "the basic values" of a "nationally important scenic and biological area."[72] Overgrazing in a national park threatened America's national heritage.

For several years in the early 1940s, officials sought, without success, to obtain funding to fence the international border.[73] In early 1947 an outbreak among Mexican cattle of *Aphtae epizooticae,* the virus that causes foot-and-mouth disease, threatened to infect the park's domestic and wild animals and heightened the fence's appeal to other federal agencies.[74] By 1949—after months of bureaucratic false starts—the National Park Service, the International Boundary and Water Commission, and the Bureau of Animal Industry collaborated to string a livestock-proof fence along nearly the entirety of the park's international border.[75] Although a 1949 report indicated that the problem of loose stock ranging over the border was resolved, a way to contain within the park the clever peccaries and nimble pronghorns and mule deer, any of which might contract the virus south of the border and bring it back across the border, continued to elude officials. The same 1949 report recommended building windmills to fill tanks with groundwater to provide the drinking sources to keep wild hooved mammals inside the park—yet more evidence of the expectation that the park must be coterminous with the national boundary. Though more sophisticated barriers—insurmountable concrete and steel walls and high-tech virtual fences that guarded the border with laser beams—would follow at the dawn of the twenty-first century, the nation-state started by demarcating the border in the 1940s with three strands of barbed wire east of Sonoyta and four strands west of it.[76]

All of this constituted an exercise of what the historian Charles Maier called "territoriality," the state power that comes from control of bordered spaces. Since the late nineteenth century, he argued, nation-states have empowered themselves by controlling the loyalties and resources (including disease-free animal populations) within enclosed borders.[77] As local islands of national control—in places where federal power is often otherwise quite weak—national parks such as Organ Pipe have been important sites at which states manifest themselves physically in order to exert sovereignty over interiors. This was intensified on the border, where protecting the park was a matter of protecting the nation, its people, and its heritage. The interagency fence in the 1940s was both the tool to accomplish that and the physical marker of the state's authority and ability to do so. Thus Organ Pipe has remained a site for establishing state authority to exert power by policing boundaries and for exemplifying the thinking that says the parks start and stop at the national boundary.

States need parks and other tools to establish sovereignty because there are so many border crossers that challenge territoriality. At Organ Pipe fencing the border was complicated in the park's southwestern corner by a Tohono O'odham man named Jim Orosco, who homesteaded on the north side of the border and took water from the park's Quitobaquito Springs to irrigate his crops on the south side. The park's fence sliced Orosco's homestead in two and made it impossible for him to cross from his house to his fields. Eventually, the park sought special permission from the International Boundary and Water Commission to install a gate to provide passage for the binational Orosco, his vehicles, and his stock across his land.[78] Meanwhile, the park's eastern boundary initially excluded the Tohono O'odham from lands on which the tribe had grazed stock for centuries, leading the Park Service to permit Indian cattle to graze inside the park boundaries—more external stock on park lands.[79] The boundaries are just as porous today. Whenever humanitarian Mike Wilson checks the water tanks he has placed to aid migrants, he checks in at the U.S.-Mexico border substation before crossing the border in exercise of his right as a Tohono O'odham member to continue the tribe's historic passage across the Sonoran Desert.[80] Additionally, the park's nonrecreational migrants from Mexico outnumber its American vacationers, and it is thought that on any given night perhaps 1,000 border crossers sleep in the park.[81] Even the grazing crisis of the 1940s, ostensibly a function of roaming livestock, was also a product of border-crossing weather, as a severe region-wide drought reduced the vegetation and therefore the carrying capacity of Sonoran Desert rangelands, setting thirsty hooved herbivores on the move.[82] In July 2008 the mesh pedestrian fence at the border dammed floodwaters that would otherwise have flowed south from Organ Pipe into Mexico, and instead deflected the torrents down the streets of Lukeville, Arizona, and Sonoyta, Mexico, inundating the international port of entry on the park's southern boundary. The episode repeated in 2011.[83] Thus as governments manifest borders physically, diseased cattle, smugglers, migrants, nonstate nations, floodwaters, climatic fluctuations, and other entities are constantly undoing the borders that nation-states have relied on to enclose their sovereignty over spaces of allegiances and resources. No fence can defend against so many different kinds of crossers. It takes something stronger.

Telling Stories

Hence people construct borders with stories just as much as with fences. And just as they do with physical territoriality, national parks lend a hand in this cultural project. One of the most powerful stories about parks goes by the heading "America's Best Idea." Attributed to British ambassador Lord James Bryce by writer Wallace Stegner, perpetuated by the National Park Service, and popularized by documentary filmmaker Ken Burns, the phrase captures the sentiment that setting aside scenic and historically significant lands to preserve a wild and democratic national heritage for the enjoyment of all people is a signal achievement of the United States and, as other nations have copied it, a gift to the entire world.[84] It is also a mechanism by which parks help build and maintain boundaries around national identity by defining who's in and who's out. One who has sung Organ Pipe's praises most beautifully is the nature writer Carol Ann Bassett, who extolled the park as "a magical realm . . . a sanctuary where I could think and dream and write—a place where I could learn to trust my instincts and find an opening for my heart." People play three roles in her sanctuary. They seek, along with Bassett, the desert's deep spiritual mysteries. They disappear, like the long-gone native custodians who once lived in harmony with natural rhythms and whose remains, when disturbed, mournfully testify to bygone primeval days. Or they defile it by grazing cattle, building surveillance towers, or cutting down mesquite trees to warm themselves on cold desert nights. "I want," Basset lamented, "to remember my desert sanctuary as it was: pristine and peaceful—a place of personal transformation that will forever be part of my deepest soul." This is a well-meaning and sincere version of the America's Best Idea narrative—after all, preserving pristine and peaceful sanctuaries lies at the heart of the NPS's mission—but in celebrating pristine nature as an expression of the nation, such narratives erase the presence of those who do not fit the story. America's Best Idea clearly counts nature revelers within the American community; twenty-first-century Indians living on impoverished federal reservations adjacent to but severed from their traditional resource base in the park and migrant mothers dropping plastic water bottles in fragile ecosystems while coaxing thirsty children toward the American dream—tragic as their plights may be—are out.[85]

In 2003 writer Tom Clynes told another version in a *National Geographic Adventure Magazine* article titled "National Park War Zone." "Organ Pipe," Clynes wrote, "is a truly unique piece of America." He cast the Michigan-born, valedictorian, track-star Baptist Eagle Scout park ranger Eggle, who grew up on a 130-year-old farm, as the park's all-American defender and Eggle's killer, Panfilo Murillo Aguila, as the face of the many who cross the border illegally, leaving in their tracks empty water jugs, discarded clothing, and sugar sacking used for wrapping marijuana, insensitive to the American botanical sanctuary they traverse. After the shooting, Clynes recounted, Murillo Aguila fled south toward the border, where fifty Mexican law enforcement officers fired a hundred rounds at the murderer, after which Eggle's park service buddy Jon Young tried in vain to resuscitate the dying fugitive. A Mexican firing squad gunning down an outlaw in the desert supplied compelling cliché for the dénouement of a drama about America's most dangerous park, and it clearly defined Mexican brutality as outside the American imagined community; a big-hearted ranger trying to save the life of his best friend's killer—definitely in. Clynes's approach is not unique. The list of recent articles employing martial and other violent metaphors to report on Organ Pipe and other border parks is nearly endless: "An Attack on the Border"; "Parks Under Siege"; "Invasion through Organ Pipe National Monument"; "Amid Cactuses, a Park's War on Smuggling"; "New Outlaws Plague Arizona Desert Refuges"; and "Park Rangers Take on Extra Duties as Scenic Lands Become Illegal Gateways to USA," to name a few.[86] Collectively, such reporting depicts nameless hordes overrunning U.S. patrimony, utterly without appreciation for the aesthetics, scientific treasure, biodiversity, recreation, and opportunity for spiritual fulfillment that far-sighted Americans have preserved. Is this really America at its best? Maybe not, but at least the park—in the hands of skilled storytellers—helps Americans tell themselves apart from Mexicans. At Organ Pipe, stories reinforce the very borders that make the park so dangerous.[87]

Scaling History

Parks, however, also have the potential to move us beyond the confines of the very racialized borders they have helped construct. Stories that make national parks coterminous with nations are at once too big and too small to contain the people's historical experiences with parks. Here's a park story for which the boundaries of the nation state are clearly too big. In the best-selling *Devil's Highway,* the writer Luis Alberto Urrea recounted the saga of twenty-six men from Veracruz and other parts of Mexico who followed an ill-prepared or ill-intentioned human smuggler—a *coyote*—across Organ Pipe and into Cabeza Prieta National Wildlife Refuge in May 2001. For several days they wandered, lost, through hundred-plus-degree heat as the cells in their bodies gave up life-sustaining water. First, headaches and thirst set in. Moisture evaporated from their bodies as fast as they could sweat it out. Pasty spit dribbled between their bleeding lips. Their fevers burned. Desperate for moisture, they drank their own urine, licking their lips—sandpaper tongues scraping across open wounds. Blood vessels in their eyes burst. Confusion. They began to hallucinate—God, the devil, the greenery of lush Veracruz, all paraded before their eyes. One by one their vital organs shut down, until fourteen of the walkers were no more of this world. A few of the survivors stumbled upon a border patrol officer, who summoned a sizable chunk of the agency's southwestern Arizona resources. Within ten minutes the rescue was under way. Helicopters droned. Trucks rumbled over the desert. "It was a mobilization worthy of a small invasion," Urrea observed.[88] And it was successful: twelve of the twenty-six survived. Focusing on bodies highlights common humanity. Mexican or American, smuggler or Eagle Scout, tourist or migrant, human bodies work the same at the cellular level. When there isn't enough water in our cells, gruesome things happen—ghastly enough to turn the cat-and-mouse relationship of the border patrol and migrants into a heroic rescue operation. The boundaries of the nation matter little when we look at the park at the cellular level; at that level history becomes universal. The nation-state and the divisions in wealth, race, and citizenship it implies are way too large a frame for understanding why strong young men hallucinate in the desert or why border patrol agents become knights in shining armor. Telling park stories at the cellular level expands the story to one of common humanity and blurs the distinctions between Americans and Mexicans that the "America's Best Idea" framing highlights so well.

For other park stories, however, the nation is much too small a frame. One example is global climate change. As the earth's average temperatures rise in many places, parks around the world—their weather, glaciers, floral and faunal populations, hydrology, visitation, and relationships with surrounding people—change, too. Even Organ Pipe, where it often seems it cannot possibly get any hotter, registers changes. The park's average temperature has risen two degrees Fahrenheit since 1949, and the elevated temperatures have aggravated the impact of recent multiyear drought. At the same time, the annual number of freezing days has dipped, a boon for certain invasive species such as Africanized honeybees and the pesky buffelgrass that is multiplying faster than the park's resources to manage it. Although these and other local climate trends have not conclusively been linked to global climate change, their continuation would dramatically change park ecology. "Determining the effects of climate change on monument resources," the *Superintendent's 2010 Report on Natural Resource Vital Signs* indicated, is a topic "to which the park is dedicating considerable effort."

Equally interesting is what it does not say about climate change. While acknowledging climate change's considerable importance, the report mentions it on only two of twenty-seven pages, devoting less than a paragraph to it in both cases. In contrast, the report devotes much more space to a far more immediate though no less transnational problem, border crossing. Illegal border-related activities occupy three full pages in the study and are mentioned on numerous others. The wildfire section, for example, notes that the frequency of burns has been increasing because of border crossing, and another section laments that backcountry security restrictions have hampered water-quality monitoring. "Unquestionably," the report concludes, "the most difficult issue faced by managers of [Organ Pipe] is the ongoing U.S./Mexico border situation. In recent years, every aspect of the park's management has been affected by border-related issues." Those issues, however, extend far beyond the immediate vicinity of the border. Rather, they stem from fluctuations in both the American and Mexican economies, which are themselves buffeted by markets, currencies, labor costs, and business models around the world. Thus at Organ Pipe, both global climate and economic changes drive management challenges. As *Vital Signs* concluded, the park's ecosystem is being "stressed by forces acting at scales beyond the park's boundary." These are the same boundaries that Superintendent Supernaugh once wanted to enclose inside a fence to defend the park from outside forces.[89]

The history of climate change exemplifies the need for a multiscaled narrative to national park scholarship that considers the interaction of local, national, and global perspectives. Until the last twenty or thirty years, the last thing on the mind of intrepid tourists attempting to climb the snow-covered peak of Mount Rainier would have been global warming. The altitude, distances, and rapidly changing weather posed enough challenges of their own. However, in recent years meteorological research conducted around the world has significantly changed perceptions of the natural world. Scientists working in places such as Antarctica have shown a strong correlation between rising temperatures and greenhouse gas emissions in the earth's past, which points toward unprecedented environmental change as a result of the world's fossil fuel–dominated economies. It is now difficult to look at snow-covered peaks such as Rainier and not wonder about the meaning of national parks when visitors might one day no longer see glaciers that helped make them famous but only the resulting landforms they carved. Organ Pipe's experience cautions against treating global climate change as a monolithic force that affects all parks the same way.

Human responses to climate change, as Mark Carey shows in chapter 12 of this volume, do not take place in a vacuum, but rather unfold in particular socioenvironmental contexts, which vary from nation to nation. Thus, the effect of global climate change cannot be predicted or managed through reductionist formulas that measure the rate of ice melt, project it into the future, and lament the impact on glaciers or other tourist attractions; rather, park historians must also take into account the capacities of people in given historical moments to respond. In this, no single nation or park can serve as the model. The impact of climate change, then, like many other facets of park histories, must be considered on both a global scale and a case-by-case basis.

Maier suggested the impulses that produced national territoriality in the twentieth century are dissipating.[90] Climate change is one of the forces overwhelming that territoriality. If he is right, we need new narratives that will simultaneously draw our attention to scales far smaller than the nation, even to the microscopic level within our own bodies, and outward toward cultural and ecological frameworks of global scope that cannot be contained within the jurisdiction of a single territorial state. Whether at home at Mount Rainier, abroad in Antarctica, or at the edge in the Sonoran Desert, national parks are embedded in both bodily and planetary ecologies and can therefore help construct new narratives. Such are the challenges and opportunities of doing national park history beyond the nation.

Notes

1. Members of the Narada chapter of the Theosophical Society of Tacoma named the falls in 1893. See Aubrey L. Haines, *Mountain Fever: Historic Conquests of Rainier* (Portland: Oregon Historical Society, 1962), 162–64; and Edmond S. Meany, ed., *Mount Rainier: A Record of Exploration* (New York: Macmillan, 1916), 315. A mystical faith with a deep interest in Asian religions, theosophy tended to attract Europeans and European Americans.

2. Alan Stein, "Osceola Mudflow from Mount Rainier Inundates the White River Valley Approximately 5,600 Years Ago," HistoryLink.org Essay No. 5095, January 23, 2003, at http://www.historylink.org/index.cfm?DisplayPage=output.cfm&file_id=5095 (accessed June 19, 2015); U.S. Geological Survey, "Significant Lahars at Mount Rainier," USGS Volcano Hazards Program—Mount Rainier Geology and History, at http://volcanoes.usgs.gov/volcanoes/mount_rainier/mount_rainier_geo_hist_79.html (accessed June 19, 2015).

3. Theodore Catton, *National Park, City Playground: Mount Rainier in the Twentieth Century* (Seattle: University of Washington Press, 2006), 55–59. Catton's work is the outstanding history, but also see Cecelia Svinth Carpenter, *Where the Waters Begin: The Traditional Nisqually Indian History of Mount Rainier* (Seattle: Northwest Interpretive Association, 1994); Bruce Barcott, *The Measure of a Mountain: Beauty and Terror on Mount Rainier* (Seattle: Sasquatch Books, 1997); and Ruth Kirk, *Sunrise to Paradise: The Story of Mount Rainier National Park* (Seattle: University of Washington Press, 1999).

4. For example, Stan Jorstad, *America's Best Idea: A Photographic Journey through Our National Parks* (New York: American Park Network, 2006). Two essays in the book, Alfred Runte, "New Terms for Conservation" (8–9), and Edwin Birnbaum, "The Spiritual and Cultural Significance of National Parks" (10–11),

depict national parks as uniquely American in origin, but also gesture to the parks' international and multiethnic elements.

5. The "our" in the title of F. W. Schmoe, *Our Greatest Mountain: A Handbook for Mount Rainier National Park* (New York: G. P. Putnam's Sons, 1925), exemplifies the well-intentioned manner in which Americans (more often than not white European Americans) gradually assimilated the mountain into a framework of modern governmental administration and national meaning.

6. Edmond S. Meany, *Vancouver's Discovery of Puget Sound: Portraits and Biographies of the Men Honored in the Naming of Geographic Features of Northwestern America* (New York: Macmillan, 1915), 1. The original account is in George Vancouver, *A Voyage of Discovery to the North Pacific Ocean and Round the World,* 3 vols. (London: G. G. and J. Robinson and J. Edwards, 1798). In addition to being a history professor, Meany was a conservationist, patron of explorers, and mountaineer. Meany Crest in Mount Rainier National Park and Mount Meany in Olympic National Park carry his name. See Alan J. Stein, "Edmond Stephen Meany (1862–1935)." HistoryLink.org Essay No. 7885, December 12, 2006, at http://www.historylink.org/index.cfm?DisplayPage=output.cfm&file_id=7885 (accessed June 19, 2015).

7. Meany, *Vancouver's Discovery of Puget Sound,* 81–82, 99–100.

8. Ibid., 13.

9. Ibid., 107–8.

10. Ibid., 13.

11. The best account of Native place names around Puget Sound is Coll Thrush, *Native Seattle: Histories from the Crossing-Over Place* (Seattle: University of Washington Press, 2007), esp. "An Atlas of Indigenous Seattle," 209–55. Additional glimpses of the cultural depth and complexity of the people can be found in Wayne Suttles, ed., *Handbook of North American Indians,* vol. 7, *Northwest Coast* (Washington, D.C.: Smithsonian, 1990). See also Barcott, *Measure of a Mountain,* 36, 40–41.

12. Catton, *National Park, City Playground,* 8–9; Barcott, *Measure of a Mountain,* 36–47.

13. Alfred Runte, *National Parks: The American Experience,* 2nd rev. ed. (Lincoln: University of Nebraska Press, 1987), 26–27.

14. C. Hart Merriam, *Report of the United States Geographic Board on the Name of Mount Rainier,* Senate Joint Resolution 64 (Washington, D.C.: Government Printing Office, 1924), 1.

15. "What Is the 'Ring of Fire'?" *Volcano World,* at http://volcano.oregonstate.edu/what-quotring-firequot (accessed November 2, 2012); "Ring of Fire," *National Geographic Education,* at http://education.nationalgeographic.com/education/encyclopedia/ring-fire/?ar_a=1 (accessed November 2, 2012); U.S. Geological Survey, Cascades Volcano Observatory, "'Ring of Fire,' Plate Tectonics, Sea-Floor Spreading, Subduction Zones, 'Hot Spots,'" at http://vulcan.wr.usgs.gov/Glossary/PlateTectonics/description_plate_tectonics.html (accessed November 2, 2012). See also Richard L. Hill, *Volcanoes of the Cascades: Their Rise and Risks* (Helena, Mont.: Falcon Press, 2004).

16. H. Byron Earhart, *Mount Fuji: Icon of Japan* (Columbia: University of South Carolina Press, 2011), 3–4.

17. Susan Shumaker, *Untold Stories from America's National Parks,* Part 8, *Mount*

Rainier and the Seattle Camera Club, 149, at http://www-tc.pbs.org/nationalparks/media/pdfs/tnp-abi-untold-stories-pt-08-mt-ranier.pdf (accessed July 22, 2015). Shoemaker's work is an offshoot of Ken Burns's documentary film *The National Parks: America's Best Idea.* See www.pbs.org/nationalparks/untold-stories/.

18. Louis Fiset, *Imprisoned Apart: The World War II Correspondence of an Issei Couple* (Seattle: University of Washington Press, 1997), 3–19; Shumaker, *Untold Stories from America's National Parks,* 141–55.

19. Shumaker, *Untold Stories from America's National Parks,* 148.

20. Ibid., 155; but see also Fiset, *Imprisoned Apart,* 11–12, 24.

21. Thomas R. H. Havens, *Parkscapes: Green Spaces in Modern Japan* (Honolulu: University of Hawai'i Press, 2011), 8–11, 54–62, 65–84. On Mount Fuji as a source of national pride, see also Earhart, *Mount Fuji.*

22. Kirk, *Sunrise to Paradise,* 8, 10; National Park Service, Mount Rainier National Park, "The Sister Mountain Project," at http://www.nps.gov/mora/forteachers/sister-mountain.htm (accessed July 22, 2015); National Park Service, Mount Rainier National Park, "Mt. Rainier National Park Centennial Timeline: The Thirties," at http://www.nps.gov/features/mora_cenn/thirties/thirties.htm (accessed July 22, 2015).

23. Fiset, *Imprisoned Apart,* 24–25; Shumaker, *Untold Stories from America's National Parks,* 149–50.

24. Shumaker, *Untold Stories from America's National Parks,* 153–61.

25. Kit Oldham, "Floyd W. Schmoe (1895–2001)," HistoryLink.org Essay No. 3876, February 25, 2010, at http://www.historylink.org/index.cfm?DisplayPage=output.cfm&file_id=3876 (accessed July 22, 2015); Shumaker, *Untold Stories from America's National Parks,* 158–59.

26. Floyd Schmoe, *A Year in Paradise* (New York: Harper & Brothers, 1959; rpt. Seattle: The Mountaineers, 1999), 159. Schmoe's best-known book, still in print, recounts the year that he and Ruth spent on the mountain but also includes additional information about their lives, Mount Rainier National Park, and his personal philosophy, which associated his love of nature with his pacifism. Schmoe mentions the Japanese Americans who traveled to the high country to hunt for mushrooms (134).

27. Oldham, "Floyd W. Schmoe"; Shumaker, *Untold Stories from America's National Parks,* 158–59.

28. Kirk, *Sunrise to Paradise,* 8, 10; National Park Service, Mount Rainier National Park, "The Sister Mountain Project."

29. Phuong Lee, "Japanese Past Displayed in International District Hotel," *Seattle Post-Intelligencer,* July 23, 1999, at http://www.panamahotelseattle.com/article2.htm (accessed April 13, 2013).

30. Jim Whittaker, *A Life on the Edge: Memoirs of Everest and Beyond* (Seattle: The Mountaineers, 1999), 107–8. For other accounts of the American Mount Everest expedition, see Thomas F. Hornbein, *Everest: The West Ridge* (San Francisco: Sierra Club, 1965); Maurice Isserman and Stewart Weaver, *Fallen Giants: A History of Himalayan Mountaineering from the Age of Empire to the Age of Extremes* (New Haven, Conn.: Yale University Press, 2008), 354–75; and Broughton Coburn, *The Vast Unknown: America's First Ascent of Everest* (New York: Crown, 2013).

31. Isserman and Weaver, *Fallen Giants,* 2, 8, 14, 16–17, 262–94. For a profusely illustrated, useful survey, see Conrad Anker, *The Call of Everest: The History, Science, and Future of the World's Tallest Peak* (Washington, D.C.: National Geographic, 2012).

32. Isserman and Weaver, *Fallen Giants,* 354–59, 514n23.

33. Dee Molenaar, *The Challenge of Rainier* (Seattle: The Mountaineers, 1971), xv, 160, 167, 177, 179, 302–5.

34. Ruth Kirk, *Exploring Mount Rainier* (Seattle: University of Washington Press, 1968), 15; Whittaker, *Life on the Edge,* 90. See also Molenaar, *Challenge of Rainier,* 303–5.

35. Molenaar, *Challenge of Rainier,* xv; Whittaker, *Life on the Edge,* 89–90; Coburn, *Vast Unknown,* 71–73. The five included Barry Bishop, Tom Hornbein, Lute Jerstad, Willi Unsoeld, and Jim Whittaker.

36. Molenaar, *Challenge of Rainier,* 83, 180, 181, 305.

37. Charlotte Austin, "Sherpas on Rainier: Nepalis Find Niche in North American Mountaineering Community," *Mountain,* August 13, 2012, at http://www.mountainonline.com/sherpas-on-rainier/ (accessed June 19, 2015).

38. Margaret Jeffries, *Mount Everest National Park: Sagarmatha, Mother of the Universe* (Seattle: The Mountaineers, 1991), 20. "We sincerely believe," Prince Gyanendra stated, in terms that went beyond the nation-state, "that this region and its surroundings in the grandeur of the Khumbu Valley are of major significance not only to us but the whole world as an ecological, cultural and geographical treasure which, we hope, should provide peace and tranquility and be a significant contribution to a better World Heritage." See also United Nations Environment Programme, World Conservation Monitoring Centre, "Sagarmatha National Park, Nepal," August 21, 2008, *Encyclopedia of Earth,* at http://www.eoearth.org/view/article/155820/ (accessed June 19, 2015); and "Sagarmatha National Park, Nepal," May 2011, United Nations Environment Programme, World Conservation Monitoring Centre, at http://www.conservation-development.net/Projekte/Nachhaltigkeit/DVD_12_WHS/Material/files/WCMC_Sagarmatha.pdf (accessed June 8, 2013).

39. Whittaker, *A Life on the Edge,* 110–11.

40. Robert K. Coote, Staff Assistant, Office of the Assistant Secretary for Public Land Management to Mr. George H. Owen, Director, Antarctica Staff, Department of State, April 19, 1961. RG 59 702.022/1–461.

41. See, for example, Assistant Director, Conservation, Interpretation and Use (L. F. Cook) to Regional Director, Western Region. Subject: Preservation of Historic Sites and Monuments in the Antarctic. June 6, 1963. RG 79 Records of the National Park Service. General Records. Administrative Files, 1949–1971 L66-L66 Box 217 L66 1-1-62–Dec 31 1963.

42. See http://parks.cityofboise.org/BoardAgendas/Board2006/0609Baggley.pdf (accessed June 8, 2013).

43. See http://www.nps.gov/yell/historyculture/upload/msc001_baggley.pdf (accessed June 8, 2013).

44. Robert Headland, *A Chronology of Antarctic Exploration: A Synopsis of Events and Activities from the Earliest Times until the International Polar Years, 2007–09* (London: Bernard Quaritch, 2009).

45. Assistant Director, Public Affairs (Daniel B. Beard) to Assistant Director, Conservation, Interpretation and Use. Subject: Antarctic Observer Program. June 6, 1963. RG 79 Records of the National Park Service. General Records. Administrative Files, 1949–1971 L66-L66 Box 217 L66 1-1-62–Dec 31 1963.

46. Ibid.

47. Ibid.

48. Acting Chief, Division of Natural History (Robert H. Rose) to Assistant Director, Conservation, Interpretation and Use. June 18, 1963. Subject: Antarctic Observer Program. RG 79 Records of the National Park Service. General Records. Administrative Files, 1949–1971 L66-L66 Box 217 L66 1-1-62–Dec 31 1963.

49. Ibid.

50. Assistant Director, Administration (Hillory A. Tolson) to Regional Director, Western Region, and Assistant Director, Design and Construction. Subject: Participation in Antarctic Resupply Activities. August 22, 1963. RG 79 Records of the National Park Service. General Records. Administrative Files, 1949–1971 L66-L66 Box 217 L66 1-1-62–Dec 31 1963.

51. Memorandum from Howard H. Eckles, Assistant to the Science Adviser to Mr. Rupert Southard, Geological Survey, January 17, 1966. Comments from the National Parks Service on the NSF Projection of the Antarctic Program. RG 79 Records of the National Park Service. General Records. Administrative Files, 1949–1971 L66 Antarctic From 1-1-66 to Dec 31 1967.

52. Associate Regional Director, Midwest Region (Mr. George Baggley) to The Director. Subject: Foreign Affairs Participation. Dec 11 1964. RG 79 Records of the National Park Service. Administrative Files, 1949–71 L66-L66. Box 2171. File: International Cooperation Yr. 1/1/64–12/31/65.

53. Hugh F. I. Elliott, *Second World Conference on National Parks* (Morges, Switzerland: International Union for Conservation of Nature and Natural Resources, 1974), 260. The debate was held during the second part of Session IX Special Park Environments, on Sunday, September 24, 2–5 P.M.

54. Ibid., 443–44. There was a ten-person recommendation committee under chairman Dr. M. E. Duncan Poore and Secretary Frank Nicholls (453). The recommendations were prepared in advance and then voted on in the final session by show of hands in a majority vote (459–60).

55. Christopher C. Joyner, *Eagle over the Ice: The U.S. in the Antarctic* (Hanover, N.H.: University Press of New England, 1977).

56. F. G. von Bellingshausen, *The Voyage of Captain Bellingshausen to the Antarctic Seas, 1819–1821,* ed. Frank Debenham (London: Hakluyt Society, 1945), 425–26.

57. Minutes of the 10th Meeting of the National Parks Centennial Commission Wednesday, March 14, 1973. RG 79 Records of the National Park Service. National Parks Centennial Commission. General Files, 1970–73.

58. James D. Hansom and John E. Gordon, *Antarctic Environments and Resources: A Geographical Perspective* (New York: Longman, 1988), 30–31.

59. See, for example, Peter Beck, "Twenty Years On: The UN and the 'Question of Antarctica,' 1983–2003," *Polar Record* 40, no. 3 (2004): 205–12.

60. Hansom and Gordon, *Antarctic Environments and Resources,* 287–88.

61. See, for example, Donald Worster, *A Passion for Nature, The Life of John Muir* (Oxford: Oxford University Press, 2008).

62. Undated Letter. FF24—1991—Antarctic treaty: correspondence, memoranda, newspaper clippings, news releases, statement, notes. Wilderness Society Archive, Conservation Collection, Denver Public Library.

63. Hansom and Gordon, *Antarctic Environments and Resources,* 255.

64. "Park Statistics," http://www.nps.gov/orpi/parkmgmt/statistics.htm (accessed May 25, 2012).

65. Tom Clynes, "National Park War Zone: Arizona's Organ Pipe Cactus National Monument," *National Geographic Adventure Magazine* (February 2003).

66. Richard S. Felger and Bill Broyles, eds., *Dry Borders: Great Natural Reserves of the Sonoran Desert* (Salt Lake City: University of Utah Press, 2007).

67. Franklin Roosevelt, "Organ Pipe Cactus National Monument—Arizona, by the President of the United States of America, a Proclamation," Proclamation No. 2232, 2 Fed. Reg.161 (April 16, 1937), emphasis added.

68. Brenna Lauren Lissoway, "An Administrative History of Organ Pipe Cactus National Monument: The First Thirty Years, 1937–1967" (MA thesis, Arizona State University, 2004), 40–43.

69. Harold M. Ratcliff, "Memorandum for the Regional Director, Region Three," July 2, 1946, folder "Grazing, 1946–1959," box 172, Correspondence and Subject Files, 1928–1959, RG 79, National Archives and Records Administration, College Park, Maryland (hereafter cited as NARAMD).

70. Newton B. Drury, "Memorandum for the Solicitor," December 23, 1947, folder "Repairs and Improvements, Fences," box 2320, Central Classified Files, 1907–1949, RG 79, NARAMD.

71. Ratcliff, "Memorandum for the Regional Director, Region Three"; E. R. Martell, Lafayette, Ind., to Conrad L. Wirth, Washington, D.C., May 2, 1955; folder "Grazing, 1946–1959," box 172, Correspondence and Subject Files, 1928–1959, RG 79, NARAMD.

72. John M. Davis, Santa Fe, to L. M. Lawson, El Paso, September 6, 1946, folder "Repairs and Improvements, Fences," box 2320, Central Classified Files, 1907–1949, RG 79, NARAMD.

73. Newton B. Drury, "Memorandum for the Solicitor," December 23, 1947, folder "Repairs and Improvements, Fences," box 2320, Central Classified Files, 1907–1949, RG 79, NARAMD.

74. William Supernaugh, "Memorandum for the Regional Director, Region Three," January 24, 1947; E. T. Scoyen, Santa Fe, to L. M. Lawson, El Paso, February, 4, 1947; E T. Scoyen to F. L. Schneider, Albuquerque, February 4, 1947; E. T. Scoyen, "Memorandum for the Director," February 4, 1947; E. T. Scoyen, Santa Fe, to Lawrence M. Lawson, El Paso, April 1, 1947; all in folder "Repairs and Improvements, Fences," box 2320, Central Classified Files, RG 79, NARAMD; L. M. Lawson, El Paso, to E. T. Scoyen, Santa Fe, February 6, 1947, folder 611–01, box 218, C1-National Archives and Records Administration, Rocky Mountain Region, Denver, Colorado (hereafter cited as NARARMR).

75. The interagency conversation leading to the fence construction is well chronicled in the Park Service papers at the National Archives branches in College Park, Maryland, and Denver, Colorado. See, for example, M. R. Tillotson, "Memorandum

for the Files," November 15, 1946, file 611–01, box 218, C1-NARARMR; and E. T. Scoyen, Santa Fe, to John C. Pace, El Centro, Calif., June 26, 1947; L. M. Lawson, El Paso, to E. T. Scoyen, Santa Fe, August 1, 1947; E. T. Scoyen, Santa Fe, to the Director, August 5, 1947; all in folder "Repairs and Improvements, Fences," box 2320, Central Classified Files, RG 79, NARAMD..

76. C. M. Aldous, "Special Report: International Boundry [*sic*] Animal Migration Investigation, Organ Pipe Cactus National Monument and Cabeza Prieta Game Refuge, Arizona," 1949, folder "Organ Pipe Rules," box 2319, NARAMD..

77. Maier, "Consigning the Twentieth Century to History," 808, 817.

78. John M. Davis, Santa Fe, to L. M. Lawson, El Paso, October 15, 1947; John M. Davis, "Memorandum for the Custodian, Organ Pipe Cactus," October 22, 1947; L. M. Lawson, El Paso, to John M. Davis, Santa Fe, October 17, 1947; Thomas Childs, Rowood, Ariz., to Carl Hayden, Washington, D.C., March 23, 1948; Newton B. Drury, Washington, D.C., to Carl Hayden, April 15, 1948; all in folder 611–01, box 218, C1-NARARMR.

79. Jerome A. Greene, "Historic Resource Study: Organ Pipe Cactus National Monument, Arizona," September 1977, 65, http://www.nps.gov/parkhistory/online_books/orpi/orpi_hrs.pdf (accessed July 22, 2015).

80. Margaret Regan, *The Death of Josseline: Immigration Stories from the Arizona Borderlands* (Boston: Beacon, 2010), 142–47.

81. Carol Ann Bassett and Michael Hyatt, *Organ Pipe: Life on the Edge* (Tucson: University of Arizona Press, 2004), 74; Clynes, "National Park War Zone."

82. Thomas J. Allen to E. R. Martell, Lafayette, Ind., n.d., folder "Grazing, 1946–1959," box 172, Correspondence and Subject Files,1928–1959, RG 79, NARAMD; Ran Bone, C. F. Dierking, and Volney M. Doublas, "Organ Pipe Cactus National Monument: Report on Land Use Conditions," November 30, 1946, folder "Grazing, 1946–1959," box 172, Correspondence and Subject Files,1928–1959, RG 79, NARAMD.

83. National Park Service, "Effects of the International Boundary Pedestrian Fence in the Vicinity of Lukeville, Arizona, on Drainage Systems and Infrastructure, Organ Pipe Cactus National Monument," August 2008, http://www.nps.gov/orpi/naturescience/upload/FloodReport_July2008_final.pdf (accessed May 31, 2012); Corey Reinig, "Flooding in Organ Pipe Cactus National Monument," National Border, National Park: A History of Organ Pipe Cactus National Monument, http://organpipehistory.wordpress.com/nature/flooding-in-organ-pipe-cactus-na-tional-monument/ (accessed May 31, 2012); Josie Kohnert, "The Environmental Consequences of a Border," National Border, National Park: A History of Organ Pipe Cactus National Monument, http://organpipehistory.wordpress.com/border/the-environmental-consequences-of-a-border/ (accessed May 31, 2012).

84. Wallace Stegner, "The Best Idea We Ever Had: An Overview," *Wilderness* (Spring 1983): 4–13.

85. Bassett, *Organ Pipe,* 4, 15–22, 32, 73–73.

86. Hugh Dellios, "Desert Park Victimized by Illegal Immigration; Designated a National Treasure in 1937, Organ Pipe Park in Arizona is No Jewel These Days," *Philadelphia Inquirer,* December 19, 2003; Tom Kenworthy, "New Outlaws Plague Arizona Desert Refuges; Park Rangers Take on Extra Duties as Scenic Lands Become Illegal Gateways to USA," *USA Today,* August 23, 2006; Gregory

McNamee, "Broken Borders: At Organ Pipe Cactus National Monument, Wildlife Suffers," Advocacy for Animals, http://advocacy.britannica.com/blog/advocacy/2009/04/broken-borders-at-organ-pipe (accessed October 23, 2013); Joshua Rhett Miller, "Five Federal Lands in Arizona Have Travel Warnings in Place," Fox News, http://www.foxnews.com/us/2010/06/18/federal-lands-arizona-travel-warnings-place/ (accessed October 22, 2013); "Pictures of Invasion through Organ Pipe National Monument," http://www.desertinvasion.us/invasion_pictures/invasion_opnm.html (accessed October 24, 2013); Tim Vanderpool, "Amid Cactuses, a Park's War on Smuggling," *Christian Science Monitor,* June 19, 2001, 3; Tim Vanderpool, "Parks Under Siege," *National Parks,* Nov.-Dec. 2002, 23–27; Traci Watson, "National Parks an Escape for Drug Smugglers," *USA Today,* December 10, 1999, 21A; Associated Press, "Illegal Border Traffic Is Wearing Out Preserve," June 19, 2006, http://www.msnbc.msn.com/id/13416072/ns/us_news-environment/t/illegal-border-traffic-wearing-out-preserve/ (accessed October 23, 2013); Burro, "Border Crossing Impact Fragile Desert Environments," Indybay, San Francisco Bay Area Independent Media Center, https://www.indybay.org/newsitems/2006/03/31/18125471.php (accessed October 22, 2013); Christopher Sharp and Randy Gimblett, "An Attack on the Border," University of Arizona School of Natural Resources and the Environment, http://www.snr.arizona.edu/node/933 (accessed October 24, 2013, link no longer active).

87. Clynes, "National Park War Zone."

88. Luis Alberto Urrea, *Devil's Highway, a True Story* (New York: Back Bay Books, 2005), 3, 120–29, 172–73.

89. Organ Pipe Cactus National Monument, *Organ Pipe Cactus National Monument: Superintendent's 2010 Report on Natural Resource Vital Signs* (Ajo, Ariz.: National Park Service, 2011), 6, 8, 9, 22, 27.

90. Maier, "Consigning the Twentieth Century to History," 808.

PART I | Origins

TWO # Canada's Best Idea?

The Canadian and American National Park Services in the 1910s

ALAN MACEACHERN

IN TRACING THE TRANSNATIONAL MOVEMENT of the national park idea, an exploration into the origin of the phrase "America's Best Idea" is as fitting a starting place as any. It is a saying that has been around for some time, but it gained immense traction when filmmaker Ken Burns used it as the title for his 2009 documentary series on the American parks.[1] The phrase's power is its economy: just three words give the parks credit for national significance and international influence. Like many before him, Burns credits western historian and nature writer Wallace Stegner with coining the phrase, but, like many before him, Burns provides no reference.[2] Some who cite Stegner quite naturally point to his "The Best Idea We Ever Had: An Overview," a 1983 *Wilderness* article about the national parks. But the phrase "An Overview" and the fact that Stegner never uses "best idea" in the essay itself suggest that he assumed readers would know he was quoting an existing expression rather than inventing a new one.[3] In 1990, perhaps because he knew he was becoming associated with the phrase, Stegner took the trouble to attribute it: "If the national park idea is, as Lord Bryce suggested, the best idea America ever had, wilderness preservation is the highest refinement of that idea."[4]

Lord James Bryce was the early-twentieth-century British ambassador to the United States, and he appears to have written only once in his career about parks: a 1912 lecture to the American Civic Association, "National Parks: The Need of the Future."[5] Bryce does indeed commend Americans parks as being an example to the world, but he never calls them the nation's "best idea"; in fact, the two words never appear in the same sentence. What's more, Bryce's address was fundamentally intended not to congratulate Americans on their parks but rather to urge them to create more parks and, in particular, to outlaw automobiles within them. No matter how sincere the ambassador was in his appraisal of the American park system, it cannot be ignored that he was also attempting to win over an audience he was about to counsel. If this lecture is indeed the ur-source of "America's Best Idea," it is ironic that Bryce's flattery has been remembered, even streamlined, while the rest of his address has been entirely forgotten. It is doubly ironic that Stegner, even after having cited Bryce, is still usually the one credited with the line—and Burns's attribution has only strengthened that. That Burns and his researchers did not come across Stegner's reference to Bryce seems unlikely. Maybe they found the ambassador's address too obscure, too nuanced, or just too thorny. (Even if he was British, Bryce's description of a love of nature as something that you couldn't have enough of, "as the old darky said about the watermelon," may have only called to mind America's worst idea.) Or maybe it just seemed more appropriate that "America's Best Idea" be an American's idea.

In this chapter, I seek to complicate the history not of the expression "America's Best Idea" itself but of the concepts that underlie it: that the U.S. national park system is fundamentally an American invention and that the subsequent international development of national parks is strictly an American export. I argue that on the contrary, parks inside the United States and out are better understood as products of the transnational movement of ideas. I do so by exploring the symbiotic and competitive relationship that developed between the American and Canadian national park systems in the 1910s. The creation in 1911 of a Dominion Parks Branch, the first agency in the world devoted to national parks,[6] meant Canada suddenly needed, and also had the capacity to request and receive, extensive advice from its American neighbors about developing a park system. The branch's existence and its early success in attracting funds and tourists to parks in turn allowed those in the United States who were lobbying for a park bureau to hold up the Canadian example as not just a model but also a source of economic threat and even national shame. With the establishment of the U.S. National Park Service in 1916, the Canadian agency took the matter full circle, quoting the American attention to validate its own existence to its political masters. The goal of this chapter, its joking title aside, is not to establish a Canadian invention of the national park idea.[7] Rather, it is to trace cross-border linkages and influences that have gone unnoticed in the histories of both nations' park systems. In doing so, I make a case that to understand fully the history of a national park system—or a nation, for that matter—one has to study it in relation to others.

In September 1911 the *New York Times* carried a full-page travel article on "America's Switzerland." "The suggestion that it is worthwhile to 'see America first' has turned comparatively few tourists from the Eastern States towards the Northwest," the author notes, but based on a recent trip to the Rockies he could affirm, "We can match travel wonders with the rest, though we have never left

America." A lengthy, admiring account of the beauties and charms of "our own Switzerland" follows. The article concludes, "So American scenery is gradually coming to its rightful place in the general holiday scheme."[8] The repeated references to America could almost make the reader forget, although it was stated plainly throughout, that the article was about the Canadian Rockies.

It is surprising to see Canada embody the best of America in a major U.S. publication—one can imagine American boosters howling that it was in fact the American Rockies that constituted America's Switzerland. But it is still more surprising to see Canada brought under the umbrella of the "See America First" campaign, which urged U.S. tourists of the era, for both economic and educational reasons, to visit their own land rather than Europe. "See America First" would seem the perfect fusion of patriotism and protectionism, and so its "America" was self-evidently shorthand for the United States, not North America. True, the 1906 conference that initiated "See America First" had included Canadian and Mexican delegates and characterized the issue in continental rather than national terms, but that fact was almost immediately forgotten.[9] So much of the ensuing weight of support for the movement came from U.S. boosters, businesses, and media, enticing U.S. tourists to spend their U.S. dollars at home, that it became associated more or less exclusively with the United States. When, in 1916, a young Cole Porter launched his first musical, *See America First*—such a colossal flop that his cowriter quit the theater and became a priest—it was not inappropriate that the title song's refrain began, "Don't leave America / Just stick around the U.S.A."[10] And that is how the campaign has been remembered. In the 2001 book *See America First,* historian Marguerite S. Shaffer treats her topic's geographical focus as so obvious that she never bothers to delineate it. Canada and Mexico do not even appear in the book's index.[11]

Although certainly most contemporary calls to "See America First" implicitly or explicitly referenced the United States alone, commentators were often willing to include Canada. A Chicago newspaper article titled "See America First" noted that "The west in the United States and Canada offers scenery that equals and in some instances surpasses what one sees in Norway and Switzerland." Another, published early in the First World War, stated, "It is hoped Americans who must travel will see their own country now if they never have before, and a neighborly visit to Canada will repay." Yet another called the Canadian Rockies "dear to the 'See America First' traveler."[12] There was even a series of "See America First" travel books that integrated Canadian destinations. Such inclusiveness—a willingness to extend a national tourism campaign beyond the nation—cannot simply be ascribed to either American cultural imperialism on the one hand or neighborliness on the other. Instead, it indicates an unconscious calculation that the potential enjoyment experienced by U.S. travelers from a trip to Canada (and in particular in this period, its Rocky Mountains) outweighed any negligible economic or philosophical damage that might come of undercutting the foundation of the national campaign. Canada was American enough, the

Canadian tourism industry insignificant enough, that the northern neighbor could be embraced. Canada, then as now, seemed pretty harmless.

But in September 1911, the same month as the *Times'* "America's Switzerland" piece, two events occurred in the United States and Canada that would ultimately transform tourism promotion in and between the two nations by transforming their national park systems. In Yellowstone National Park the first-ever parks conference was held, with government officials, railroad companies, and private sector advocates meeting to discuss how to better administer the parks and attract more visitors to them. The U.S. parks had until then been managed among the Departments of Interior, War, and Agriculture, with the result being little communication, let alone coordination. A campaign to bring the parks under the control of one bureau had begun a year earlier and found its first concrete formulation at Yellowstone.

Meanwhile, in Ottawa, a Dominion Parks Branch within the Department of the Interior actually began operation. That Canada established an agency devoted to national parks five years before the United States does not mean its park system was better developed. Beginning when the 1885 act making Banff Canada's first national park mimicked the wording of the 1872 Yellowstone act, Canada's park system had tended to imitate the American one, including its shortcomings. Canada established a small series of national parks in the Rockies in the late nineteenth and early twentieth centuries by a variety of mechanisms, under a variety of regulations, and under no central control. The parks were generally treated with benign neglect. In 1908, at the height of the conservation movement, they were placed under the care of the Forest Branch, but the change spoke to the growing importance not of parks but rather of forest reserves, which had also sprung up in the late nineteenth century. To conservationists, forest reserves seemed to be places where the conservation of forests, wildlife, and water could all be achieved together, and so the reserves were taking on many of the resource management goals that would later be associated with parks. When in 1911 the relationship between the two government properties was formalized in a Dominion Forest Reserves and Parks Act, the creation of a separate Parks Branch seemed almost an afterthought. The new bureau would have been quite justified to have defined its responsibilities conservatively, as being whatever the Forest Branch was not already doing, in whatever parks already existed.[13]

The Parks Branch that set up shop in Ottawa in September 1911 had a staff of just seven employees (1 percent of the overall Department of the Interior), a budget of $200,000 (4 percent), and a mandate to systematize and manage a half dozen national parks 3,000 miles away. Heading the new agency was Commissioner J. B. Harkin, a former newspaperman who had served for the previous ten years as private secretary to successive ministers of the interior in Sir Wilfrid Laurier's government.[14] That Laurier's government was swept from office within days undoubtedly made Harkin's new job more difficult. Harkin is attributed with saying that he became commissioner knowing nothing of parks, a claim that has had a surprisingly important role in turning him into one of Canada's environmental heroes: his initial ignorance only magnifies the significance of his conversion to conservation, demonstrating the transformative power of parks. While Harkin may have been speaking with excessive modesty—he had been active at a senior level in Interior for a decade, after all—his statement was undoubtedly fundamentally true, because no one anywhere knew much about national parks. There were individuals who knew individual parks, of course, but there had never before been an agency overseeing an entire park system. In the words of Mabel "M. B." Williams, who arrived as a clerk in 1911 but went on to become the park system's principal guidebook author and in retirement its first historian, "There was little in the new office at Ottawa to serve for guide or inspiration."[15]

J. B. Harkin directed his staff to find out as much as possible about matters relating to national parks. What was the purpose of parks and how could they be justified? What should their policies and regulations be? Relevant Canadian literature was pored over and other Canadian government departments contacted, but it was clear from the outset that the U.S. national parks would serve as the best comparison for Canada's. The United States had the world's first parks and the largest number of them, and its proximity to Canada made comparison easy and apt. Although Canada was most assuredly a British nation, the mother country was not seen as offering a relevant model in terms of parks—or environmental concerns generally, for that matter. One would be hard-pressed to find mention of Great Britain in Canadian park correspondence of this era.

When the Dominion Parks Branch was just two months old, a letter was sent under Harkin's signature[16] to the U.S. Department of Interior saying that the branch was "greatly interested in the work being done by your Government in the administration and development of the National Parks of the United States. Our information in this regard, however, is somewhat indefinite and I write to ask you if you would send us copies of any publications or reports issued by your Department which contain some account of your work."[17] In effect, could you tell us everything about your park system and how to run one? And could you include a map? The response was encouraging enough that a regular stream of requests followed, with the letters soon being channeled to the U.S. Interior Department's chief clerk, Clement S. Ucker. For example, on October 17, 1912, Harkin wrote Ucker asking for a blank copy of the superintendents' report form. Two days later he wrote seeking budget information about the parks, especially Yellowstone and Glacier. Two days later he wrote yet again, this time for information on the economic value of tourism.[18] Harkin also contacted the superintendents of individual U.S. parks for advice on matters such as predator control.[19] The American authorities responded politely and obligingly to these early requests—and on rare occasions with requests of their own, as when Ucker asked for and received the parliamentary debates leading up to the 1911 park act.[20] But in general the Americans shied away from anything that might suggest parity between the two park systems. When Ucker asked his superior about having the two park systems share photo collections for public exhibition, he was told to let the matter drop. "A large portion of the tourist travel to the Canadian national parks comes from the United States," he was told, "and, while I do not consider it advisable for this office to make any concerted movement in opposition to the Canadian parks, I consider it inadvisable for this office to have any part in advertising the Canadian reserves."[21] That Canada had the first and, for the moment, only park service did not lead inevitably to a sense that the two park systems were kindred.

The Dominion Parks Branch set about in its first years creating a coherent philosophy, one that would serve as the basis for descriptions and defenses of parks for decades to come. Knowing that parks appropriations had to be justified to Parliament and people, the agency fixed upon tourism as the prime economic justification for parks, and spiritual, mental, and physical betterment as the justification for tourists coming in the first place. This general theme, always bearing Harkin's name and so putting a focus to the branch's efforts, was then communicated in newspaper columns, magazine articles, and memos to the minister and prime minister. Most imaginatively, the commissioner's annual reports became the agency's key forum to lobby the government for more money and to reach the broader public. The reports were sent to every member of Parliament and newspaper in the country—effectively turning a mandatory accounting into a publicity campaign—and earned favorable responses in both the House and editorial pages.[22]

The United States figured heavily in these reports. In just the second paragraph of his first report Harkin stated, "It is interesting to note that the United States is following Canada's example in the matter of specializing in regard to National Parks administration," and cited U.S. president William Howard Taft on the need for a national park service.[23] In this and subsequent reports Harkin quoted extensively from American sources, such as John Muir on the national park idea, J. Horace McFarland on the value of parks, or "an American writer" on the nature of modern existence.[24] British sources were quoted, too, but the American example plainly had special significance because of the proximity and similarity of the two North American nations. Harkin noted when quoting McFarland, "His remarks, of course, referred to Americans, but change the word 'American' to 'Canadian,' and the concluding portion of his address crystallizes a thought of equal application to Canada."[25] The United States was genuinely a model but it was also a shortcut, allowing the Canadian park system to develop a philosophy more easily, without having to build one entirely from scratch.

The commissioner's reports also occasionally offered a sense of the Americans as rivals. Harkin quoted a University of Toronto geologist and alpinist at length on why the Canadian Rockies outshone the American ones. In another instance, the commissioner went so far as to say, "In the United States there has been an active movement whose slogan is: 'See America First.' The slogan which Canada's outstanding advantages in the way of natural scenic and other attractions justify using in regard to parks' development in Canada is: 'See America's Best.'"[26] This latter statement was made in October 1914, by which point Harkin may have felt some justification for swagger. Just three years after coming into being, the branch was experiencing remarkable growth. It was finding considerable success in drawing public attention to the parks' existence, and as a result attendance was climbing. But it was government attention that was really skyrocketing. In fact, as early as 1913 Canada was devoting more funds to its national park system than the United States was to its. The Dominion Parks Branch's budget grew from $200,000 in 1911–12 to $680,000 in 1914–15—a rise from 4 to 9.5 percent of the departmental budget—while the appropriation for parks in the United States increased from $375,000 in 1911 to $540,000 in 1914. Staffing in the Canadian bureau climbed from seven to thirteen in the same period; it would be twenty-one a year later.[27] New parks had already been created, with more on the way.

The new Parks Branch's successes were not entirely due to its own efforts: it benefited greatly from two factors for which it could not take credit. First, because the parks had been established in part to draw traffic on the Canadian Pacific Railway, they were well integrated with it and its thriving marketing campaign for the "Canadian Pacific Rockies." Although the Canadian park system was not as well developed as the American one, the CPR was every bit the match for railway companies to the south.[28] At the 1911 Yellowstone Conference, for example, Louis W. Hill, president of the Great Northern Railway Company, complained, "Thousands of Americans go to Canada every year for things they might just as well get in the United States. . . . The reason for it is the advertising which is being done by the Canadians."[29] Second, whereas the American parks stretched all across the West, from Mount Rainier, Washington, to Hot Springs, Arkansas, the Canadian system was smaller and relatively compact, with the core parks—Banff, Glacier, Yoho, Waterton Lakes, Jasper—all in the Rockies, and so easier for tourists to tour and easier for the managers to manage.[30] Nevertheless, what the Harkin-led Dominion Parks Branch was starting to achieve could not help but draw notice in the United States. Late in 1914 U.S. Secretary of the Interior Franklin K. Lane asked Stephen Mather to run the American park system. Lane is quoted as saying: "I'm looking for a new kind of public official, one who will go out in the field and sell the public on conservation, then work with Congress to get laws passed to protect the national parks. The job calls for a man with vision." He might well have been describing J. B. Harkin. Given that Lane was born in Canada and was well aware of happenings there, perhaps he was.[31]

"Canada Points Way." So the *Christian Science Monitor* stated in a 1913 article about a bill being introduced in Washington to have the national parks administered by a single agency. Supporters of Utah senator Reed Smoot's bill were calling attention to Commissioner Harkin's first annual report and commending Canada for recognizing national parks' potential, "thus opening the way to the many tourists who are quite ready to 'see America first,' once the necessary conveniences and comforts are provided." The newspaper then quoted a long passage from Harkin's report that offered evidence of the economic value of tourism, and therefore parks.[32] That Harkin's examples were largely American went unmentioned.

Even as the Canadian park authorities sought advice from their American counterparts throughout the early to mid-1910s, the American media began regularly citing the Canadian example when discussing what their own nation's park system needed. It seemed so clear-cut: an agency solely devoted to parks meant more attention to parks, which meant better advertising and services, which meant more tourists, which meant more revenue. The *New York Times* ended an article on the American parks simply, "By the way, our Canadian neighbors have a Dominions Parks Commission that works excellently. One result of it is millions of money from tourists."[33] The details were not all that important, and so not always accurate, as when a 1915 *Forest and Stream* article stated that Canada was spending more than twice as much that year as the United States on parks.[34] While the authors of such pieces were admiring of what Canada was up to, they also revealed a competitive streak—and evidence that they were not above shaming Americans for inactivity. After expounding on the Canadian model, the American Civic Association stated, "And while this has been done by Canada, the American parks as a rule have been all but neglected, notwithstanding they offer as fine scenery as is to be found in Canada and much wider diversity than Canada affords."[35] Similarly, when the *Oregonian* maintained that "Even Canada has gone far ahead of the United States in the improvement of its national parks, and for that reason Canada is getting a large share of the tourist travel that normally would be ours," the opening "Even" can be read as a slur on both countries.[36]

By 1916 bills to create a U.S. park agency had failed to make it out of committee five years running, defeated by opposition from the Forest Service, but the need for such an agency seemed greater than ever with the campaign for one coming to a head. The war in Europe was keeping Americans at home, and as a result the parks were enjoying unprecedented popularity: attendance rose from 240,000 to 334,000 between 1914 and 1915 alone.[37] There were bills put forward to create sixteen more parks in 1916. Stephen Mather, tasked by Interior Minister Lane to run the parks, and Robert Sterling Yard, the journalist hired by Mather (largely with his own money) to promote the parks, launched an all-out publicity offensive on behalf of a park agency early in the year.[38] Canadian competition figured heavily in that campaign, particularly because of the prevalent view that Canada had poached tourists from the United States the summer before. With Europe closed to travelers, promoters of the 1915 Panama-Pacific Exposition in San Francisco had seized upon the "See America First" idea to a degree and with a success unprecedented; in the words of historian

Marguerite S. Shaffer, "If one had to pick a moment at which touring the United States became feasible and fashionable, this was it."[39] But the Canadian Pacific Railway had also seen the opportunity and promoted its line as the best route for traveling to or from the exposition, with stops in the Canadian national parks along the way. The Canadian railway and parks profited hugely.[40] Even to the limited degree to which Canada actually had siphoned off American tourism, this could well have been interpreted as a shrewd but isolated business decision. Instead, it was held up as a sign of growing Canadian dominance in the tourism field. It was repeatedly stated, as quoted in *Outlook,* "Canada, through its Department of Parks, has so successfully exploited its possessions that during the season of 1915, when there was such a large volume of travel through the West, Canadian parks attracted in the aggregate more visitors than the parks of the United States."[41] The fact that no commentator suggested the flourishing Canadian park system was actually a threat to the long-term health or viability of the American one, however, implies that the cross-border comparison was being made fundamentally as a way to raise awareness about the lack of a coordinated system in the United States. Disappointment at opportunity untapped remained the dominant trope. Robert Sterling Yard, in a prominent article in *The Nation's Business,* described how Americans had long gone to Europe, in particular Switzerland, not knowing about the beauty of the scenery at home. He then told of how Canada had "entered the scenery business," adopting and improving on Swiss methods. "And the results?" he asked. "These countries, Switzerland and Canada, today share the scenic reputation of the world. . . . Abroad, the Canadian Rockies are supposed to be the only scenery in North America worth looking at, and before last summer (and perhaps today) it was the fixed belief of most Americans that they excelled anything in the United States." Yard never mentioned Canada's national parks per se—why give them free advertising, after all. But he compared the treatment accorded Canadian scenery to that of American national parks, lamenting how underpublicized those parks have been. "What has the United States been doing with them? Conserving them. Just conserving them."[42]

With the Dominion Parks Branch thriving and the Americans taking notice, Commissioner J. B. Harkin saw a chance in the winter of 1916 to bring the two park systems closer together. He had received a letter from Stephen Mather, who a year into the job of running the U.S. parks was making a study of parks around the world. Harkin responded to Mather with an intense and moving statement as to the meaning of parks. He argued that national parks must move away from being managed largely as scenery, and instead focus on how they can improve human welfare. "It seems to me that national parks work will be a failure unless it contributes very largely to the development of our human units with respect to their physical, mental and moral well-being." He concluded by committing to cooperate fully with Mather: "I know that every new development that you may succeed in bringing about with respect to your parks will undoubtedly help me in bringing on the same or other developments with respect to ours. I assume that the same is true with regard to your case. There is such a tremendous field in relation to human welfare which could be developed in association with national parks work that personally I sometimes feel appalled at the prospect."[43] Harkin's letter was arguably the most complex expression of parks that he would ever offer, one simultaneously idealistic and pessimistic, and one that perhaps could only have been formulated at the height of a world war. That it was not written for a public audience indicates it may also have been his most heartfelt. Its exposed nature suggests, too, how valuable Harkin felt it would be to bond personally with Mather. But rather than take up Harkin's implicit invitation to discuss parks philosophically, Mather steered his response to the more prosaic matter of having the two nations' contiguous parks, the U.S.'s Glacier and Canada's Waterton Lakes, share a road.[44] While American park authorities were happy to work with the Canadian ones, they showed no interest in forming a special relationship.

The April 1916 hearing before the House Committee on Public Lands was the final push for those advocating the establishment of a U.S. national park service. As in recent years, the existence and accomplishment of the new Dominion Parks Branch was an extremely useful precedent to cite. The hearing's first speaker was Richard W. Watrous, secretary of the American Civic Association, and within minutes he embarked upon an almost 2,000-word analysis of the Canadian case by stating, "Canada has been ahead of us on the national park proposition in every respect—in almost every respect." As several more speakers would, he cited the case of the Panama-Pacific Exposition, claiming "the Canadian national parks, because of their exploitation, and because of the things that had been done to make them ready for the comfort and convenience and safety of the tourists, drew the great, wholesale travel—I learned on very good authority that of the travel which went west about 75 per cent was routed, either going or returning, by Canadian railroad systems. . . . That meant thousands upon thousands of dollars of cold American cash for Canada, to be credited to its parks." Watrous then quoted Commissioner Harkin at length, both from a recent speech in Washington and from his first annual report—a "beautiful little report"—as to the economic value of parks. Those who followed Watrous backed him up. Stephen Mather spoke admiringly of Canada's advertising campaigns. P. S. Eustis of the Burlington Railroad claimed that in respect to the integration between railways and parks, "They seem to have better park arrangements in Canada than we have in the United States. We are rather scattered." Publicity agent Robert Sterling Yard focused on Canada's publicity effort, describing how "a few years ago Canada went into the scenery business" and stating somewhat sullenly, "Till then in this country every man, woman, and child had been brought up to the belief that the greatest scenery of the world was in Switzerland; and now, in the last few years, they have also added the Canadian Rockies. That is the great word in this country to-day—the Canadian Rockies."

American Civic Association president J. Horace McFarland concluded bluntly that in terms of developing the park system, "Canada beat us to it."[45]

The bill to create a U.S. national park service was passed into law in August 1916, not without some of the opposition that had caused it die out on the committee floor the previous five years, but with a growing acceptance, thanks in large part to the Canadian example, that the benefit of having a single agency to harmonize, coordinate, and develop the park system was self-evident. It hardly mattered that the statistics that had been repeatedly bandied about concerning the rapid growth of the Canadian national park system were short-lived or just plain wrong. While it was true that the Canadian parks received more funds than the American ones did in 1913–14 and 1914–15, this advantage disappeared almost at the moment Americans started citing it. With Canada's first wartime budget in 1915–16, the appropriation for parks dipped from $681,000 to $349,000, even as the United States increased its support for parks to $610,000. It would not be until 1920–21 that the Canadian park system returned to the funding level it had enjoyed six years earlier, at which point the American parks budget had left it far behind.[46] Likewise, the widely reported statement that Canadian national parks received more visitors than the American ones was mistaken. The U.S. parks had begun the 1910s with systemwide attendance of 200,000; it would take Canada more than a decade to reach that number, at which point the U.S. parks had reached one million. Commissioner Harkin himself provided the most specific and trustworthy comparative figures when he claimed that in 1915 the U.S. parks welcomed 278,000 visitors while the Canadian ones welcomed 121,000.[47]

Regardless of the underlying reality, both national park systems had been well served by the appearance that they were becoming well-matched rivals. The Americans could use this as ammunition to help bring an agency into existence, and the Canadians could use it to prove the worth of the agency they already had. J. B. Harkin included in his 1917 annual report a section titled "Canada's Success," in which he quoted all five American speakers who had praised Canada during the U.S. hearing into the national park service bill the year before; it was a case of Canada using the United States using Canada.[48] But whether this temporarily symbiotic relationship would turn into a permanent, shared sense that they were equals was another matter. Perhaps to express thanks for the role the Dominion Parks Branch had played in its creation, the new U.S. National Park Service invited Harkin to speak at its introductory conference in Washington early in 1917. Harkin dedicated his talk to outlining why the two park systems, which on first sight might appear to be competitors, were in fact mutually beneficial. "I hold, sir, that every man who visits one national park, whether it is in Canada or the United States, will thereby get a taste for what national parks alone can give him and that he will never be satisfied till he has visited other parks." Harkin thus rejected American magazine articles that spoke in catastrophic terms about Americans visiting Canadian parks. "To me it seemed that they should in reality rejoice that so many Americans were getting a taste for national parks and, if you will pardon me for saying it, getting such a good start on the right line. . . . Personally, I rejoice to see the figures regarding visitors to your parks jump up, because I then feel perfectly certain that a considerable proportion of those people are going to visit our parks eventually." Harkin dearly wanted his American counterparts to take the Canadian park system seriously. It must have seemed a bad sign that when the conference proceedings were published, his name was given as "Harker."[49]

In the spring of 1918, the new U.S. National Park Service produced "the Lane letter," a major statement on policy objectives; Horace Albright would later call it "our basic creed." The Lane letter at one point advised the service to work with groups and associations involved in parks, tourism, and outdoor living. In its only reference to a specific international body, it stated, "In particular, you should maintain close working relationships with the Dominion parks branch of the Canadian department of the interior and assist in the solution of park problems of an international character."[50]

The U.S. bureau stayed true to that dictum in the years that followed, but the relationship that developed between the American and Canadian agencies was never one of equality. Although technically younger, the U.S. National Park Service had more experience, more staff, and more funding, and it oversaw a larger park system. The Dominion Parks Branch came to rely on it greatly when formulating policy. Predator control policy is a good example. The Americans supplied the Canadians with evidence that predators were a threat to park wildlife populations; the Canadian agency's archival files are bulging with Yellowstone Nature Notes, Bureau of Biological Survey Monthly News, USDA bulletins, and other material provided by the National Park Service. The Americans also supplied the Canadians with testimonials about spring traps and on at least one occasion even sold them poison.[51] As a result of such reliance, park policy in the two nations grew quite similar, even more so than their similar environmental and cultural conditions would anticipate.[52] But by essentially subcontracting expertise to the United States, Canada failed to develop sufficient homegrown expertise; by the time the Canadian Wildlife Service was finally created in the late 1940s, for example, Canadian wildlife science was significantly behind that of the United States. This likely only reinforced a preexisting feeling within the American service that it had little to learn from the park system to the north. Whereas there were many requests over the years for information and advice from the Canadian park agency to the American one, there were few traveling in the opposite direction. It is surely telling that the Dominion Parks Branch came to file its correspondence with other park agencies as either "foreign" or "American," while the U.S. National Park Service adopted the single category "foreign."

But one can read the American-Canadian relationship that developed in another fashion: maybe the Americans were not just ignoring, but studiously ignoring, Canada's park program. Over the first half of the 1910s, American interests had moved from thinking Canada so harmless that it could be included within "See America First" to considering it a genuine rival, one whose park service and whose acumen in attracting funds and tourists had to be respected. In 1916 Robert Sterling Yard had spoken enviously in magazines and at the park service hearings about Canada's parks and publicity efforts. But in 1917 he published *The Top of the Continent: The Story of a Cheerful Journey through Our National Parks*—an unusual hybrid, a children's guidebook, that from its title onward seemed to forget Canada's very existence. Canada appears only once in the entire book, in passing. Yard's protagonists are hiking in Montana's Glacier National Park, which borders not just Canada but also its Waterton Lakes National Park. Yard writes, "Thousands of feet below them lay the broad green Waterton Valley, dotted with lakes and backed by splendid glacier-shrouded heights. To the north lay the lesser Canadian Rockies."[53]

"Lesser." For an American booster such as Yard to belittle Canada's scenery was indication of how much Canada had achieved over the previous few years. It was, in its way, the highest of compliments.

Notes

1. In this chapter I use "America" to refer to the United States of America, never North America. It would seem either imperialistic or lazy to refer to the continent and a nation on that continent interchangeably; Africa is not synonymous with South Africa. Given that the essay involves discussion of the somewhat plastic historical use of the term "America," it might seem preferable that whenever possible I simply avoid the term altogether. However, that would mean continually employing awkward constructions such as "that of the United States."

   A version of this chapter's introductory section appeared on the NiCHE: Network in Canadian History & Environment blog *The Otter* in October 2011: http://niche-canada.org/2011/10/23/who-had-americas-best-idea/ (accessed August 4, 2015).

2. Ken Burns, director, *The National Parks: America's Best Idea,* 2009. See also the accompanying Dayton Duncan and Ken Burns, *The National Parks: America's Best Idea* (New York: Knopf, 2009), xxii, 252, 353, 386.

3. Wallace Stegner, "The Best Idea We Ever Had: An Overview," *Wilderness* 46 (Spring 1983): 4–13. There is more evidence that the phrase already existed. In *The National Park Service* (Boulder, Co.: Westview, 1983), William C. Everhart recalls conservationist Mardy Murie "a few years ago" saying of national parks, "I wonder if it is not the best idea the U.S.A. ever gave the world" (157).

4. Stegner, "It All Began with Conservation," *Smithsonian* 21, no. 1 (April 1990): 34–46, reprinted in *Where the Bluebird Sings to the Lemonade Springs: Living and Writing in the West* (New York: Random House, 1992), 128.

5. Rt. Hon. James Bryce, "National Parks: The Need of the Future," American Civic Association series 11, no. 6 (December 1912), 6–13. The speech was reprinted in both *The Outlook,* December 14, 1912, 811–15, and Bryce, *University and Historical Addresses* (New York: Macmillan, 1913), 389–406.

6. See, for example, John Sheail, *Nature's Spectacle: The World's First National Parks and Protected Places* (London: Earthscan, 2010), 71, or the website of Parks Canada itself, http://news.gc.ca/web/article-en.do?nid=653759 (accessed August 29, 2015). Sweden lays claim to legislating the first national park *system* in 1909. See Tom Mels, *Wild Landscapes: The Cultural Nature of Swedish National Parks* (Lund, Sweden: Lund University Press, 1999), 81. Elsewhere in this volume, Paul Sutter discusses the problems, if not the ridiculousness, of attempting to establish primacy.

   The Canadian national park agency went through a number of names over the years, becoming known as its present-day Parks Canada in 1973. I use "Dominion Parks Branch" throughout, the name it held for its first decade.

7. Only the most rabid of Canadian nationalists would at this point mention Stegner's Saskatchewan roots.

8. "America's Switzerland," *New York Times,* September 17, 1911.

9. The conference concluded, "First—This continent is annually drained by an immense sum of money spent on foreign travel. Second—That it would mean much to the United States, Mexico, and Canada if even a portion of this money could be diverted into home channels of circulation. . . . Fifth—That it is possible, by concerted action of some sort, to correct this condition and to accomplish thereby a vast amount of good for the entire continent of North America." Marguerite S. Shaffer, "'See America First': Re-Envisioning Nation and Region through Western Tourism," *Pacific Historical Review* 65 (November 1996): 571. See also Shaffer, *See America First: Tourism and National Identity, 1880–1940* (Washington, D.C.: Smithsonian Institution Press, 2001).

10. The lyric continues, "Cheer for America / And get that grand old strain of Yankee Doodle / In your noodle; / Yell for America, / Altho' your vocal cords may burst, / And if you ever take an outing, / Leave the station shouting: / 'See America First!'" Robert Kimball, ed., *The Complete Lyrics of Cole Porter* (New York: Knopf, 1983), 43.

11. Perhaps more remarkable, her book does not mention the Cole Porter play.

12. "See America First," *Chicago Defender,* May 10, 1913, 7; "Seeing All America," *Chicago Daily Tribune,* May 3, 1915, 8; "A New Gateway to the Pacific," *The Independent,* January 3, 1916, 18. See also, for example, "See America First," *New York Times,* October 8, 1911; "The Banner Year of Tours Begins," *The Independent,* April 5, 1919, 36.

13. I discuss the branch's establishment in greater detail in "M. B. Williams and the Early Years of Parks Canada," *A Century of Parks Canada, 1911–2011,* ed. Claire Elizabeth Campbell (Calgary: University of Calgary Press, 2011), 21–52, especially 27–29. See also C. J. Taylor, "Legislating Nature: The National Parks Act of 1930," *To See Ourselves/To Save Ourselves: Ecology and Culture in Canada,* ed. Rowland Lorimer et al. (Montreal: Association for Canadian Studies, 1991), 125–38; R. Peter Gillis and Thomas R. Roach, *Lost Initiatives: Canada's Forest Industries, Forest Policy, and Forest Conservation* (Westport, Conn.: Greenwood, 1986), especially 62–70; Peter J. Murphy, "'Following the Base of the Foothills': Tracing the Boundaries of Jasper Park and Its Adjacent Rocky Mountains Forest Reserve," *Culturing Wilderness:*

Studies in Two Centuries of Human History in the Upper Athabasca River Watershed, ed. I. S. MacLaren (Edmonton: University of Alberta Press, 2007), 71–121.

14. On Harkin, see E. J. (Ted) Hart, *J. B. Harkin: Father of Canada's National Parks* (Edmonton: University of Alberta Press, 2010); but also Alan MacEachern, *Natural Selections: National Parks in Atlantic Canada* (Montreal-Kingston: McGill–Queen's University Press, 2001), chap. 2.

15. M. B. Williams, *Guardians of the Wild* (London: Thomas Nelson & Sons, 1936), 7. On Williams, see my "M.B.: Living and Writing the Early Years of Parks Canada," online exhibit and archive, http://mbwilliams.academic-news.org/ (accessed August 29, 2015).

16. It is in most cases impossible to know who actually wrote such letters. Staff drafted a considerable amount of the correspondence sent out under Harkin's name, and certainly letters to U.S. park authorities would have carried more weight if bearing the commissioner's signature. But for the same reason, Harkin may well have written many of these letters himself. Given that this chapter began by exploring how complicated stories become simplified, and given that I have argued elsewhere against attributing the Parks Branch's early development to Harkin alone (see "M. B. Williams"), I am tempted to repeatedly qualify the authorship of this correspondence. However, for the sake of simplicity, I will give credit throughout to Harkin, as the sender.

17. Harkin to U.S. Secretary of the Interior, December 18, 1911, Record Group 79, Box 630, File 0–30 Pt 1, General, Foreign Parks, Canada; National Archives and Records Administration, College Park, Maryland (hereafter RG 79). Historian Janet Foster goes so far as to call Harkin's attempts to open channels of communications with Washington his "first efforts as Commissioner." Foster, *Working for Wildlife: The Beginning of Preservation in Canada,* 2nd ed. (Toronto: University of Toronto Press, 1998 [1978]), 82.

18. Harkin to Clement S. Ucker, Chief Clerk, U.S. Secretary of the Interior, October 17, 19, and 21, 1912, RG 79.

19. RG 84 vol. 35, file U300 vol. 1, Library and Archives Canada, Ottawa, Ontario.

20. See Harkin to Ucker, January 20, 1913, RG 79.

21. Laurence F. Schmeckebier, Clerk in Charge of Publications, Department of Interior to Ucker, February 17, 1913, RG 79.

22. MacEachern, "M. B. Williams," 30–33.

23. Harkin, July 4, 1912, "Report of the Commissioner of Dominion Parks" [henceforth "Report"] in Canada, *Sessional Papers* vol. 25, 1913, 3, http://eco.canadiana.ca/view/oocihm.9_08052 (accessed August 5, 2015).

24. Harkin, September 30, 1913, "Report," in Canada, *Sessional Papers* vol. 25, 1914, 5, 7–9; Harkin, October 1, 1914, "Report," in Canada, *Sessional Papers* vol. 25, 1915, 6.

25. Harkin, September 30, 1913, "Report," in Canada, *Sessional Papers* vol. 25, 1914, 8. Likewise, "While Ambassador Bryce's words were addressed to Americans, it is obvious that they apply with equal force to Canadians." Ibid., 3.

26. A. P. Coleman cited in ibid., 9; Harkin, October 1, 1914, "Report," in Canada, *Sessional Papers* vol. 25, 1915, 4.

27. For Canadian appropriation and staffing figures, see Auditor-General Annual Reports, Canada, *Sessional Papers,* 1911–15, compiled and cited in C. J. Taylor, "A History of National Parks Administration," unpublished manuscript, 1989, Table 1 (no pagination). The American appropriation figures were determined by adding War Department and Interior Department spending on parks in *Reports of the Department of Interior,* 1918, vol. 1 (Washington, D.C.: Government Printing Office, 1919), 1011. But note that the Canadian and American figures are not fully comparable, both because the U.S. appropriations also take in national monuments and because the United States used calendar years and Canada fiscal years.

28. On the relationship between the Canadian national parks and the CPR, see E. J. Hart, *The Selling of Canada: The C.P.R. and the Beginnings of Canadian Tourism* (Banff: Altitude Press, 1983), especially chap. 7.

29. Some U.S. railway companies were conflicted because they traveled into Canada. For example, under the "See America First" motto the Northern Pacific ran advertisements that included British Columbia destinations. *Grand Forks Herald,* June 8, 1906.

30. Prior to 1911, only Saint Lawrence Islands National Park had been established east of the Rockies.

31. Lane, quoted in Horace M. Albright, as told to Robert Cahn, *The Birth of the National Park Service: The Founding Years, 1913–33* (Salt Lake City, Utah: Howe Brothers, 1985), 16.

32. "Federal Bureau of Public Parks Bill Advocated—Canada Points Way," *Christian Science Monitor,* March 26, 1913, 6.

33. "The National Parks," *New York Times,* December 18, 1915, 10.

34. "Conservation—or Conversation?" *Forest and Stream,* March 1915, 160.

35. "National Parks Greatly Need Improving," *Idaho Statesman,* December 2, 1915, 5. See also *Washington Evening Star,* cited in United States, *Congressional Record,* January 18, 1915, 1793.

36. "Park Appeals Denied," *Oregonian,* December 11, 1915, 4.

37. Jenks Cameron, *The National Park Service: Its History, Activities, and Organization* (New York and London: D. Appleton & Co., 1922), 137; *Reports of the Department of Interior,* 1918, vol. 1 (Washington, D.C.: Government Printing Office, 1919), 1004.

38. On the run-up to the creation of the National Park Service, see Albright, *Birth of the National Park Service,* chap. 3.

39. Shaffer, *See America First,* 33–36, quotation on 36.

40. See Hart, *Selling of Canada,* 97–98. Like many across the Western world assuming that the war would be over quickly, Canadians were not fully in war mode in 1915. It is nevertheless remarkable to read a Canada Steamship Lines representative tell the *Washington Post,* in encouraging travel to Canada, "Contrary to general belief, Canada is not at war, except, of course, in so far as active cooperation with Great Britain in making certain of a successful issue to the great struggle can be called being at war." *Washington Post,* February 13, 1915, 6.

41. "Wanted, a National Parks Service," *Outlook,* March 1, 1916, 491. Very similar phrasing appears in, for example, "A National Park Service," *The Independent,* May 29, 1916, 321, and *Kansas City Star,* June 1, 1916, 2.

42. Robert Sterling Yard, "Making a Business of Scenery," *Nation's Business,* June 1916, 10–11.

43. Harkin to Mather, February 8, 1916, RG 79.

44. Mather to Harkin, February 16, 1916, RG 79. As Karen Routledge makes clear in her chapter on Waterton Lakes and Glacier in this volume, Canada–U.S. parks cooperation happened more readily at the local than the national level.

45. All quotations from *National Park Service Hearing before the U.S. Congress House Committee on the Public Lands, House of Representatives, 64th Congress First Session, on H.R. 434 and H.R. 8668 Bills to Establish a National Park Service and for Other Purposes, 5–6 April 1916* (Washington, D.C.: Government Printing Office, 1916), 5–9, 35, 66, 61–62, 54. Available at http://www.archive.org/details/nationalparkser00unkngoog (accessed August 5, 2015).

46. For Canadian figures, see Canada, House of Commons, *Debates,* June 8, 1920, 3283. For American figures, see *Reports of the Department of Interior,* 1918, 1011.

47. Harkin, "Report," in Canada, *Sessional Papers* vol. 25, 1918, 4. For the Canadian figures, see the Branch's 1911–16 and 1922 annual reports. However, attendance is often listed only for Rocky Mountains National Park, or simply "Banff." For American figures, see Cameron, *National Park Service,* 137–38.

48. Harkin, "Report," in Canada, *Sessional Papers* vol. 25, 1917, 4.

49. Hon. J. B. Harker, "Canadian National Parks," delivered at the National Parks Conference at Washington, D.C., January 5, 1917 (Washington, D.C.: Government Printing Office, 1917). http://www.biodiversitylibrary.org/item/83097#page/3 (accessed August 5, 2015). Actually, the name was wrong in two ways, in that Harkin was neither Honorable nor a Harker.

50. Albright himself was in fact the initial and primary author of the statement, and after feedback was provided by other senior administrators and prominent conservationists, it was issued as a letter from Secretary of the Interior Lane to Director Mather. See Albright, *Birth of the National Park Service,* 68–73.

51. RG 84 vol. 162, file W266, Library and Archives Canada.

52. This could be a problem, as for example in the 1920s when staff of the U.S. Biological Survey defended the local extermination of large predators by maintaining that the targeted series would continue to thrive in Canada and Mexico. Since so much of Canadian wildlife management policy grew out of the national parks, in effect Canadian policy was being formulated by imitating American policy, which was itself being defended on the assumption of independent Canadian policy making. Alan MacEachern, "Rationality and Rationalization in Canadian National Parks Predator Policy," *Consuming Canada: Readings in Environmental History,* ed. Chad Gaffield and Pam Gaffield (Toronto: Copp Clark, 1995), 197–212.

53. Robert Sterling Yard, *The Top of the Continent: The Story of a Cheerful Journey through Our National Parks* (New York: Charles Scribner's Sons, 1917), 101. For the record, the Canadian and American Rockies actually have quite similar prominence, the distance from base to summit, but the Canadian mountains begin and end at a lower altitude.

| # A Short History of the New Zealand National Park System

THEODORE CATTON

NEW ZEALAND IS A SMALL COUNTRY with world-renowned scenery. Its natural beauty is associated with peaceful pastoral landscapes and empty beaches, with fantasy (especially as brought to the silver screen in *The Lord of the Rings* and *Narnia* movies), and perhaps most of all with rugged mountains. Snowcapped peaks, massive glaciers, and majestic fjords characterize the mountain topography in South Island, while volcanoes are the most prominent mountain feature in North Island. No fewer than fourteen national parks showcase New Zealand's superlative mountain scenery. Two of these encompass North Island's high volcanoes, two more take in areas of North Island's rugged, forest-clad hill country, nine are located in the mountain cordillera known as the Southern Alps that forms the spine of South Island, and the last one covers three-quarters of Stewart Island, which is located off South Island's southern tip.[1]

New Zealand's national parks, like the nation's superlative scenery, have world stature. Not only are some of the areas internationally famous as recreation and tourist attractions, but they contain a wealth of natural and cultural resources under government protection. New Zealand's first national park and the world's fourth, Tongariro, which centers on North Island's main volcanoes, is now a World Heritage Site. A grouping of four other national parks in the southwest corner of South Island forms a second World Heritage Site, Te Wahipounamu. Altogether, New Zealand's fourteen national parks cover an area of 11,841 square miles, or more than 11 percent of the nation. By world standards, this is a very generous allocation of the national domain for the preservation of nature.

As in the United States, where the first national park was established in 1872, the purpose of New Zealand's national parks is complex and layered, deriving from many sources. The conception of these special places owes much to world culture: to the romantic age of the eighteenth and nineteenth centuries and its sanctification of nature; to the wilderness heritage that New Zealand shares with other settler societies in North America and around the globe; to the remarkable growth of tourism since the Victorian age, and more recently, the growth of ecotourism; to the rise of ecology and environmentalism; and to considerable technical guidance provided by the U.S. National Park Service. At the same time, New Zealand national parks are fundamentally an expression of New Zealand nationalism, as the very term *national park* would indicate. As such, the New Zealand national park idea can be seen as an outgrowth of New Zealand's history, political economy, and peculiar ecology.

This chapter considers how a small British Commonwealth nation in the South Pacific adapted the national park idea to its unique national and even local circumstances, and how the New Zealand national park idea evolved from the late nineteenth century to the present. I find that New Zealanders quite deliberately modeled their national parks after the U.S. national park system, but they also adapted the national park idea to serve their needs as a distinctive settler society within the British Empire and Commonwealth and, more recently, to suit their aspirations as a bicultural nation of Polynesian Māori and non-Māori, or Pākehā.

A Weird and Unsettled Ecology

Unlike other Pacific Islands, New Zealand is a landmass with ancient origins in the supercontinent of Gondwanaland. The islands of New Zealand ride on a tectonic plate that detached from Gondwanaland some 80 million years ago. When this plate moved off into Oceania, most of it lay beneath the surface of the sea; all but a sliver of New Zealand's present land area was uplifted from the bed of the ocean millions of years later. Separated from other landmasses by thousands of miles of water, New Zealand's assemblage of plant and animal species coevolved through millions of years in relative isolation from terrestrial life on the continents. As a result, New Zealand is now home to a native terrestrial flora and fauna containing many relic forms. A large proportion of its 50,000 plant and animal species are found nowhere else in the world.[2]

Before the arrival of humans, New Zealand was practically devoid of mammals. The land's only native mammals are three species of bat. New Zealand's birds, reptiles, amphibians, and insects came to fill many niches usually dominated by mammals. New Zealand's diverse birdlife once included the great Haast eagle, which occupied the niche of top predator, and many flightless birds such as the giant moa, a plant eater. Many other flightless birds, including the football-size kiwi, persist today in smaller numbers. In the absence of mammals, other critters such as frogs, lizards, and snails were relatively abundant. A family of large, crawling insects called weta took the place of small rodents. New Zealand's oceanic isolation and lack of mammals also gave rise to an unusual native flora. Its trees, shrubs, and grasses evolved without the evolutionary pressure of grazing and browsing animals. New Zealand forests or "bush" were peculiarly adapted to a rugged landscape swept by frequent rainstorms and beset by sometimes phenomenal rates of soil erosion.

New Zealand was the last, large land area in the world to be inhabited by humans. The Māori arrived around 1000 A.D., the Pākehā some eight hundred years later. There are living trees in New Zealand that started life before our species was part of the ecosystem. The human impact was profound. Since the advent of humankind in New Zealand, no fewer than thirty exotic species of mammals have become established, with two-thirds of these having major impacts on native species. Furthermore, both Māori and Pākehā set to work burning the forest. Though they had different cultural reasons for clearing the land, both peoples in turn contributed significantly to New Zealand's progressive deforestation. Both peoples were implicated as well in the extinction of numerous species—especially the disappearance of flightless birds—through overhunting, habitat degradation, and the unintended consequences of exotic mammal introductions. As stated in an anthology of scientific writings on New Zealand's biological invasions, "No other island nation has sustained that level of alien impact."[3]

Although the processes of Māori and Pākehā settlement were very different from one another, both were unsettling to the native ecology. In the modern view put forward by archaeologists and ecologists, Māori settlement constituted a first wave of human colonization that was essentially destructive toward New Zealand's indigenous flora and fauna. Even traditional Māori consider themselves colonizers of the land, albeit with a long tenure compared to that of the Pākehā. New Zealand's indigenous people do not possess the same sense of "deep time" with relation to the land as do Australia's Aborigines or East Africa's tribes. Nor do New Zealanders of European descent think of the Māori as a people of the land in the same way that Americans often regard Native American tribes. When Americans imagine nature in its pristine condition—when they idealize America as a "virgin land"—the place in their imagination is populated with Indians. When New Zealanders do the same for their country, it is a primordial world without humans.[4]

Still, the indigenous Māori of New Zealand occupy a place in modern environmentalism that is somewhat akin to Native Americans in the United States and Canada. That is, they have a claim to environmental stewardship based on a blend of moral and legal factors. Their moral claim is grounded in their aboriginal forms of land stewardship and the fact that their contemporary culture stands somewhat apart from the Western tradition of industrialized resource exploitation. Traditional Māori culture contains certain concepts of land stewardship such as *tapu,* which describes an object or area as sacred and inviolable, and *tāonga,* which describes a thing or a piece of land that is so precious it must be passed from generation to generation with great care and respect.[5] Their legal claim, meanwhile, rests on their relationship to the Crown as set down in the Treaty of Waitangi of 1840. Like the treaties made with North American Indian tribes, the Treaty of Waitangi opened their lands to purchase and settlement while reserving to them certain rights as the indigenous people. As in North America, the European settler society abused those rights in its headlong rush to secure the land and resources for its own use. As will be seen later in this chapter, modern Māori seek restitution for the historical abuse of their treaty rights as well as revitalization of their cultural connection to the land.

If prehistoric Māori were responsible for hunting the moa to extinction and for introducing New Zealand's first two exotic mammals (the Pacific rat and the dog), Europeans went much farther in disturbing the native ecology. When Europeans arrived toward the end of the eighteenth century, they initiated a second wave of human colonization and biological invasion. Captain Cook's circumnavigation of New Zealand in 1769–70 was followed by further explorations of the coastline from the 1770s through the 1790s and then the coming of whalers, sealers, and missionaries around 1800. These first European visitors introduced the cat, now a widespread feral carnivore, as well as the Norway rat, which eventually replaced the Pacific rat. Pākehā settlement of New Zealand began in the 1840s and from the beginning it was oriented toward clearing the native bush to create pastureland for sheep and other livestock. From an estimated 28 million acres in 1847 (about 42 percent of the land area) the amount of New Zealand forestland shrank to around 17 million acres in 1909.[6] As Pākehā settlers cleared forest and drained wetlands, they sowed the land with grass seed imported from Britain, creating monocultures where there had been diverse forest habitats. As the process of land conversion gained momentum, the amount of grassland doubled during the 1880s alone.[7] This was a landscape transformation without parallel anywhere in the American West over the same time period, except perhaps in California's Central Valley.

Besides importing livestock and grass seed, Pākehā settlers introduced game animals, songbirds, horticultural plants and freshwater fish in an effort to stock New Zealand with species that they found familiar or useful. As individual efforts to transport and release exotic species often met with failure, New Zealand's early colonists formed "acclimatization societies" to organize and sustain these introductions. The idea of propagating wild species in new environments was not unique to New Zealand; acclimatization societies were active in this period from Paris and London to California and Australia. However, they reached their apogee in the New Zealand setting. Among dozens of different species that the acclimatization societies introduced, the most important were the European red deer, the Australian brushtail possum, and the rabbit. The red deer eventually spread throughout New Zealand's mountain lands, while the possum found a new home in New Zealand's most vulnerable rain forests. In their new environments both animals established food habits that hinder forest reproduction. The rabbit, meanwhile, became a threat to the sheep industry, which led to the introduction of three predators—the ferret, the weasel, and the stoat—with devastating consequences for New Zealand's flightless birds.[8]

Ironically, the acclimatization societies called forth New Zealand's first conservation law, which aimed at prohibiting the hunting of game until such time as the exotic wildlife populations became firmly established. Subsequently, when the wildlife law was amended to permit hunting, it allowed for the registration of acclimatization societies to administer hunting licenses.[9] By the end of the nineteenth century, the acclimatization societies had changed their focus from importation of exotic species to management of the wildlife resource. In effect, they became local fish and game departments, collecting license fees, maintaining staff, and, as the decades passed, taking a role in habitat improvement projects. A senior wildlife administrator with the U.S. Forest Service who visited New Zealand in 1972 tried to fit the role of the acclimatization societies into the American context. "In a way," he observed, "it would be similar to a sportsman's club acting as a little state fish and game department for each county in a state."[10] This form of local control in New Zealand wildlife management set the pattern for New Zealand national park administration, as local park boards made up of private citizens were established to guide the development and management of each national park.

It is no exaggeration to say that New Zealand's national park system developed against a backdrop of biological invasion, faunal extinctions, and landscape transformation. Although settler societies around the world experienced similar profound changes in biota and landscape, the circumstances in New Zealand were more pronounced than elsewhere because the South Pacific archipelago was so remote from the rest of the world. Moreover, the Pākehā settlers' late-nineteenth-century awakening to their new home's unique fauna and flora coincided with the first stirrings of New Zealand nationalism. The movement to preserve threatened native bird and plant species was integral to New Zealand's transition from colony to commonwealth.[11] Indeed, New Zealanders derived their nickname "Kiwis" from the odd-looking flightless bird. Both the country's remoteness and the historical timing of European colonization ensured that the national park idea would develop there in a way that was peculiar to that nation.

Early Establishment of National Parks

In 1887, amid contentious proceedings of the Native Land Court in central North Island, a Māori chief named Te HeuHeu Tukino offered the summits of three volcanoes to the Crown for the purpose of making a national park in perpetuity for all the people of New Zealand. The government passed the Tongariro National Park Act, which held the gifted peaks in reserve until it could purchase enough surrounding land to make a national park. Twenty years later the government completed its task and proclaimed the national park. The park was enlarged in 1916, and expanded again in 1983. Conceived in 1887, it was only the fourth national park in the world.[12]

Tongariro National Park encompasses North Island's three central volcanoes and the high desert ecosystem surrounding them. Early proponents of the national park idea, including Te HeuHeu's son-in-law Lawrence Grace, a member of the colony's House of Representatives, and John Balance, the author of the first Tongariro National Park bill, saw the opportunity for attracting tourists. A central trunk railroad was then under construction that would pass nearby. New Zealand was already noted for its scenery, and a national park would make the beauty spot in the center of North Island famous around the world.[13]

New Zealanders established their second national park, Egmont, in 1900. Mount Egmont, or Taranaki as it is known today, is a nearly symmetrical volcano that rises 7,500 feet above the surrounding plain. The original purpose of this national park was to preserve the bush on the mountain's slopes as a catchment for the area's heavy rainfall. For this reason the farmers on the surrounding Taranaki Plain strongly supported the national park. The Egmont National Park Board, composed of representatives of the farming districts encircling the mountain, quickly went to work to develop the park with "tracks" (hiking trails) and simple visitor accommodations. Despite its utilitarian premise, Egmont National Park soon became the most visited mountain resort in New Zealand, demonstrating that national parks could stimulate tourism. Its 2,500 visitors in the summer of 1902–1903 amounted to about ten times the number of hotel guests who registered at the Hermitage near the base of Mount Cook in the heart of the Southern Alps that same year. Egmont's popularity was due mainly to its accessibility by rail from New Zealand's capital city of Wellington.[14] Although the visitor numbers might seem insignificant by today's standards, they were on par with those for Mount Rainier National Park in Washington State at that time, which was established one year before Egmont and like Egmont was served by a railroad from a nearby city.[15]

While the Māori concept of *tapu* formed the germ of New Zealand's first national park, and forest conservation and tourism development were the two main purposes served by New Zealand's second national park, the nation's third national park was conceived for still another purpose: to preserve a sample of New Zealand's native ecology. Arthur's Pass National Park straddles the Main Divide in the Southern Alps and centers on Arthur's Pass, the primary transportation route through the mountains. Although the national park was not officially proclaimed until 1929, the government set aside most of the area for national park purposes in 1901. The person principally responsible for the 1901 reservation was a New Zealand ecologist by the name of Leonard Cockayne. He was then the world expert on New Zealand's unique flora and was held in high esteem by European botanists. Following his success in getting a reserve established around Arthur's Pass, he became a leading voice for preservation of New Zealand's native ecology from the 1900s through the 1920s.[16]

Cockayne challenged one of the fundamental tenets of New Zealand ecology at that time, an idea known as displacement theory. Displacement theory held that New Zealand plant and animal forms were intrinsically inferior and doomed to give way to exotic species. In time, this theory predicted, exotic species would take hold and thrive in their new setting. Cockayne warned that this was fallacy. Exotic species might do well in the short term but in the long run they would not flourish in New Zealand's unique soil and climate conditions. New Zealanders should protect their natural heritage, he argued, for it was the land's native stock that would eventually show staying power. Cockayne's attack on displacement theory was especially pertinent for New Zealand forestry as the nation began to experiment on a large scale with plantation forests—mainly composed of exotic Monterey pine—to secure its future timber supply.[17]

Cockayne gave scientific substance to a sentiment already shared by many lay New Zealanders by the turn of the twentieth century. Most New Zealanders were by this time either native born or thoroughly accustomed to their new home; consequently, they showed less interest in reproducing the English countryside and more interest in protecting what was native. The idea of preserving nature began to gain a significant following in New Zealand in the 1890s. Early efforts to protect native birds focused on New Zealand's offshore islands, which were still ecologically nearly pristine. The government established four different island reserves during the decade. Meanwhile, efforts to protect remnants of native bush began with the formation of numerous scenery preservation societies. These local groups attracted such a popular following that the government passed the Scenery Preservation Act 1903.[18] The scenic reserves that sprang from this law varied greatly in size and purpose: most were quite small and protected no more than a roadside stand of native bush or wetland, while others covered extensive areas and formed the basis for future national parks. Notably, in 1905 the government established a vast reserve in the southwest corner of New Zealand that eventually became known as Fiordland National Park. The reserve encompassed 2,326,200 acres—an area slightly bigger than Yellowstone.[19]

The first four national parks established a pattern. They encompassed mountainous areas that were rugged, remote, and subject to harsh weather. Though the lands possessed scenic grandeur, they were not representative of the variety of New Zealand's landscapes or ecosystems, and therefore the nascent national park system did not encompass all of New Zealand's biodiversity. New Zealanders were hardly the only people in the world to skew their preservationist impulses toward so-called wastelands where the potential for alternative forms of economic development were severely limited; indeed, their emphasis on mountain parks paralleled both the American and Canadian examples. As in the United States and Canada, the national park idea in New Zealand combined nature preservation and tourism promotion, with the latter purpose often sustaining and circumscribing the former.[20]

The Search for Unified Administration

Administration of the national parks and scenic reserves remained rudimentary through the first half of the twentieth century. National park boards guided the administration of each area with only nominal government support. Responsibility for the parks and reserves was divided between the Department of Tourist and Health Resorts and the Department of Lands, both of which were driven by wider agendas emphasizing economic development. The Department of Tourist and Health Resorts, for example, continued to support introductions of exotic game species in the Southern Alps in the early 1900s in spite of mounting evidence that the introductions were harmful to many native species of birds and trees. New introductions made under government auspices included chamois from the Swiss Alps, thar from the Himalayas, and moose and elk from North America. The Department of Lands, for its part, allowed some scenic reserves such as the large reserve around Mount Cook to be used for sheep pasture.[21]

The Federated Mountain Clubs of New Zealand (FMC) played a central role in persuading the government to form a unified national parks authority. The FMC membership was composed of climbers and "trampers" (backpackers). New Zealand tramping began to acquire popularity after World War I. The first purely tramping club formed in Wellington in 1919. Another formed in Christchurch in 1925, one year after that city acquired railroad access to Arthur's Pass. A distinctive feature of New Zealand's emerging tramping culture was its enthusiasm for developing backcountry huts for overnight shelter, and Christchurch's club began immediately to erect huts in the Arthur's Pass area. With or without huts, tramping through New Zealand's steep, primitive, rain-soaked backcountry was no picnic, and the growing popularity of this recreational activity in the 1920s and 1930s was an indication of how New Zealanders took increasing pride and interest in their country's natural landscapes and mountain scenery.[22]

In 1938 the FMC adopted a formal policy on national parks. The FMC's thrust was for the government to create a national board that would work with the existing park boards on matters of general policy and oversee the allocation of government moneys for upkeep and development of the national parks. This new authority would have sole power to make leases and issue licenses in national parks. The FMC wanted representation on the National Parks Authority, and it believed that mountain clubs should have the right to build huts where they were not provided by the government. The FMC further recommended two basic principles for national park administration: that the public should enjoy free and unfettered access to these areas subject only to reasonable safeguards for the protection of national park values, and that native species should be encouraged and exotic species suppressed as far as possible.[23]

The FMC presented its policy to the minister of lands. The government promised to respond, but World War II intervened and the government postponed action. After World War II, the FMC reaffirmed its policy on national parks and in 1946 joined with another grassroots conservation group, the Forest and Bird Protection Society, in renewing its campaign. In 1950 both groups joined with the Royal Society of New Zealand, the Forestry League, and other organizations in forming the Nature Protection Council. The government drafted legislation and circulated it to the park boards and the several member organizations of the Nature Protection Council. The minister of lands then introduced the bill in the New Zealand legislature, and the National Parks Act became law in 1952.[24]

The National Parks Act 1952 established the National Parks Authority, roughly equivalent to the National Park Service in the United States. The mission of the National Parks Authority was to provide for the use of the national parks for recreation and enjoyment while preserving their natural state. The law defined preserving a natural state according to the native/exotic dichotomy that New Zealanders found so important: "Native plant and animal life should as far as possible be preserved, and introduced plant and animal life should as far as possible be exterminated."[25] The law also affirmed New Zealanders' commitment to local citizen control, as each park would remain under the administration of a park board whose responsibilities were to plan and implement development, appoint ranger staff to protect and interpret park resources, and license private enterprise to provide visitor amenities.[26]

In a little more than a decade after the National Parks Act 1952, the New Zealand government proclaimed six more national parks. Together with one other that had been established in 1940, this brought the number of national parks to ten. At this point the New Zealand national park system had a solid foundation in law and a robust land base (about 7 percent of the nation's land area) but very little money. Compared to the U.S. National Park Service, the New Zealand National Parks Authority was very lean indeed. In 1953 it oversaw a staff of just five rangers, or one ranger per park; a dozen years later the number had risen to thirty-eight rangers for all ten parks, which was still a miniscule force for the protection of some 5 million acres.[27] As one former official of the National Parks Authority describes the situation at the end of the 1960s, the New Zealand national parks were in transition from a maintenance phase to a management phase, a new era "notable for the increasing professionalism in parks and reserve management."[28] It was in this transition period that New Zealand looked keenly to foreign examples, especially the example of the U.S. National Park Service, for the future direction of its national parks.

Lessons from North America

The U.S. National Park Service was equally keen to share its expertise with New Zealand. At the start of the 1960s, Secretary of the Interior Stewart Udall strongly supported global advancement of the national park idea and pushed the agency to expand its international outreach. The initiative was part of a U.S. foreign policy aimed at bolstering American influence around the world to forestall communist movements. The U.S. National Park Service established an International Affairs Office with funding provided by the U.S. Foreign Assistance Act of 1961. While the overarching goal of containment had more relevance in other parts of the world, national park officials felt a natural affinity toward their English-speaking counterparts in New Zealand, as well as in countries such as Canada, Australia, and South Africa. The U.S. effort began with the First World Conference on National Parks held in Seattle, Washington, where Secretary Udall delivered the keynote address. The conference drew 145 delegates from sixty-three nations, including two delegates from the New Zealand National Parks Authority.[29]

In the wake of the conference, the service joined with Parks Canada and the University of Michigan in developing an "international short course" on administration of national parks and equivalent reserves to expose park administrators from other nations to the important principles and policies of the U.S. and Canadian national park systems. Three New Zealanders enrolled in the course in three successive years in the late 1960s, each one assisted by a Winston Churchill Fellowship awarded by the New Zealand government. The first recipient was Gordon Nichols, then the chief administrator of Westland National Park, who was soon to occupy the position of supervisor of national parks and assume the hands-on role in the New Zealand National Parks Authority, similar to Horace M. Albright's when he was western field director and the number two man in the National Park Service in its early days. The second recipient was John Mazey, chief administrator of Tongariro National Park. The third recipient was P. H. C. "Bing" Lucas, who immediately on returning to New Zealand was appointed director of the National Parks Authority. Promoting and modernizing the New Zealand national park system in the 1970s, Lucas's role was analogous to Stephen Mather's in the United States in the 1920s. Through the international short course, the National Park Service was able to provide a strong influence on these three senior officials during a crucial time in the development of New Zealand's national parks.[30]

Lucas spent a total of four months in 1969 touring the United States and Canada. Arriving in mid-July, he spent the first month traveling from San Francisco to various parks and regional offices in a camper van, staying in park campgrounds whenever possible. In the second month he attended the international short course with forty other people from thirty-two nations. The course included field visits to Jasper, Banff, Mesa Verde, and Grand Canyon. Following the course, Lucas traveled to many more national parks, historic sites, and recreation areas in the United States and Canada, as well as pertinent government offices in Washington, Ottawa, and Denver. In those two months he logged a total of 16,000 miles. Upon returning to New Zealand, Lucas described what he had learned in a ninety-four-page booklet, *Conserving New Zealand's Heritage*.[31]

Lucas's strongest impressions from his time in North America related to the sheer visitor load on the U.S. national parks and how the National Park

Service was managing it. He was appalled to find a jail in Yosemite and a traffic light over a busy intersection of roads in the Great Smoky Mountains, but he generally admired what he saw. In his view, he was not just visiting a much more populous society but peering metaphorically into his own country's future. While he welcomed New Zealanders' growing enthusiasm for their national parks, he recognized that the National Parks Authority and park boards had limited time and resources to prepare these areas to meet rising demands.[32] His colleagues generally agreed that New Zealand national parks would soon face pressures similar to those in the United States, and that New Zealanders had an opportunity to learn from the U.S. experience, choose what they admired most in the U.S. national park system, and chart their own course.[33] Lucas's tour was the most significant among many professional exchanges between the U.S. and New Zealand in the 1960s.[34]

The New Zealand National Parks Authority consciously molded New Zealand national park administration after the U.S. example in numerous ways, three of which stand out. First, it professionalized the ranger workforce by establishing ranger training schools and integrating the ranger job with career service in government. Second, it made interpretation one of the central purposes of national parks. Visitor centers became the anchor points for interpretive programs in each New Zealand national park just as they were in the United States. And third, it emulated the National Park Service's commitment to the master plan. The master plan had been a cornerstone of U.S. national park administration since the 1920s. Forty years on, the master plan combined two potent ideas. The first was to establish a comprehensive vision for development and management of a park based on interdisciplinary exchange of ideas among all concerned professional staff. The second was to use the planning process as a way to marshal public support. By the 1960s the U.S. National Park Service was making public hearings and public review of the draft document an integral part of producing a master plan. The interplay of government technical expertise and citizen involvement appealed to New Zealanders, and they devised a park planning system that closely resembled that of the National Park Service.

In other areas of national park administration, New Zealanders took stock of the U.S. model and fashioned their own approaches. Regarding visitor accommodations, they wanted their national parks to serve families with modest incomes rather than wealthy tourists from abroad. Lance McCaskill, who preceded Lucas as head of the National Parks Authority, bluntly stated in 1965: "It is necessary to accept that part of a park will be a rural slum." Park administrators sought to locate such sacrifice areas away from the principal attractions and keep them as confined as possible.[35] Thus, the National Parks Authority supported construction of camping areas with cinderblock washrooms and other basic amenities but did not seek development of grand lodges, despite their potential to draw foreign tourists and generate tourism. "Facilities should never become primary attractions and they should be appropriate to national parks," wrote Ray Cleland, supervisor of national parks.[36] At the same time, the National Parks Authority took a more lenient stand than the U.S. National Park Service on backcountry accommodations. In New Zealand, park users would continue to have access to primitive backcountry huts (some built by tramping clubs, others by the New Zealand government).[37]

Some aspects of the U.S. and Canadian national park systems simply did not translate to the New Zealand context. The strong prohibition against hunting in North American national parks exemplified such differences. Through the middle decades of the twentieth century, New Zealanders wrestled with the problem of controlling deer and other exotic wildlife populations. The government carried on a war of extermination, convinced that these animals were destroying native bush and indirectly causing an increase in soil erosion injurious to agriculture. Scientific studies in the 1950s and 1960s muddied this picture, and it began to appear that all-out extermination was neither achievable nor justifiable on purely economic grounds. Meanwhile, in a further twist of public opinion on the relative merits of native and exotic species, New Zealand's sport hunters argued that the time had come for New Zealanders to accept deer and other wild ungulates as part of the natural fauna; the deer, after all, had been in the country for as many generations as Pākehā. But even as the government scaled back efforts to control animal pests, the National Parks Authority insisted that it was still its policy to exterminate exotic species in the national parks as far as possible. To this end, the National Parks Authority encouraged recreational hunting in the national parks.[38] Without question, invasive species posed the biggest threat to the ecological integrity of New Zealand's national parks.

In contrast, the U.S. national parks faced mounting threats from myriad adverse land uses happening outside of park boundaries, including hydroelectric projects, commercial logging, mining, air pollution, and residential subdivisions. New Zealand national parks did not yet face those external threats to the same degree, but far-sighted New Zealanders nevertheless found the U.S. example instructive. At the First World Conference on National Parks, Secretary of the Interior Udall stated that national parks derived more and more significance from being "nature islands for the world."[39] A statement by the National Parks Authority in 1970 echoed that sentiment and recast the powerful metaphor: "In a world of growing population pressure national parks have become international islands of nature in a sea of development."[40]

The first external threat to a New Zealand national park posed by a development scheme surfaced in the early 1960s when the government announced that it had authorized an Australian mining company to build a hydroelectric dam that would raise the level of a large natural lake in Fiordland National Park. The controversy over Lake Manapouri slowly gained public attention through the course of the decade. When the issue came to a head in the early 1970s it gave rise to the first nationwide environmental campaign, culminating in a government decision to preserve the natural lake level. The "Save Manapouri" campaign is viewed as the awakening of New Zealand's environmental movement, analogous to Earth Day in the United States.[41]

During the 1970s New Zealand's environmental movement focused attention on South Island's West Coast, where commercial logging of native bush increased as it dropped to almost nil throughout the rest of the nation. In response to the public's growing sentiment for protecting what was left of New Zealand's native forest, the government established fourteen forest parks in the 1970s. Tracts of West Coast forest were protected through establishment of Victoria Forest Park and additions to Westland National Park. At the same time, the government took steps to permit clear-cutting of native beech species over large areas of West Coast lowlands and foothills. The government's so-called Beech Scheme contemplated forest conversion to eucalyptus in some areas and regeneration of native beech for sustained-yield production in other areas. A citizen group called the Native Forest Action Council led opposition to these moves, and in 1977 about one in ten New Zealanders signed the Maruia Declaration, a petition to Parliament to phase out logging of native forests on the West Coast. The movement to protect native bush on the West Coast culminated in the establishment of Paparoa National Park in 1987, New Zealand's eleventh national park and the first addition to the system in twenty-three years.[42]

The Department of Conservation and the Politics of Retrenchment

New Zealand's national parks prospered from the early 1970s through the mid-1980s. With more and more New Zealanders taking an interest in outdoor recreation, and with international tourism contributing an increasing amount to the national economy, the government lavished money on them. New visitor centers were built. Park roads were improved. Backcountry huts were upgraded. A ranger training program was instituted at New Zealand's leading agricultural college, and the National Parks Authority oversaw the professionalization of the ranger service. Park managers implemented a more science-based form of resource management. By the Third World Congress on National Parks in 1982, New Zealand's national park system had acquired international stature. New Zealanders provided technical assistance to national parks and conservation programs in Nepal and elsewhere, and they offered ranger training opportunities to foreign nationals from numerous countries in South Asia, the Pacific, and South America.[43]

While the national parks prospered, other parts of the environment were getting hammered by new energy development schemes being promoted by New Zealand's conservative National Party government. Prime Minister Robert Muldoon's "Think Big" program of foreign-capitalized megaprojects came to include aluminum smelting, converting natural gas to petroleum, oil refining, and building a suite of hydroelectric dams. Most of these projects depended on a relaxation of environmental protections. But even as the people of New Zealand appeared to support the energy development schemes, they also demanded greater protection of their national parks, indigenous forests, threatened native fauna, and remaining free-flowing rivers. Environmental organizations grew in strength, and a new Values Party arose, presaging the worldwide Green movement.[44]

In 1984, the opposition Labour Party was returned to power partly on the basis of its promise to protect New Zealand's environment from further "State vandalism." However, shortly after assuming power, the Labour government had to contend with a debt crisis, the severity of which had been concealed by the previous government. Under strained and confused circumstances, the left-wing Labour Party proceeded to move the country in the direction of New Right monetarist economic theory, smaller government, greater transparency in public administration, and "market-led" environmental policy. Its signature reform in the environmental arena was the Conservation Act 1987. This law had far-reaching consequences for the administration of national parks.[45]

The Conservation Act abolished the National Parks Authority and its parent organization, the Department of Lands and Survey, and established a new Department of Conservation (DOC) to manage New Zealand's entire conservation estate. The basic idea behind the establishment of DOC was to differentiate "protection lands" from "production lands" and place all of the former under a single department that would be clearly accountable to environmental interests. It answered a long-standing complaint by the environmental community that bureaucrats too often answered to obfuscating dual mandates that defeated good environmental stewardship. This criticism was loudest with reference to the Forest Service, which managed state forests (mostly exotic plantation forests) for maximum sustained-yield production while administering forest parks (mostly native bush) according to multiple-use principles. Similar charges were leveled at the Department of Lands and Survey, which controlled the national parks. Since the Department of Lands and Survey also administered Crown pastoral lands, it had strong ties to New Zealand's powerful agricultural interests. By creating DOC, the Labour government sought to sweep away dual mandates and establish a cabinet-level department concerned solely with environmental protection. Under the new regime, roughly 30 percent of the nation's land area was dedicated to preserving biodiversity. In that regard, the Conservation Act 1987 was a big win for the environmental lobby and an important step forward in implementing a national conservation strategy.

Yet, as it was implemented during a time of sharp retrenchment in the public sector, the Conservation Act had somewhat mixed implications for New Zealand's national parks. Compared with its predecessor, the National Parks Authority, DOC had to do far more with less. The reorganization of government agencies initiated a scramble for limited financial and human resources. DOC came into being with fewer personnel than the combined staffs of the agencies that it replaced, and in its first decade of existence it was reduced in size from around 2,000 to 1,200 permanent employees.[46] The contraction of the public sector and the shift toward New Right economic thinking forced national park administrators to adopt, in the words of one former member of the National Parks Authority, "business-like approaches to conservation in an increasingly commercialized economic and social environment."[47]

The way in which the Conservation Act bifurcated the national domain into "protection lands" and "production lands" had further mixed results for the national parks. On one hand, the law buttressed the national park system by enfolding it within a strengthened conservation estate. The law established two new national parks, made additions to existing national parks, and transferred native forest lands to DOC control. Since DOC was given a strict conservation mandate, the new agency took major elements of national park administration and applied them to the whole conservation estate. Most of the national parks thereby acquired large buffer areas around them. But, on the other hand, these same changes partially effaced the national park system. While the national parks and various other types of reserves kept their previous names, henceforward they were under the administration of district conservation officers. National park rangers became conservation officers. A few chief rangers (equivalent to park superintendents in the United States) became district conservation officers while most took reassignments. The national park boards were abolished and in their place the government established a network of seventeen regional conservation boards, each having oversight of all protection lands in its sector. Each conservation board was commissioned to obtain public input and produce a citizen-based conservation management strategy (CMS) for its region. Each CMS was to be essentially a statement of resources, conservation values, and desired future conditions for all protection lands in that region. To a certain degree, the national parks faded into a larger framework of biodiversity conservation and lost some of their potency.[48]

After the National Party swept back into power in 1990, weaknesses in the Conservation Act became more apparent. Under National Party rule, the bifurcation of government's conservation and development functions only served to marginalize DOC. Whereas DOC's predecessor agencies had had a direct line to powerful ministers of government representing farming and forestry, DOC had no such cabinet-level support. National Party leaders regarded the Ministry of Conservation as such a slim portfolio that they did not even accord it a place in the cabinet. DOC suffered attrition accordingly.

Under National Party leadership, DOC was required to develop more private-sector sources of funding to fulfill its conservation mandate. In the early 1990s DOC turned to national park concession and commercial use fees to help pay its bills. (Interestingly, DOC staunchly opposed collecting park entrance fees the way the U.S. National Park Service did, insisting that New Zealanders would not abide it.) Some concerned citizens decried the move toward private-sector funding, suggesting that the government was squeezing DOC in order to undermine its conservation mission and force commercialization of the national parks. They protested when DOC began to put some of its limited resources toward developing a "Visitor Strategy" to abet the tourism industry.[49]

New Zealand's tourism industry was primed to exploit the national parks more fulsomely. Indeed, the contraction of New Zealand's public sector in the late 1980s and early 1990s happened to coincide with a period of mounting pressure for more national park development. Between 1985 and 1995 the number of foreign visitors to New Zealand doubled. National park visitor surveys in the mid-1990s revealed that 60 to 70 percent of visitors at such prime locations as Mount Cook and Fox and Franz Josef Glaciers came from overseas. Even at Arthur's Pass National Park, long known as a backcountry user's park, the proportion of foreign visitors rose from 25 percent in 1980 to 45 percent in 1995. Affluent overseas visitors actually displaced New Zealanders in some instances, such as on famous tramping routes where DOC imposed hefty user fees and a reservation system. With adventure travel booming, the tourism industry brought pressure on DOC to allow sightseeing flights and build roads through a number of South Island national parks.[50]

Māori Land Claims

Although most New Zealanders had long taken pride in the perception that their nation had a history of relatively enlightened race relations, Māori activism and New Left ideology in the 1970s finally challenged that orthodoxy. In the census of 1980, Māori numbered about 270,000 in a nation of not quite 3.4 million people. Suffering from a high rate of poverty, Māori pressed for recognition of their *tangata whenua,* their rightful place as an indigenous people. In response, the government formed a claims court, the Waitangi Tribunal, which it named after the Treaty of Waitangi of 1840, a seminal document in New Zealand history. In 1985 the government took a significant step when it expanded the purview of the Waitangi Tribunal so that Māori could bring claims based on past, as well as present-day, infractions of the treaty. Most of the historical claims revolved around land loss. The magnitude of Māori land claims surprised a large number of Pākehā and raised concerns that the Waitangi Tribunal had opened a Pandora's box. Some of the backlash against Māori land claims came from environmentalists who feared that the cash-strapped government would settle claims by giving away valuable chunks of the conservation estate—maybe even one or more national parks.[51]

Through the 1990s, environmentalists' concerns focused on the claims of one Māori *iwi* or tribe, Ngāi Tahu. To understand the significance of this tribe's land claims, it is helpful to observe that the Māori population in 1840 was concentrated in the far north of New Zealand and was quite sparse in the cooler latitudes of South Island. Ngāi Tahu, as the southernmost tribe in New Zealand, had a disproportionately large tribal territory covering roughly the southern three-fourths of South Island. Being widely dispersed, with habitations confined mainly to coastlines, the tribe was easily dispossessed of its lands. In ten land sales from 1844 to 1864, the Crown divested Ngāi Tahu of its entire estate, some 34 million acres.

The gist of the Ngāi Tahu claim before the Waitangi Tribunal was that the Crown had defaulted on its terms, in particular its promises to set aside one-tenth of all land sales as tribal reserves and protect the tribe's *mahinga kai* or essential resource areas. Ngāi Tahu sought restitution both in the form of money and land. The tribe's interest in land was both economic and cultural. On the economic side, it wanted some land for tribally owned for-profit enterprises, and it wanted other land for traditional economic activities such as fishing, hunting, and gathering. On the cultural side, it wanted land for the purpose of restoring *mana* or the tribe's guardian power (in modern terms, authority or stewardship responsibility), and it wanted recognition of the tribe's *tāonga* or natural treasures and sacred sites.[52] When Ngāi Tahu and the government began to discuss a settlement, it was not clear how national parks might figure into it. Environmentalists' worst fears were that national parks would be brought into the negotiations as economic assets. They might be conveyed to the tribe and then leased back to the Crown for an annual fee, for example.[53]

DOC became closely involved in the Ngāi Tahu land claims settlement talks from the late 1980s through the 1990s. Some environmentalists complained that DOC's involvement in Māori land claims caused it to stray from its intended role as a single-purpose conservation agency. To their way of thinking, the reason for DOC's involvement was obvious and simple: the government had pledged that no private lands would be used to make restitution for Māori land claims, so the conservation estate would serve as the government's land bank, and DOC would serve as its realtor. FMC president Hugh Barr made a rhetorical jab when he stated, "DOC sees itself as a Maori land agency."[54]

But viewed another way, the Māori land claims settlement talks simply provided a timely venue for DOC and Māori to work out how the national parks would change as New Zealand strove to become a more bicultural nation. DOC had a mandate to go down that very road after the New Zealand government passed the Resource Management Act (RMA) in 1991. An important objective of the law was to integrate "Māori interests and values and the Treaty of Waitangi" into the nation's whole conservation program. The RMA specifically required the government to recognize "the relationship of Māori and their culture and traditions with their ancestral lands, water, sites, wáhi tapu and other tāonga." It also required the government to give due consideration to Māori *kaitiakitanga,* which was defined in the act as "the exercise of guardianship; and, in relation to a resource, includes the ethic of stewardship based on the nature of the resource itself."[55]

The Ngāi Tahu settlement, signed in October 1996 and ratified in September 1998, was a significant step toward realizing those objectives. Besides a $180 million package aimed at securing a better economic future for the roughly

8,000-member tribe, the settlement contained several elements of significance to the national parks. It dedicated a number of seats on various local and regional boards to Ngāi Tahu, including two seats on each regional conservation board within the Ngāi Tahu tribal area. It obligated the national parks to ensure that Ngāi Tahu were actively involved in the protection of their *tāonga* species found within each park, and to work with Ngāi Tahu in developing protocols for customary use of traditional materials and indigenous species.[56]

With the settlement money, Ngāi Tahu invested heavily in ecotourism ventures and quickly became a major force in New Zealand's tourism industry. During the early 2000s Ngāi Tahu pushed for control of national park concessions, claiming that it should receive preference rights on the basis of customary use. It did not succeed with its claim, but it nonetheless acquired several national park concessions by dint of its strong capitalization and business organization. By the mid-2000s, Ngāi Tahu owned Franz Josef Glacier Guides in Westland National Park, Hollyford Valley Walks in Fiordland National Park, and Kaiteriteri Kayaks in Abel Tasman National Park, among other concessions.[57]

In 2013 the government settled with another Māori tribe, Ngāi Tūhoe, whose homeland, Te Urewera, once centered in the rugged hills of North Island's East Cape. Toward the end of the nineteenth century, Te Urewera became legendary as the last stronghold of Māori resistance to British control. After the iwi's remaining lands were confiscated and enfolded into Te Urewera National Park in the 1950s, many came to believe the government had wrongly punished the iwi for past bloody acts that the iwi had never actually committed. The settlement with Ngāi Tūhoe included a $170 million commercial redress package, an acknowledgment by the government of breach of the Treaty of Waitangi, an official apology, and an agreed historical account of the government-iwi relationship. Most significantly for the New Zealand national park system, it called for a new management board for Te Urewera National Park composed of equal numbers of government and Ngāi Tūhoe appointees.

As the Waitangi Tribunal nears the end of the land claim cases, it is clear that the Māori have effectively used the land claim process to regain a strong voice in the environmental stewardship of New Zealand. Ngāi Tahu and other iwi now have solid representation on regional conservation boards. Many Māori now work for DOC. Through extensive staff training, DOC employees are far better informed about Māori rights and culture than they were twenty years ago. To what extent the indigenous people's perspective will reshape the national park idea in New Zealand remains to be seen. Certainly New Zealanders have positioned their national park system to be a bellwether in that regard.

Conclusion

The New Zealand national park system bears many similarities to its U.S. counterpart, but it is also distinctive in ways that reflect New Zealand's unique ecology, history, and political economy. New Zealand's fourteen national parks cover a generous portion of the national domain, yet they are practically all confined to mountain and steep-land topography, reflecting the fact that most low-lying parts of the country were dedicated to production of livestock pasturage and exotic plantation forests. The protection of natural conditions in New Zealand national parks places heavy emphasis on the preservation of native species and the extermination of exotic species. This stark juxtaposition of native and exotic species developed in response to New Zealand's history of biological invasions and reflects a combination of popular attitudes and state-directed policies. New Zealand national parks have followed the U.S. lead in some areas of park administration such as interpretation and planning, but New Zealanders have shown an enduring attachment to the use of citizen-staffed boards for guiding park management. New Zealanders' concepts of wilderness closely parallel American concepts, but they also show the markings of New Zealand's unique tramping culture. New Zealand national parks are now in the process of giving more recognition to Māori. This change comes amid growing sensitivity to indigenous people's rights around the world; however, it is also linked to a major initiative by New Zealanders to make their society more bicultural. For all these reasons, New Zealand's national parks, despite their obvious connections to the international arena, are best understood as an outgrowth of the nation-state.

Notes

1. Department of Conservation/Te Papa Atawhai, "National Parks," http://doc.govt .nz/parks (accessed June 23, 2011).
2. George Gibbs, *Ghosts of Gondwana: The History of Life in New Zealand* (Nelson, New Zealand: Craig Potton, 2006), 49–58.
3. I. A. E. Atkinson, "Introduced Mammals in a New Environment," in *Biological Invasions of New Zealand,* ed. R. B. Allen and W. G. Lee (Berlin: Springer-Verlag, 2006), 49. Also see R. N. Holdaway, "New Zealand's Pre-Human Avifauna and Its Vulnerability," *New Zealand Journal of Ecology* 12 (1989): 12.
4. Bronwyn M. Newton, John R. Fairweather, and Simon R. Swaffield, "Public Perceptions of Natural Character in New Zealand: Wild Nature versus Cultured Nature," *New Zealand Geographer* 58, no. 2 (2002): 17–18; Alan Mulgan, "Literature and Landscape in New Zealand," *New Zealand Geographer* 2, no. 1 (1946): 189–90, 202–3.
5. Atholl Anderson, "A Fragile Plenty: Pre-European Māori and the New Zealand Environment," in *Environmental Histories of New Zealand,* ed. Eric Pawson and Tom Brooking (Melbourne: Oxford University Press, 2002), 19–34; R. J. Cameron, "Destruction of the Indigenous Forests for Maori Agriculture during the Nineteenth Century," *New Zealand Journal of Forestry* 9, no. 1 (1964): 98–109; M. S. McGlone, "Polynesian Deforestation of New Zealand: A Preliminary Synthesis," *Archaeology of Oceania* 18 (1983): 11–25; M. S. McGlone, "The

Polynesian Settlement of New Zealand in Relation to Environmental and Biotic Changes," *New Zealand Journal of Ecology* 12 (1989): 115–29.

6. S. E. Masters, J. T. Holloway, and P. J. McKelvey, *The National Forest Survey of New Zealand, 1955, Volume 1: The Indigenous Forest Resources of New Zealand* (Wellington: R. E. Owen, Government Printer, 1957), 3.

7. Tom Brooking and Eric Pawson, "The Contours of Transformation," in *Seeds of Empire: The Environmental Transformation of New Zealand,* ed. Tom Brooking and Eric Pawson (London: I. B. Tauris, 2011), 13.

8. Nigel Pears, "Familiar Aliens: The Acclimatisation Societies' Role in New Zealand's Biogeography," *Scottish Geographical Magazine* 98, no. 1 (1982): 23–34; Andrew Hill Clark, *The Invasion of New Zealand by People, Plants, and Animals: South Island* (New Brunswick, N.J.: Rutgers University Press, 1949), 259–82.

9. Graeme Caughley, *The Deer Wars: The Story of Deer in New Zealand* (Auckland: Heinemann, 1983), 144–45.

10. Lloyd W. Swift, "Exotic New Zealand," *American Forests* 78 (March 1972): 51. Swift was director of the U.S. Forest Service's Division of Wildlife Management from 1944 to 1963.

11. Paul Star, "Native Bird Protection, National Identity and the Rise of Preservation in New Zealand to 1914," *New Zealand Journal of History* 36, no. 2 (2002): 123–24.

12. David Thom, *Heritage: The Parks of the People* (Auckland: Landsdowne, 1987), 91–97, 101–2.

13. Ibid., 97.

14. Eric Pawson, "The Meanings of Mountains," in Pawson and Brooking, *Environmental Histories of New Zealand,* 141.

15. Theodore Catton, *Wonderland: An Administrative History of Mount Rainier National Park,* appendix B (May 1996), www.npshistory.com/park_histories. htm#mora (accessed August 16, 2015).

16. Thom, *Heritage: The Parks of the People,* 106; Paul Star and Lynne Lochhead, "Children of the Burnt Bush: New Zealanders and the Indigenous Remnant, 1880–1930," in Pawson and Brooking, *Environmental Histories of New Zealand,* 128–29.

17. Paul Star, "Ecology: A Science of Nation? The Utilization of Plant Ecology in New Zealand, 1896–1930," *Historical Records of Australian Science* 17 (2006): 198–99.

18. Star and Lochhead, "Children of the Burnt Bush," 121–27.

19. Thom, *Heritage: The Parks of the People,* 132.

20. Pawson, "The Meanings of Mountains," 143.

21. Ibid., 116–21.

22. Mavis M. Davidson, "Trampers, Climbers, and Forest Recreation," *New Zealand Journal of Forestry* 10, no. 2 (1965): 194–96.

23. "The Federation and National Parks," *FMC Bulletin* 27 (April 1967): 1–2.

24. Ibid., 3–4; Thom, *Heritage: The Parks of the People,* 151; Tom Bührs and Robert V. Bartlett, *Environmental Policy in New Zealand: The Politics of Clean and Green?* (Auckland: Oxford University Press, 1993), 41.

25. Quoted in Thom, *Heritage: The Parks of the People,* 150.

26. *Encyclopedia of New Zealand* (1966) at http://www.teara.govt.nz (accessed June 21, 2011).

27. Thom, *Heritage: The Parks of the People,* 161.

28. Paul R. Dingwall, "Antarctica/New Zealand," in *Protecting Nature: Regional Reviews of Protected Areas,* ed. J. A. McNeely, J. Harrison, and P. Dingwall (Gland, Switzerland: International Union for Conservation of Nature and Natural Resources, 1994), 233–34.

29. Terence Young and Lary M. Dilsaver, "Collecting and Diffusing 'the World's Best Thought': International Cooperation by the National Park Service," *George Wright Forum* 28, no. 3 (2011): 273–74.

30. For the Albright and Mather analogy I am indebted to my father (William R. Catton Jr., personal communication, June 2, 2011). As professor of sociology at the University of Canterbury from 1970 to 1972, he knew both Nichols and Lucas and had insight into their leadership roles and how they worked together. For more on his impressions of New Zealand national parks at this stage in their evolution, see William R. Catton Jr., "The Wildland Recreation Boom and Sociology," *Pacific Sociological Review* 14, no. 3 (July 1971): 339–59.

31. P. H. C. Lucas, *Conserving New Zealand's Heritage* (Wellington: A. R. Shearer, Government Printer, 1970).

32. Ibid., 10–15; P. H. C. Lucas to Director, National Park Service, November 8, 1965, Record Group 79 (Records of the National Park Service), Administrative Files 1949–71, Box 2181, Folder "Foreign Parks and Historic Sites, New Zealand, 1949–69, Part 5," National Archives and Records Administration, College Park, Maryland.

33. L. W. McCaskill, "General Aspects of Recreation in National Parks," *New Zealand Journal of Forestry* 10, no. 2 (1965): 155–57; R. W. Cleland, "National Parks—How the People Use Them," *New Zealand Journal of Forestry* 10, no. 2 (1965): 159–69.

34. There were also visits in the other direction. Samuel P. Weems, parks adviser in the International Affairs Division of the National Park Service, made a six-week study tour of the New Zealand national parks the same year that Lucas visited the United States. Lance McCaskill, who headed the National Parks Authority through the 1950s and 1960s, credited Olaus Murie with introducing the "wilderness area" concept to New Zealanders in 1949 when he went as part of a National Geographic Society expedition to Fiordland. See McCaskill, "General Aspects of Recreation in National Parks," 157, and correspondence from 1941 to 1969 in Box 2181 (see n. 32).

35. McCaskill, "General Aspects of Recreation in National Parks," 156.

36. Cleland, "National Parks—How the People Use Them," 163.

37. Lucas, *Conserving New Zealand's Heritage,* 26.

38. Caughley, *Deer Wars*; Ross Galbreath, *Working for Wildlife: A History of the New Zealand Wildlife Service* (Wellington: Bridget Williams Books, 1993); Cleland, "National Parks—How the People Use Them," 168.

39. Stewart L. Udall, "Nature Islands for the World," in *First World Conference on National Parks* (Washington, D.C.: Government Printing Office, 1962), 3.

40. National Parks Authority, "People, Parks and Fire," *FMC Bulletin* 37 (October 1970): 3–4.

41. Nicola Wheen, "A History of New Zealand Environmental Law," in Pawson and Brooking, *Environmental Histories of New Zealand,* 263–64.

42. A. J. Tilling, "Indigenous Forest Management in New Zealand: From Interventionist to Monetarist Policies and the Special Case of the South Island's West Coast," *New Zealand Journal of Forestry* 36, no. 4 (February 1992): 8–13.

43. Paul R. Dingwall, interview by the author, April 13, 2012.

44. W. Keith Jackson, "New Zealand in the 1970s," *Current History* 58 (April 1970): 217–22; Henry S. Albinski, "Australia and New Zealand in the 1980s," *Current History* 85 (April 1986): 149–56; Tilling, "Indigenous Forest Management in New Zealand," 8–13; Hugh Barr, "Establishing a Wilderness Preservation System in New Zealand: A User's Perspective," *International Journal of Wilderness* 3, no. 2 (June 1997): 9.

45. Bührs and Bartlett, *Environmental Policy in New Zealand,* 90–91; Hugh Logan, interview by the author, March 15, 2012.

46. Dingwall interview.

47. Dingwall, "Antarctica/New Zealand," 234–37, quotation on 234.

48. J. C. Halkett, "Transparency of Purpose: An Improvement on the Forest Service? (A Consideration of Government's Environmental Administration Restructuring)," *New Zealand Journal of Forestry* 32, no. 1 (May 1987): 20–23; Robert Cahn, "Reorganizing Conservation Efforts in New Zealand," *Environment* 31, no. 3 (April 1989): 19–20, 40–45; Les Molloy, "Wilderness in New Zealand: A Policy Searching for Someone to Implement It," *International Journal of Wilderness* 3, no. 2 (June 1997): 11–13.

49. Hugh Barr, "DOC Advocates Treaty Land Grab," *FMC Bulletin* 115 (March 1994): 4–5; Barr, "Establishing a Wilderness Preservation System in New Zealand," 9–10.

50. Stephen R. Espiner and David G. Simmons, "A National Park Revisited: Assessing Change in Recreational Use of Arthur's Pass National Park," *New Zealand Geographer* 54, no. 1 (1998): 37–45; Molloy, "Wilderness in New Zealand," 12.

51. Dora Alves, *The Maori and the Crown: An Indigenous People's Struggle for Self-Determination* (Westport, Conn.: Greenwood, 1999), 39–41, 57–58, 67–68; Hugh Barr, "Treaty Claim Legislation Threatens National Parks," *FMC Bulletin* 113 (June 1993): 19–22.

52. Te Rūnanga o Ngāi Tahu, "The Settlement" (1996), ngaitahu.iwi.nz/ngai-tahu/the-settlement (accessed June 27, 2011).

53. David Henson, "FMC Submission to the Waitangi Tribunal re the Ngai Tahu Trust Board Claim," *FMC Bulletin* 99 (September 1989): 24–25; Hugh Barr, "Tribunal Rulings on Ngai Tahu Claims," *FMC Bulletin* 109 (March 1992): 23; Hugh Barr, "Under-Valuing Our Parks," *FMC Bulletin* 116 (May 1994): 3.

54. Barr, "DOC Advocates Treaty Land Grab," 4.

55. Quoted in Alexander Gillespie, "Environmental Politics in New Zealand/Aotearoa: Clashes and Commonality between Māoridom and Environmentalists," *New Zealand Geographer* 54, no. 1 (1998): 20. The first quote is by the minister of environment referring to the bill in 1988; the second and third quotes are from the act.

56. Te Rūnanga o Ngāi Tahu, "The Settlement"; Department of Conservation/Te Papa Atawhai/Canterbury Conservancy, *Arthur's Pass National Park Management*

Plan (Christchurch: Department of Conservation/Te Papa Atawhai/Canterbury Conservancy, 2007), 15–16, 20, 46–47, 123.

57. Poma Palmer, interview by the author, March 13, 2012; Heather Zeppel, "National Parks as Cultural Landscapes: Indigenous Peoples, Conservation, and Tourism," in *Tourism and National Parks: International Perspectives on Development, Histories, and Change,* ed. Warwick Frost and C. Michael Hall (New York: Routledge, 2009), 277.

EMILY WAKILD

FOUR | # Conservation on Tour
Comparing Nations, Scientists, and Parks in the Americas

IN THE LATE NINETEENTH AND EARLY TWENTIETH CENTURIES, citizens throughout the Americas organized frontier expeditions, sent forth scientific surveyors,
and began to compile information in new ways. Such strategic journeys into
the wilderness—with all of the weighty symbolism that phrase implies—produced new knowledge about vast extensions of territories and their inhabitants
far from political centers of power. The expeditions underwrote a platform for
state expansion, promoted the often-forced incorporation of indigenous peoples, and ambitiously mapped locations for economic development. Scientists
at the forefront of these changes simultaneously made some of the staunchest
arguments for nature conservation, laying the groundwork for the vast national
parks created by the United States, Argentina, and Brazil. Of course, the historical contexts in each nation formed unique crucibles in which evolving notions
of citizenship, nationhood, and conservation grew into their own tropes. To
compare the contexts is not to conflate the experiences. Yet, by comparing these
processes, a vibrant scientific network emerges to provide a more encompassing
view of the motivations behind nationally protected nature.

The contours of this network appeared in high relief when former U.S. president Theodore Roosevelt toured South America in 1913 and 1914. On this trip,
Roosevelt gave speeches and met dignitaries in the largest cities in the Southern
Cone, but his stated purpose was to record scientific observations and collect
specimens of mammals and birds for the American Museum of Natural History
in New York. Much to his delight, he found similar processes of exploration and
observation already under way.[1] Roosevelt's role could be overstated; his trip
was neither the origin of scientific expeditions nor the mechanism for spreading

national parks. Roosevelt could no more spread science than force political change.[2] Instead, his trip is illustrative precisely because Roosevelt encountered men who would soon become national heroes engaged in frontier expeditions and fledgling conservation projects.

During the trip, he found kindred souls in two men. In Argentina, Francisco Pascasio Moreno personally guided Roosevelt on a tour across the icy extremes of the Andes Mountains and through the desolate plains of Patagonia to South America's first national park.[3] Moreno was more than qualified to host the former president since the parklands, territory along the shores of Nahuel Huapi Lake, had been given to Moreno by the federal government as compensation for his work negotiating the boundary line with Chile. In 1903 Moreno returned three leagues (nearly 23,000 acres) to the Argentine nation with the explicit demand that they be incorporated as Parque Nacional del Sur. The government agreed in principle, but Moreno would not live to see the creation of the federal parks commission or his former property's official designation and expansion into Parque Nacional Nahuel Huapi in 1934. Nevertheless, his vision would become the world's first privately donated national park, one that remains a jewel of the Argentine patrimony.[4]

After his journey across Patagonia, Roosevelt joined a telegraph survey commission already under way to form the Brazilian-American Expedição Científica Roosevelt-Rondon. Among other accomplishments, this "zoogeographic reconnaissance" charted a new (to them) river that pushed water northward through dense tropical rain forest before reaching the Madeira, and then the Amazon River.[5] It was, in Roosevelt's words, "the biggest tributary of the biggest tributary of the mightiest river in the world."[6] Brazilians subsequently renamed the river the Rio Roosevelt after the famed "wilderness warrior."[7] But the true expedition chief was the frontiersman Cândido Mariano da Silva Rondon. Descended from the Terena and Borôro peoples and orphaned at a young age, Rondon grew to demonstrate fierce military discipline and adherence to positivism.[8] Rondon had spent nearly all of his life walking, charting, scouting, and surveying his native Amazonian woodlands in federal employ. Roosevelt's enthusiastic participation in the river navigation and other components of his South American tour illuminate a network of scientists engaged in the type of investigation that laid the groundwork for national park creation from the U.S. West to Argentina and Brazil.

Popular interest in Roosevelt's trip to South America has resurged in recent years, and a parallel academic literature exists investigating the role of science in displaying, conveying, and even constructing the wilds of South America.[9] Several important articles on the history of nature conservation have begun to reexamine the legacies of parks in the region.[10] But the overall picture of parks' past in South America remains unformed. Roosevelt's tour captures a moment in time that illuminates a wide process of exchange and action throughout the hemisphere. The men of science who accompanied and astonished Roosevelt played formative roles creating the conservation networks these nations administer today. Rondon and Moreno assured the success of Roosevelt's collecting objectives, mediated the extremes of South American nature, and conveyed the solidarity of a network of naturalist-scientist-explorers who would chart the path of nature conservation in their respective nations. Roosevelt, Rondon, and Moreno, and the parks that bear their legacies, are prominent—but certainly not solitary—examples of the material manifestations resulting from the interplay between science and society. These men grew out of a broad scientific tradition (based largely but loosely on the work of Alexander von Humboldt) not yet freighted with separate disciplines and specialties.[11]

Rather than a return to a "great man" version of history, this interpretation of these men intends to let the few stand for the many.[12] Women scientists did play important roles in each country, but women rarely gained the profiles necessary to advocate for conservation areas until much later. Furthermore, the continual constitution of manhood figured into the adventures and advocacy, continually exemplifying and challenging appropriate behaviors and powers.[13] Men such as Rondon, Moreno, and Roosevelt reveal the ways personalities, relationships, and expeditions shaped conservation. The actions and lives of individuals help explain how national parks came to inhabit the world. State development was never faceless; instead, it reflected highly personal projects. Historians have fruitfully discussed the rise of ideas about wild nature, landscape aesthetics, and relationships with nature in Latin America under colonial rule.[14] Building on these works, this chapter offers three minibiographies of Roosevelt, Moreno, and Rondon to compare the distinct but analogous trajectories of the development of conservation ideas in the U.S. West, Amazonian Brazil, and Patagonian Argentina. The characteristics these men displayed individually and in their relationships with each other, the regions they represented and knew through vast personal experience, and the ways in which they reached similar conclusions about the role of parks in protecting the contents of these regions illuminate national scientific agendas and nature protection schemes. This chapter is neither an administrative history of three parks nor a trinational comparison of state conservation strategies. Instead, it is an attempt to examine some of the adventures of three "great men" and read those escapades against the transnational lens of science in practice. This will give a fuller picture of how the networks of natural field science formed the basis for early efforts at creating national parks in three distinct nations. American ideas about conservation (symbolized by Roosevelt) influenced, intertwined, and inspired park creation but did not dislocate processes set in motion by internal agendas.[15]

Roosevelt: A Naturalist Roughly Riding

As a conservationist, national figure, and scientific explorer, Roosevelt was shaped by intense personal experiences in nature. Born in New York City in 1858, during his early life he explored the Adirondacks and Maine woods of New

England accumulating observations of animals in various habitats. Roosevelt's encyclopedic knowledge of birds—their songs, markings, and habits—acquired by close, careful, and repeated study, was outmatched only by his eagerness for large mammals to study and hunt. As a young man, Roosevelt's westward trip to the Badlands of the Dakota Territory stirred in him a sense of openness and freedom he'd not felt before. He desperately yearned for a buffalo to hang on his library wall, and the completion of the Northern Pacific Railroad offered him the opportunity to reach the desolate Badlands relatively quickly.[16] Despite his thirst for the hunt, Theodore Roosevelt was a notoriously poor shot. Cursed with poor eyesight, he wore spectacles that often got in the way, and he rarely took down an animal with grace.[17] In September 1883 he arrived on the western edge of the Dakota Territory just days after the railroad's completion and soon became acquainted with the infinite horizon and open range in personally transformative ways.[18] Between bouts of asthma and cholera, Roosevelt stayed himself physically to obtain the buffalo. He employed a guide, rented a mare, acquired a proper gun, and rode into the foreboding geography of the rugged plains with men suspicious of his aristocratic air. Despite the massive slaughter of buffalo that occurred in the wake of railroads, hunting the animal was no easy endeavor. Through rising creeks and merciless downpours, Roosevelt's party rode over intriguing geological features until Roosevelt finally shot a bull. The bullet made contact but the buffalo got away. The poor weather was not the only obstacle—Roosevelt fell into a cactus patch, pricking his hands and causing them to swell, and in another instance he banged his rifle on his own forehead, causing a deep bloody gash. Finally, as the group rode into Montana, Roosevelt peered over a ridge at a buffalo and pierced its shoulder with a shot. After the animal staggered into a ditch, Roosevelt "encircled the buffalo, whooping and chanting as if he were White Bull or Two Moons in an effort to pay this 'lordly buffalo' due reverence."[19]

The buffalo-hunting expedition highlights the contradictions between Roosevelt's naturalist tendencies, his ideas about native peoples, and his approach to conservation. The paradoxes have not been lost on historians. He was an aristocrat who immersed himself in physical trials; a bombastic president not afraid to burn bridges even in his own party; a man of privilege who suffered great tragedy (he lost his wife and mother on the same day); a tireless champion of U.S. imperialism in the Caribbean basin who softened his stance as he traveled farther south in the hemisphere; a Progressive's progressive who broke up monopolies and busted trusts; a Harvard-educated military man and avid birder who wrote books of history; and a Nobel Peace Prize winner who championed conservation in parks and forest reserves. Although his bombastic political style gained him a negative reputation in Latin America for instituting the "Roosevelt Corollary" to the Monroe Doctrine, and his arrogance contributed to wresting Panama apart from Colombia, his scientific leanings reveal a different agenda. Roosevelt had a lifelong appreciation for the type of knowledge gained by natural history, noting late in life that "while my interest in natural history has added very little to my sum of achievement, it has added immeasurably to my sum of enjoyment of life."[20] He was eager to participate in scientific data accumulation and did so regularly for natural history museums.

Roosevelt was fascinated with the frontier and undertook lengthy expeditions at home and abroad. In 1901, in the months before he took office as vice president, he traveled to Colorado to hunt cougar and describe their habits for a biological survey. In 1903, as president, he made the "Great Loop Tour" across the continent presenting conservation policies and visiting Yellowstone, Yosemite, and the Grand Canyon.[21] In addition to his expansive domestic travels, which helped craft his state-making agendas at home, Roosevelt was the first sitting U.S. president to travel abroad (by visiting the Panama Canal in 1906). After leaving office, he spent a year (1909) in northern and eastern Africa hunting and acquiring the biggest collection of large game specimens in the world for the National Museum in Washington; Roosevelt himself shot nine lions.[22] His South American tour was an extension of a wide variety of collecting expeditions.

The "Far West" that Roosevelt adored inspired his conflicting ideas about native inhabitants. As Thomas Dyer has shown, broad generalizations in Roosevelt's writings reveal his contempt for native people and his belief in their inferior status.[23] While he occasionally celebrated attributes he believed they possessed (such as strong fighting skills), he characterized them as "savage" and found them unfavorable to the lauded image of heroic frontiersman. But in certain ways Roosevelt's approach was more measured. Roosevelt did not promote allotment, although his sympathies had more to do with romanticizing Native American ways of life than defending their sovereignty. The U.S. Congress passed the Dawes Act in 1887 to reduce reservations and concentrate Indians into individual plots of private land.[24] The act remained in effect until 1934, destructively stripping native people of millions of acres of land. Roosevelt argued that allotment was "a vast pulverizing engine to break up the tribal mass,"[25] and he adhered to the belief that Indians should be allotted land in severalty while also allowing for white settlers to gain from "surplus" land.[26] Roosevelt hardly sympathized with native peoples, yet he recognized vast differences among them (on both individual and cultural levels). He stressed the necessity of assimilation largely through intermarriage and the necessity for obedience to state-building projects including the Indian Service.[27] His self-cultivated cowboy image belied a nostalgic view of Indian life.[28] Once in the presidential office, Roosevelt's personal encounters softened his stance and made him more sympathetic—calling for both the study of Indian societies and the maintenance of aspects of their life—in opposition of those who preached total assimilation.[29] He set a precedent that made Indian affairs the purview of westerners.[30] Historian Frederick Hoxie has shown that in part under the leadership of Roosevelt, Indian policy in the first decade of the twentieth century shifted from a push for rapid incorporation to "racism, nostalgia, and disinterest."[31]

Of the three men under consideration, Roosevelt was the only one who held elected political office providing a position to directly influence park creation. As president he encouraged the establishment of six national parks, but parks were a small part of his conservation legacy. He supported Gifford Pinchot and the creation or enlargement of 150 national forests, named fifty-one bird reservations and four game preserves (most of which became part of the National Wildlife Refuge System), and set aside eighteen national monuments, many using the Antiquities Act, which he signed in 1906.[32] These conservation areas took seriously the continuum of connections between humans and nature as they included Spanish missions, caves, canyons, and cliff dwellings. In the national forests alone, Roosevelt added more than 150 million acres to the national patrimony, an area larger than France, Belgium, and the Netherlands combined.[33] In a position of power and with a personal affinity for conservation, Roosevelt created an enormous footprint for conservation, largely in the western states.

An unexpected, yet revealing, context for his escapades is against a backdrop of his peers in nations to the south. Roosevelt, no stranger to travels in little-known regions, did not design or plan the 1913 expedition but undertook it under the sponsorship of the American Museum of Natural History. He was hardly alone in the wild; in addition to dozens of muleteers, paddlers, guards, porters, and cooks, the Brazilian staff alone boasted no fewer than ten highly trained scientists—ranging from astronomers and surveyors to botanists and taxidermists. Neither was the initial route new, although the serendipity of planning an encounter led them into uncharted regions. Heading to Patagonia before Amazonia, Roosevelt's South American tour (which included the countries of Brazil, Uruguay, Argentina, and Chile) provides glimpses of similar scientific ideas, state-building enterprises, and conservationist agendas circulating through the hemisphere.

Moreno: An Expert on Patagonia

One frigid, blustery morning just before sunrise in March 1877, a young man not yet twenty-five years old rose from his camp on the southern edge of the glacier-fed and iceberg-filled Lago Viedma to make a topographical sketch of the lake.[34] Armed only with a prismatic compass in a leather case strapped to his chest, he bundled himself in a poncho made of guanaco (a wild camelid, cousin of the llama) hides and began to walk. Scarcely outside camp, a female puma sprang from behind and knocked the scientist to the ground, lacerating his face and back with her powerful paw. She tumbled with him then lost her balance before she could sink her teeth into him. In a flash, he sprung up, dropped off the poncho (mitigating the chance the cat thought it was going to feast on a meaty camelid and not a man), and positioned the compass case as the puma pounced again. He managed to stun her in the head with the case and she slunk off, puzzled. The frightened scientist made his way back to camp periodically yelling and shaking the poncho as she followed in slow pursuit. The puma took refuge in a patch of bushes as the man reached camp and raised an alarm. Shortly after, the man's companions tracked down and "brained" the cougar with a set of *bolas,* traditional weapons made of two wooden balls joined by a braided leather cord. The man, Francisco Moreno, then rested, relieved that, despite the pain, his injuries would not be fatal.

At the time of this attack Moreno had spent more than six years in the wilds of Patagonia. Born in Buenos Aires in 1852 to a British immigrant's daughter and an Argentine father, he was on his second southern expedition, this one to the Santa Cruz River.[35] Having already surveyed the paleontology of the littoral in Buenos Aires province, Moreno's first southern expedition carried the flag of field science. In 1872 (the same year the U.S. Congress designated Yellowstone a national park) Argentines founded the Sociedad Científica Argentina with the objective of creating a museum and funding an expedition to fill that museum. Moreno was trusted with the expedition and with what would become the destiny of Patagonia.[36] Besides collections, expedition objectives included identifying a pass to cross the Andean mountain range through territory hardly subject to laws originating in Buenos Aires. Each trip begot further trips until Moreno spent nearly twenty years exploring the region. He, more than any other single person, is credited with affirming Argentine national sovereignty over thousands of miles of land in Patagonia.[37]

Argentina was a nation dominated by the eastern riverine capital and province of Buenos Aires. Its transition from colonial rule to become a unified country came with many fits and starts, and frontier exploration played heavily into this process.[38] Despite a declaration of independence in 1816, it remained a loose confederation of provinces facing internal power struggles, including the secession of Buenos Aires. Bartolomé Mitre was finally elected president of a unified country in 1862, and new forms of state making took shape. Visions of culture and progress that embraced European liberalism were debated amid tensions between the supposedly backward provinces and the metropolitan city.[39]

Moreno was part of this process; he was one of many scientists and explorers sent south from Buenos Aires (and Santiago in Chile) to bring back information about the natural resources, native populations, mining prospects, watercourses, and settlement possibilities in the sparsely settled lands of Patagonia.[40] Moreno compiled a series of important firsts on these expeditions through the interior, including hoisting a national flag at Nahuel Huapi in 1876, renaming today's climbing mecca of a peak after the British captain of the *Beagle*, Robert FitzRoy (it was known to the Tehuelche people as El Chaltén), and "discovering" the vibrant blue lakes he would name Lago Argentino and Lago San Martin after the independence fighter. Moreno took his pen to so much topography "naming and claiming" the region that one article reasoned, "Moreno was not only the

Christopher Columbus but the St. John the Baptist of all the lakes and hills of Patagonia."[41] He did more than patriotically rearrange the names; he amassed a sizeable collection of artifacts, including fossils, Pleistocene animal skins, ancient skulls, and ethnographies of the peoples he encountered. The objects became the basis of the La Plata Museum of Natural Sciences, founded in 1886, and the linguistic abilities came in handy when he was kidnapped by a band of Tehuelche in 1880, barely escaping with his life.

A staunch champion of scientific inquiry, Moreno was neither a persecutor nor a champion of indigenous peoples. He had more than his share of contact during his expeditions, with mixed results. Moreno's aim of establishing a route through the cordillera—where Chilean liquor was smuggled in and Argentine cattle rustled out—was postponed when Chief Shaihueque did not permit him to cross.[42] Seen in the context of the day, Moreno's actions were more moderate than the military campaigns of his contemporaries in the government and military. Julio Argentino Roca's famed Desert Campaign that began in 1879 showed how quickly patience with submitting native peoples to state rule turned to violence.[43] Moreno criticized government policy that provided the Indians with rations but not with education. He lamented the overuse of weapons in the "Remington argument" that needlessly destroyed thousands of lives rather than employing "good Indians" in a commission to work out ways to have them yield to national authority.[44] A proponent of gradual assimilation and pacification, Moreno harshly criticized the mindless slaughter so-called civilizing campaigns caused, arguing they "lost the *patria* [fatherland] thousands of sons." He noted distinctions among groups he encountered, learned many customs, trusted Indians in his parties, and at times relied on their expertise and skill. He also remarked on native peoples' generosity, rationality, and eagerness for education.[45] He recognized their forms of governance and that these applied to him, especially that he needed permission to conduct his surveys.[46]

Yet Moreno discussed the nearing extinction of native peoples in Patagonia as a foregone conclusion, remarking that "the Gennaken, like the Ahonekenes or Patagones, are destined to rapid extinction; their character, their customs are completely primitive and can't resist rapid changes, and you will see them languish and disappear without assimilation with the invading races."[47] His explorations served as indirect catalysts for the Desert Campaign despite his refusal to support it. As a scientist, he collected skulls as specimens, but he did forcibly create them himself. Moreno remained curious about Indian lives and their antecedents yet refrained from political action or personal participation in their defense. He instead focused his energies on science and development, with the hope that answers to the struggles with native peoples would follow.

Moreno's pursuit of knowledge caused him to link ignorance about Patagonia to its lack of development. He remained firm in his commitment that the federal government should play a role in developing basic infrastructure—roads, railways, water systems—that would support settlement, leaving the rest to industrious colonists.[48] In the 1880s proceedings took place to determine the boundary line between Chile and Argentina. The nations signed a treaty in 1881 establishing the preliminary determination of the boundary that would follow the flow of water—the point at which water flowed to the Pacific would be Chilean territory, the Atlantic, Argentine.[49] This simple solution had a difficult application that resulted in negotiations in Santiago in 1898 and final arbitration in London in 1903. Moreno served as the primary Argentine expert during the proceedings, and his reputation as a faithful defender of Argentina's territorial sovereignty was further validated.[50]

In payment for his commission work, in August 1903 the government rewarded him with twenty-five leagues of property (some 100,000 acres)—a mere sliver of the more than 40,000 leagues or nearly 200 million acres he retained for the nation during the dispute.[51] Three months later, he donated three leagues (well over 10,000 acres) back to the federal government under the covenant that it would form the nucleus of a national park. Moreno argued that the land's conversion to public property would allow it to serve as "a center of great intellectual and social activity, and even more, an excellent instrument of human progress."[52] On February 1, 1904, the congressional body accepted the land and agreed to his conditions—including prohibitions on hunting, cutting trees, and altering watercourses—by publishing Decree 4192 as the official designation.

Although private benefactors have sustained parks in other countries, Moreno's donation of land surpassed, at least in terms of personal generosity, that of more famous conservation philanthropists such as the Rockefellers, because he died poor, leaving no inheritance for his children. Scientific study was not a profit-making profession, and Moreno gave all his energies to museums and publishing with little remuneration. He returned portions of the land donation and hoped to restrict its use before he had a chance to make any personal fortune. In one of his final diary entries, he lamented, "I am 66 years old and haven't a cent! I have given eighteen hundred leagues to my country [in the dispute negotiation] and a National Park for the benefit of future citizens so that they may find solace and renewed strength to serve this country. Yet I have not so much as a square meter of land to give my children to bury my ashes!"[53] In fact, in his final wishes he asked his children to lay his remains inside the national park, an honor finally arranged by his son nearly twenty-five years after his father's death.[54] Driven by faith in fair development and perpetual warnings about the problems with land concentration, Moreno's commitments shifted quietly throughout his life from the rugged frontiersman fending off wild animals to an advocate for the conservation of wild places for his countrymen.[55]

Moreno met Roosevelt at a point in his life where he had already achieved much. On his tour, Roosevelt decided to explore the Andean and lake regions of Chile and Argentina despite his wife Edith's objections. Having crossed over the mountain range in the northern city of Mendoza, Roosevelt returned via a more remote southern route by car, horseback, and train while Edith sailed home. Roosevelt arrived in Patagonia during early summer, in December 1913, with his entourage of more than a dozen people, and they spent a single night in the town of San Carlos de Bariloche on the shores of Nahuel Huapi Lake. The party included Moreno, whom he pronounced "as devoted a friend as if he had been my aide." They dined with education reformers and immigrant pioneers who, since Moreno had first planted the flag less than ten years before, had turned the town into "the best of Swiss Argentine villages."[56]

Roosevelt enjoyed the sophisticated companionship and was similarly impressed with the landscape, likening the region to the Alps or the parks already lining the Rockies and Cascades of the American West. Roosevelt explained the area in comparative superlatives: "We had been through a stretch of scenery as lovely as can be found anywhere in the world—a stretch that in parts suggested the Swiss lakes and mountains, and in other parts Yellowstone Park or the Yosemite or the mountains near Puget Sound." The comparisons to existing and proposed U.S. parks were not incidental; Roosevelt recognized the park designation around Bariloche and applauded the action. He noted, "Thanks to Doctor Moreno, the Argentine end of it is already a national park; I trust the Chilean end soon will be."[57] By the 1930s an image of Roosevelt in the national park graced the inside of a Spanish language Nahuel Huapi park brochure, and the town's reputation as a national park worthy of visitation solidified after construction of the Hotel Llao Llao, which drew trekkers in the summer and skiers in the winter.[58]

Roosevelt gave Moreno publicity but not necessarily political leverage; Moreno acquired that on his own. He participated in a global network of investigation, exchanging his results across the Atlantic and throughout the hemisphere. Moreno held the U.S. park model in great esteem, as did his successors such as Exequiel Bustillo, the first director of the Dirección de Parques Nacionales created in 1934.[59] Bustillo drew much from the U.S. system, not the least of which was his belief that parks should be supported by the state especially with regard to infrastructure.[60] But the parks also drew heavily from European influences, turning to Germany for attention to rural landscapes and Italy as a source of inspiration for ski hills and cable cars.[61] Contact among park managers was frequent—if erratic—but never hegemonic.[62] Moreno, Bustillo, and other Argentines picked and chose from the array of strategies to protect nature in a pattern that fit their national development trajectory, political situation, strategies for social integration, and temperate (rarely desert or tropical) landscapes. Never static, ideas about conservation and its benefits drew from the changing scientific understandings of both nature and society. Eager to share the wealth and wonder of Patagonia with fellow Argentines, immigrants, and visitors, Moreno's desire to create a park came from his wealth of personal experience and deeply nationalist affinities. Other mestizo men of science developed parallel affinities.

Rondon: Through the Amazonian Wilderness

Choosing a shallow pool on the edge of the river, Rondon carefully inspected the site until satisfied there were no piranhas in the area. Yet, as soon as he put his foot into the water, one of them attacked, biting off a toe completely. Hardly the savage showdown of man versus top predator, this incident—the aggression of a small but ferocious fish toward a sacrificial bodily appendage—captures the contradictions of the tropical landscape and the role of scientist-adventurers within it. After hearing what seemed the hundredth story about piranhas, Roosevelt explained, "These man-eating fish are a veritable scourge in the waters they frequent."[63] The ferocity of the tropical wilderness was not measured in the size of the foe. Caimans, jaguars, and piranhas had their place, but Roosevelt was not the first to note that the true dangers of the Brazilian wilderness were "the torment and menace of attacks by the swarming insects, by mosquitoes and the even more intolerable tiny gnats, by the ticks, and by the vicious poisonous ants which occasionally cause villages and even whole districts to be deserted by human beings."[64] In the midst of these angry, insect-laden woods, the casual piranha strike on Rondon underscored his long-earned knowledge, skill, and tenacity in ways as revealing as the puma assault on Moreno or Roosevelt's buffalo hunt.

If Moreno provides a counterpoint to Roosevelt for Argentina, Rondon serves that role for Brazil. Born in the far western state of Mato Grosso in 1865 to a mixed race Indian mother and a cattle-ranching father, he lost both parents as an infant and was raised by an uncle.[65] After excelling at the Cuyaba government school, Rondon briefly served in the army in Rio de Janeiro then attended military school. After nearly ten years in the capital, he declined a magisterial position and returned to Mato Grosso to construct telegraph lines across the Amazon basin.[66] The Strategic Telegraph Commission of Mato Grosso to Amazonas (later called the Rondon Commission) constructed the first telegraph line to the farthest reaches of the basin and explored, surveyed, and mapped vast extensions of the western territory.[67] Between 1900 and 1906, Rondon oversaw and participated in the building of telegraph stations, bridges, and more than 1,000 miles of telegraph line.[68]

To say Rondon revised the nation's geography understates his contributions: he surveyed more than 50,000 linear kilometers of land, adding twelve new rivers to the map of Brazil. By 1956 a new state (the size of Italy) bore his name, as did a tiny marmoset monkey.[69] He, and the scientists he supervised, contributed 8,837 plants, 5,676 zoological specimens, and 3,380 cultural artifacts to the National Museum.[70] He was honored by geographical societies in New York, Paris, Munich, the Netherlands, Peru, Colombia, and more.[71] His contributions to ethnography were no less impressive. He wrote three volumes of the series *Indios do Brasil* and grammatical guides to the Borôro and Ariti (Parici) Indian languages.[72] He refused a government subvention for his boundary dispute work so the funds could be used to build a school.[73]

When Roosevelt arrived in Brazil, Rondon was less than halfway through this extremely productive life after having spent twenty-four years exploring. Deeply engaged in the telegraph project, he did not offer to host the visiting former president. Instead, the journey down the river was suggested by Lauro Müller, Brazilian minister of foreign affairs and a former classmate of Rondon's, who hoped to make the best of Roosevelt's South American tour. The Roosevelt party intended to visit the Amazon for collections, but the expedition his American colleagues had planned was much less spectacular than the river expedition Müller presented in his first meeting with Roosevelt. In Rondon's explanation, five expeditions were offered to Roosevelt, and "of these five alternatives, the one which offered the greatest unforeseen difficulties was that relative to the river Dúvida. This was the one chosen by Mr. Roosevelt."[74] This route included the exploration of an uncharted river Rondon had encountered while installing telegraph stations.

Roosevelt's enthusiasm for potentially placing a new river on the map was matched by Rondon's capacity to make it happen. Roosevelt described him as "all, and more than all, that could be desired," explaining this to mean Rondon was "a peculiarly hardy and competent explorer, a good field naturalist and scientific man, a student and a philosopher."[75] Rondon continued to defer the leadership title, calling himself the assistant, but Rondon's demeanor, experience, and prior work on the telegraph commission impressed Roosevelt.[76] Explaining the attributes of the soft-spoken Brazilian, Roosevelt noted, "With him the conversation ranged from jaguar-hunting and the perils of exploration in the 'Matto Grosso,' the great wilderness, to Indian anthropology, to the dangers of a purely materialistic industrial civilization, and to Positivist morality."[77] The geographic objective gave Roosevelt reason to elevate the trip beyond a hunting expedition and thus avoid some of the criticism he had received for his earlier African safari.[78] Indeed, the trip was not a bloodthirsty slaughter. Not only were large mammals fewer and harder to acquire, Rondon noted places they chose not to hunt after adequate specimens—of jaguars, tapirs, giant anteaters, capybaras, and others—had been collected, "as there was no need to get more."[79]

After nearly four months of arduous travel through treacherous rapids, lost dugout canoes, and dwindling rations, the commission emerged from the wilderness having surveyed the river and renamed it the Rio Roosevelt. It cost them the lives of two commission members (one to drowning, another to murder) as well as Rondon's favorite dog (to Nambiquara arrows), and nearly Roosevelt's life as well (to fever and an infected leg wound). Roosevelt returned to Manaus and then New York where he unveiled the Rio Roosevelt to geographic societies worldwide. Rondon went back to work on the telegraph lines until radios made them obsolete.

Along with the line came colonization and an expansion of state authority. Through his years in the swamps, forests, and plains of Amazonia, Rondon became increasingly aware of the nation-building and territorial implications of his work for those already living there. After the telegraph commission, Rondon's crowning achievement was his founding of the Indian Protection Service (SPI, 1910) guided by the philosophy he exercised toward native peoples throughout his career: "To die if necessary; to kill, never." The service functioned to protect Indians from persecution by settlers by providing education, land titles, and other forms of state-directed patronage.

Rondon's work with indigenous peoples is both his most praised and most critiqued endeavor. Official publications and histories hail him as a national hero, while recent scholarship argues that his actions constituted a different shade of paternalistic conquest.[80] Todd Diacon's work on the telegraph commission argues for a middle ground between these divisive extremes. Like Moreno, Rondon sought to understand the native peoples he encountered. His guiding principles for the state's relationship with indigenous peoples were radical given the context. For instance, he saw indigenous groups as sovereign peoples and strove to guarantee them rights to the lands they needed to survive. At a time when throughout the hemisphere Indians were increasingly evicted from their territories through a range of schemes, the assurance of territorial sovereignty marked a different path.[81] Yet Rondon's adherence to positivist models of development shaped his philosophy toward native peoples, and he endeavored to turn Indians into "Brazilians" as a specific end to the course of state tutelage. Perhaps the most biting critique of Rondon's relationship with indigenous peoples was the line itself, which became the path for BR-230, the Trans-Amazon Highway. The road, like the telegraph line before it, brought people into the interior, facilitating vast social and environmental transformations. New settlers took Indian land and spread disease as they cleared the forests. Despite the fact that he did perhaps more than any other single person to shelter indigenous peoples from change, during the last half of Rondon's life, Indian populations declined from an estimated 1 million to less than 200,000.[82]

In his later years, Rondon deepened his work with native peoples, including national park developments in the Amazonian region. In one of their many early forms, some national parks in the Brazilian Amazon were intended to set aside land for indigenous people to remain living there. One of the more interesting examples of the manifestation of the idea of territorial protection alongside the protection of indigenous peoples is the case of Xingu National and Indigenous Park. Rondon played a role on the committee proposing a park at Xingu in 1952 among a cadre of federal administrators pursuing an alliance between native people and nature protection. Unlike many U.S. reservations that removed and resettled peoples, this park allowed—indeed, was designed specifically to facilitate—native peoples to remain living on traditional lands. Xingu, an isolated area around the river by the same name, in the northern portion of Mato Grosso due east of the Rio Roosevelt, is in a transitional ecological zone between tropical savannah and rain forest hosting native peoples from all four major aboriginal language groups in Brazil (Tupi, Arawak, Carib, and Gê.[83] The park, created in 1961 along with five other national parks, represented a pathbreaking ideological design for conservation. Park proponents included Rondon, his close friend anthropologist Darcy Ribeiro, Museu Nacional scientist José Cândido de Melo Carvalho, and the park's future champions and first administrators, Orlando, Leonardo, and Claudio Villas Bôas. Making the case to President Getulio Vargas in 1952, Ribeiro argued that only the Indians were "capable of preserving a living sample of the original nature of Brazil, which everywhere is being destroyed."[84]

The promoters advocated two objectives for the park: to build a natural reservation to safeguard flora and fauna for the future, and to extend immediate protection to the indigenous peoples of the region, including defense against settlers. The Villas Bôas brothers viewed protection of the Indians as the best insurance against the "white invaders, lured by riches in diamonds, gold, pelts, and land, [who] spread disease and at times even machine-gunned the tribes."[85] The first objective resonated with nature protection elsewhere in the hemisphere. The second marked a radical extension of the idea of preservation to native peoples. To be sure, constructing indigenous cultures as "another endangered species" did not grant native peoples administrative or decision-making rights, but it assured them a place inside a national park, unlike that of their peers elsewhere.[86] This was more sophisticated than the romantic notion of a "noble savage" living in nature because Rondon and others worked to assure native sovereignty through land rights. Because of native activism, new policies, and a dynamic evolution of conservation areas, additional people-centered parks including extractive reserves recognizing traditional people's rights to use natural resources were added to Brazil's portfolio of parks by 1987.[87]

The radical foresight of Xingu's design reveals the specific national context of park creation.[88] Some park promoters, including Rondon and the Villas Bôas brothers, argued for the indigenous residents of the region as the most appropriate custodians of the natural environment in a language reminiscent of more recent calls for native stewardship. For instance, journalist Mark Dowie has recently lamented the loss of people in protected areas worldwide, claiming, "If we really want people to live in harmony with nature, history is showing us that the dumbest thing we can do is kick them [native peoples] out of it."[89] History in this case shows that people were not kicked out; the national park was designed with them in mind. However, that parks and peoples were merged did not stop the rapaciousness of development around them. The lack of historical introspection or context in recent denunciations such as Dowie's begs a reevaluation of the early conservation actions in some of South America's most paradigmatic landscapes. Rondon advocated Brazil's (and possibly the world's) first combined indigenous reserve *and* national park (although the park is today managed solely as an indigenous area). It is unlikely the political room to create a park such as Xingu would have opened without his prior work trying to understand native peoples and administering the SPI, and his status as a national hero for doing so. Competing local and state interests fought vociferously against the park and what it signified as an early courtship between protecting the environment and indigenous peoples.[90]

Comparative Hemispheric Conservation

Threads of similarity weave throughout the lives of these three men. Their scientific activities found support from research institutes, museums, and federal governments. Private institutions the men collected for in Rio and Buenos Aires were not so different from those in New York. Located in metropolitan areas and dedicated to enlightenment questions, learned, wealthy patrons paid for these institutions and liked the modernity science implied. But many of the actual scientists were of a more humble nationalist stripe. Rondon was mestizo, born in the frontier. Moreno was born to an immigrant mother in the city. Both spent the formative years of their adulthood in wild country in constant interaction with native peoples, learning their languages and practices and conducting projects of great scientific merit. Rondon could converse with many native peoples he encountered but not with Roosevelt in English.[91] Like their American peer, they became experts in a language of science that relied on a suite of technical skills including mapping and charting rivers, surveying landscapes, describing geological and biological attributes of specific areas, and identifying birds, plants, and other flora and fauna. In contrast to Roosevelt, both men were excellent marksmen and had the marks of men—scars from their encounters with dangerous animals. Diligent note-takers and prolific writers, these men also had an unassuming air about them that allowed them to move between high society in the seats of power and remote outposts staffed by rugged settlers.

The lives of Roosevelt, Moreno, and Rondon offer conservationists a more rounded picture of how and where nature becomes part of a national patrimony. The Amazonian and Patagonian regions today protect some of the largest contiguous tracts of intact ecosystems on the planet. These are not unpeopled landscapes, they are not exclusive enclaves, but they are reserves in the public sphere as a compromise with the future. And people in Argentina and Brazil recognized their worth up to a century ago. Ian Tyrell has recently argued that U.S. historians have focused too narrowly on parks at the expense of seeing a more expansive view of concerns over the impact of civilization on the natural world.[92] In his letter donating the land for the park, Moreno mentions the United States but also the work of Australia and New Zealand.[93] That the United States had parks was not enough of an argument for Argentina to create them—men such as Moreno recognized it was a global trend and a national necessity.

For some, that Moreno and Rondon knew Roosevelt would be enough to dismiss their intentions as at best borrowed, at worst tools of the long arm of U.S. imperialism. In such an interpretation, South America might be one more place where conservationists used protected areas as "an instrument of capitalism" in the "century-old conflict between global conservation and native peoples" that resulted in "thousands of unmanageable protected areas."[94] But the historical record does not support this conclusion; Rondon and Moreno were well-established men of science *before* they met Roosevelt. Their work has been overlooked or lost—along with that of scores of scientists, educators, and tireless champions of wild spaces—by the generations in between.[95] Much attention has been given to the role of scientific travelers as appropriative forces distorting alternative ways of evincing knowledge. Such a reading overlooks national politicians and domestic leaders who played a much larger role in park administration than did external organizations.[96]

While postcolonial theorists may see scientists as agents of empire representing the heart of the colonial project, there is much evidence to suggest that the men themselves inhabited the margins of this enterprise.[97] Here Roosevelt proves an imperfect comparison; other U.S. conservationists, such as John Muir, John Wesley Powell, or Bob Marshall, might serve better parallels, but Moreno and Rondon's national recognition was more similar to Roosevelt's, and they knew him personally. Candace Millard characterizes Rondon's life, unlike Roosevelt's, as "spent at the edge of Brazil's frontier—and at the margins of its society."[98] Moreno died broke and forgotten. That these men and their accomplishments infiltrated mainstream national life speaks to the wealth of their accomplishments, their risk taking, constant questioning, and years of often grueling personal experiences.

Like Humboldt before them, these men saw nature as both an intellectual and a social pursuit. Their actions were not driven by the personal profit schemes that motivated many of their compatriots. That these men were collecting for science, not for the sale of what they collected, matters. Make no mistake—Moreno and Rondon were concerned about development more than conservation in its traditional sense. For them, the tragedy was that resources were not being used to benefit more people. In this sense they had concern over what today might be called social justice. Moreno lamented concentration of land among few owners.[99] Rondon's positivist view drove his commitment to "protect" indigenous peoples from those he saw as cruel and to work for national incorporation of distant regions. That these men were developmentalists does not cancel out their passion for nature, their understanding of it, and their wish to see it remain as part of the natural patrimony. Often conservation is seen in stark contrasts rather than many shades; it has been too frequently subject to false dichotomies.[100] Moreno, Rondon, and Roosevelt show us that strict binaries such as development *or* conservation, parks *or* indigenous peoples are more recent constructs.

Notes

1. Theodore Roosevelt, "Buenos Aires: A Fine Modern Capital," *The Outlook,* March 28, 1914, http://www.theodorerooseveltcenter.org/en/Research/Digital-Library/Record/ImageViewer.aspx?libID=o279305&imageNo=1 (accessed August 27, 2015).

2. Roosevelt's former actions, especially in Cuba and Panama, provided important context for his reception; he was criticized heavily in Santiago, Chile.

3. The literature on Moreno is immense. For an introduction, see Francisco P. Moreno, *Viaje a la Patagonia Austral* (Buenos Aires: Imprenta de la Nación, 1879); Eduardo V. Moreno, *Reminisciencias del Perito Moreno* (Buenos Aires: Elefante Blanco, 1997); Jose Liebermann, *"Francisco P. Moreno: Precursor Argentino," Anales de la Sociedad Científica* Argentina (1945): 396–427; Carlos A. Bertomeu, *El Perito Moreno, centinela de la Patagonia* (Buenos Aires: Editorial El Ateneo, 1949); Miranda I. Márquez, "Francisco P. Moreno y las 'ciencias del hombre' en la Argentina," *Ciencia e Investigación* 8 (1952): 484–92; Aquiles Ygobone, *Viajeros científicos de la Patagonia* (Buenos Aires: Editorial Galerna, 1977); idem, *Francisco P. Moreno: Arquetipo de la Argentinidad* (Buenos Aires: Editorial, 1954); Antonio Requeni, *Francisco P. Moreno: Perito en Argentinidad* (Buenos Aires: Artes Gráficas Yerbal, 1998); Roberto Hosne, *Francisco Moreno: Una herencia patagónica desperdiciada* (Buenos Aires: Émece, 2004). There is debate as to whether or not Nahuel Huapi is the *first* park—it was the first piece of land accepted in a government document as a national park, although it was not formally established for management until 1934. This was neither the first proposal for a park (which was made by André Rebouças in Brazil in 1876, but the recommended parks were not created until 1959) nor was it the first formally designated park (which was created by Chile at Vicente Pérez Rosales National Park in 1926). See discussion in Gary Wetterberg, "The History and Status of South American Parks and an Evaluation of Selected Management Options" (PhD diss., University of Washington, 1974), 37–40.

4. Mary and Laurance Rockefeller, "How South America Guards Her Green Legacy: Parks, Plans, and People," *National Geographic Magazine* 131, no. 1 (1968): 74–119.

5. Theodore Roosevelt, *Through the Brazilian Wilderness* (New York: Charles Scribner & Sons, 1914), preface.

6. Edmund Morris, *Colonel Roosevelt* (New York: Random House, 2010), 347; Joseph R. Ornig, *My Last Chance to Be a Boy: Theodore Roosevelt's South American Expedition* (Mechanicsburg, Penn.: Stackpole, 1994), vii.

7. Rio da Dúvida (River of Doubt) was renamed the Rio Roosevelt. Candace Millard, *River of Doubt: Theodore Roosevelt's Darkest Journey* (New York: Doubleday, 2005); Morris, *Colonel Roosevelt.* For an introduction to literature on Roosevelt, see David McCullough, *Mornings on Horseback: The Story of an Extraordinary Family, a Vanished Way of Life, and the Unique Child Who Became Theodore Roosevelt* (New York: Simon & Schuster, 1982); Edmund Morris's three volumes, *The Rise of Theodore Roosevelt* (New York: Modern Library, 2001), *Theodore Rex* (New York: Random House, 2002), and *Colonel Roosevelt*; Kathleen Dalton, *Theodore Roosevelt: A Strenuous Life* (New York: Vintage, 2004); and Douglas

Brinkley, *Wilderness Warrior: Theodore Roosevelt and the Crusade for America* (New York: Harper Perennial, 2009).

8. On Rondon, see Todd A. Diacon, *Stringing Together a Nation: Cândido Mariano da Silva Rondon and the Construction of a Modern Brazil, 1906–1930* (Durham, N.C.: Duke University Press, 2004); Darcy Ribeiro, *O Indigenista Rondon* (Rio de Janeiro: Ministério da Educação e Cultura, 1958); Mário Garcia de Paiva, *A grande aventura de Rondon* (Rio de Janeiro: Instituto Nacional do Livro, Ministério da Educação e Cultura, 1971); Antônio Carlos de Souza Lima, *Um grande cerco de paz: poder tutelar indianidade e formação do Estado no Brasil* (Petrópolis: Vozes, 1995).

9. Diacon, *Stringing Together a Nation;* Nancy Leys Stepan, *Picturing Tropical Nature* (New York: Reaktion Books, 2006); Aaron Sachs, "The Ultimate 'Other': Post-Colonialism and Alexander von Humboldt's Ecological Relationship with Nature," *History and Theory* 42, no. 4 (2003): 111–35; Mark Carey, *In the Shadow of Melting Glaciers: Climate Change and Andean Society* (New York: Oxford University Press, 2010); Carolina Marotta Capanema, "A natureza no projeto de construção de um Brasil Moderno e a obra de Alberto José de Sampaio" (PhD diss., Universidade Federal de Minas Gerais, 2006). Works within history of science pay more attention to the colonial era: Antonio Barrera Osorio, *Experiencing Nature: The Spanish American Empire and the Early Scientific Revolution* (Austin: University of Texas Press, 2006); José Augusto Pádua, *Um sopro de destruição: Pensamento político e crítica ambiental no Brazil escravista (1786–1888)* (Rio de Janeiro: Jorge Zahar, 2004).

10. José Luiz de Andrade Franco and José Augusto Drummond, "Wilderness and the Brazilian Mind (I): Nation and Nature in Brazil from the 1920s to the 1940s," *Environmental History* 13, no. 4 (2008): 724–50, and "Wilderness and the Brazilian Mind (II): The First Brazilian Conference on Nature Protection (Rio de Janeiro, 1934)," *Environmental History* 14, no. 1 (2009): 82–102. For Patagonia, see Eugenia Scarzanella, "Las bellezas naturales y la nación: Los parques nacionales en Argentina en la primera mitad del siglo XX," *Revista Europea de Estudios Latinamericanos y del Caribe* 73 (2002): 5–21; Eduardo Miguel E. Bessera, "Politicas de estado en la Norpatagonia Andina, Parques Nacionales, desarrollo turístico y consolidación de la frontera. El caso de San Carlos de Bariloche (1934–1955)" (PhD diss., Universidad Nacional del Comahue, 2008). For a review, see Sterling Evans, "Historiografía verde: Estado de la historia sobre la conservación de la naturaleza e América Latina," in,*Naturaleza en declive: miradas a la historia ambiental de América Latina y el Caribe,* ed. Reinaldo Funes Monzote (Valencia, Spain: Centro Francisco Tomás y Valiente UNED Alzira-Valencia/Fundación Instituto de Historia Social, 2008).

11. Mary Louise Pratt, *Imperial Eyes: Travel Writing and Transculturation* (London: Routledge, 1992); Karl S. Zimmerer, "Humboldt's Nodes and Modes of Interdisciplinary Environmental Science in the Andean World," *Geographical Review* 96, no. 3 (2006): 335–60; Aaron Sachs, *Humboldt Current: Nineteenth-Century Exploration and the Roots of American Environmentalism* (New York: Viking, 2006); Laura Dassow Walls, *Passage to Cosmos: Alexander*

Humboldt and the Shaping of America (Chicago: University of Chicago Press, 2009).

12. Argentine forestry engineer Carlos Gallardo pushed for the legislation expanding Nahuel Huapi (see editorial series in *La Nacion* [Buenos Aires], 1913). The Villas Bôas brothers were as formative as Rondon in creating Xingu National Park in the 1960s. See the discussion later in this chapter as well as Seth Garfield, "A Nationalist Environment: Indians, Nature, and the Construction of the Xingu National Park in Brazil," *Luso-Brazilian Review* 41, no. 1 (2004): 139–67.

13. Sarah Watts, *Rough Rider in the White House: Theodore Roosevelt and the Politics of Desire* (Chicago: University of Chicago Press, 2003); Gail Bederman, *Manliness and Civilization: A Cultural History of Gender and Race in the United States, 1880–1917* (Chicago: University of Chicago Press, 1996), 7.

14. Sérgio Buarque de Holanda, *Visão do paraíso: Os motivos edênicos no descobrimento e colonização do Brasil* (São Paulo: Editora Nacional, [1959] 1977); Cynthia Radding, *Landscapes of Power and Identity: Comparative Histories in the Sonoran Desert and the Forests of Amazonia from Colony to Republic* (Durham, N.C.: Duke University Press, 2005); Pádua, *Um sopro*; Margarota Gascón, *Naturaleza e imperio: Araucanía, Patagonia, Pampas (1598–1740)* (Buenos Aires: Editorial Dunken, 2007); Barrera Osorio, *Experiencing Nature;* Neil Safier, *Measuring the New World: Enlightenment Science and South America* (Chicago: University of Chicago Press, 2008); Susan Deans-Smith, "Nature and Scientific Knowledge in the Spanish Empire: Introduction," *Colonial Latin American Review* 15, no. 1 (2006): 29–38.

15. Identifying the appropriate descriptor to talk about actions undertaken by people of the United States is tricky in a discussion of South America. "American" is not self-evident as a reference to the United States, as many Latin American nations employ it. Neither is "North American" precise. Canada and Mexico have distinct histories of conservation that cannot be mindlessly grouped with that of the United States. "United States" does not have a conventional possessive term (United Statesian is awkward) and "United States American" is long and unwieldy. For these reasons, I use "American" to distinguish the United States or actions of those citizens of the United States hereafter. For reflections on these issues, see Felipe Fernandez Armesto, *The Americas: A Hemispheric History* (New York: Random House, 2003), and Greg Grandin, "Your Americanism and Mine: Americanism and Anti-Americanism in the Americas," *American Historical Review* (2006): 1042–66.

16. Brinkley, *Wilderness Warrior,* 147.

17. Theodore Roosevelt, "My Life as a Naturalist," *American Museum Journal,* May 1918, reprinted in *The Works of Theodore Roosevelt,* National Edition, Vol. 5 (New York: Charles Scribner & Sons, 1926; hereafter cited as *Works*), 385; Brinkley, *Wilderness Warrior,* 161.

18. Brinkley, *Wilderness Warrior,* 152; Theodore Roosevelt, *Hunting Trips of a Ranchman* (New York: G. P. Putnam's Sons, 1885).

19. Brinkley, *Wilderness Warrior,* 163; letter from Theodore Roosevelt to Alice Lee Roosevelt, 1883–09–20, Theodore Roosevelt Collection, MS Am 1541.9 (103), Houghton Library, Harvard University.

20. Roosevelt, "My Life as a Naturalist," 393.

21. Brinkley, *Wilderness Warrior,* 502.

22. Theodore Roosevelt, "My Trip in Africa," address to Union League Club, February 22, 1911, reprinted in *Works,* 5:394, 398.

23. Thomas G. Dyer, *Theodore Roosevelt and the Idea of Race* (Baton Rouge: Louisiana State University Press, 1980), 71.

24. Colin G. Calloway, *First Peoples: A Documentary Survey of American Indian History* (Boston: Bedford/St. Martin's, 2012), 420.

25. Ibid.

26. Dyer, *Roosevelt and Idea of Race,* 85.

27. Ibid., 84; Frederick E. Hoxie, *A Final Promise: The Campaign to Assimilate the Indians, 1880–1920* (Lincoln: University of Nebraska Press, 1984), 107.

28. Hoxie, *Final Promise,* 102.

29. Ibid., 104; Dyer, *Roosevelt and Idea of Race,* 87.

30. Hoxie, *Final Promise,* 108.

31. Ibid., 113.

32. Totals derived from appendices in Brinkley, *Wilderness Warrior,* 818–30, and *Theodore Roosevelt and Conservation, 1858–1958.* Theodore Roosevelt Centennial Symposium. Dickinson State University, 1958, http://www.theodore rooseveltcenter.org/Research/Digital-Library/Record.aspx?libID=o275771 (accessed August 5, 2015).

33. Brinkley, *Wilderness Warrior,* 406.

34. Description of attack derived from composite of what Moreno reports in *Viaje,* 375–80; what he told Roosevelt in *Through the Brazilian Wilderness,* 19–21; Hosne's rendering in *Francisco Moreno,* 86–87; and Requeni, *Moreno,* 18–19.

35. The initial P stood for Pascasio, the saint on whose day he was born. Hosne, *Francisco Moreno,* 11. Later the P was taken to mean "Perito" or "the expert," which became an affectionate nickname.

36. Alejandro Winograd, *Patagonia: Mitos y certezas* (Buenos Aires: Edhasa, 2008), 174.

37. Eduardo Moreno, *Reminiscencias,* 10; Ygobone, *Moreno,* xii.

38. Many works examine this transition: Ariel de la Fuente, *Children of Facundo: Caudillo and Gaucho Insurgency during the Argentine State-Formation Process* (Durham, N.C.: Duke University Press, 2000); Jeffrey M. Shumway, *The Case of the Ugly Suitor and Other Histories of Love, Gender, and Nation in Buenos Aires, 1776–1870* (Lincoln: University of Nebraska Press, 2005).

39. Tulio Halperín Donghi, Iván Jacsic, Gwen Kirkpatrick, and Francine Masiello, eds., *Sarmiento: Author of a Nation* (Berkeley: University of California Press, 1994).

40. Chris Moss, *Patagonia: A Cultural History* (New York: Oxford University Press, 2008), 181; Susana Bandieri, *Historia de la Patagonia* (Buenos Aires: Editorial Sudamericana, 2005), 114.

41. Moss, *Patagonia,* quotation on p. 182.

42. Requeni, *Moreno,* 15; Winograd, *Patagonia,* 174; Francisco Moreno, *Viaje,* 11.

43. Bandieri, *Historia de la Patagonia,* 125; Juan C. Walter, *La Conquista del*

Desierto: Lucha de frontera con el indio (Buenos Aires: Eudeba, 1970); Eduardo Moreno, *Reminiscencias*, 34.

44. Eduardo Moreno, *Reminiscencias,* 122–26.

45. Francisco Moreno, *Viaje,* 9–10.

46. Eduardo Moreno, *Reminiscencias*, 36.

47. Ibid., 146.

48. Winograd, *Patagonia,* 177; Francisco Moreno, "Reservas nacionales: La acción parlamentaria," *La Nación* (Buenos Aires)*,* October 3, 1912.

49. Francisco Moreno, *Records of the Proceedings of the Argentine and Chilian* [*sic*] *Experts Concerning the Demarcation of the Boundary-Line between the Argentine Republic and Chili* [*sic*] (Buenos Aires: Imprenta de M. Biedma e hijo, 1898).

50. Ygobone, *Moreno,* x.

51. Ibid., xii.

52. "El parque Nahuel-Huapi: Aceptación de la donación Moreno. Una importante obra internacional," *El Diario* (Buenos Aires), February 8, 1904. Archivo del Museo de la Patagonia, San Carlos de Bariloche, Argentina.

53. Moss, *Patagonia*, 185.

54. Eduardo Moreno to Emilio Frey, May 21, 1934, Colección Emilio Frey, Archivo del Museo de la Patagonia, San Carlos de Bariloche, Argentina (hereafter CEF); Liebermann, "Moreno," 202.

55. Francisco Moreno, *Viaje*; Eduardo Moreno, *Reminiscencias.*

56. Primo Caprario to Emilio Frey, December 1, 1913, CEF.

57. Roosevelt, *Works*, v. 3, 273. Chile created the Benjamin Vicuña Mackenna National Park in 1925 and Vicente Pérez Rosales National Park in 1926.

58. Nahuel Huapi brochure, National Archives and Records Administration (U.S.), Record Group 79, Entry 10: Central Classified Files, 1907–49, Box 2913, Folder: Proposed Foreign Parks, Argentina, 1934–47; Emilio Frey to Eduardo Moreno, July 14, 1934, CEF.

59. Scarzanella, "Las bellezas naturales," 10; Exequiel Bustillo, *El despertar de Bariloche* (Buenos Aires: Casa Pardo, 1968).

60. Bailey Willis, "El parque nacional del sud," *Dirección General de Agricultura y Defensa Agrícola Boletín* (Buenos Aires), no. 2, 1913. Explicit comparisons were made to Yosemite and Yellowstone by Gallardo. See editorial in *La Nacion,* March 4, 1913.

61. Scarzanella, "Las belleza naturales," 10.

62. See correspondence by Schleimar of Buenos Aires in NARA, Record Group 79, Entry 10: Central Classified Files, 1907–49, Box 630, Foreign Parks Brazil 1919–31.

63. Roosevelt, *Through the Brazilian Wilderness,* 31.

64. Ibid., 32.

65. Todd Diacon, "Cândido Mariano da Silva Rondon and the Politics of Indian Protection in Brazil," *Past and Present* 158 (2002):158; Roosevelt, *Through the Brazilian Wilderness,* 125.

66. Ribeiro, *O Indigenista,* 15.

67. Comissão de Linhas Telegráficas Estratégicas de Mato Grosso ao Amazonas (CLTEMTA). See Diacon, *Stringing Together a Nation*, 4.

68. Ibid., 17.

69. Ribeiro, *O Indigenista,* 18, 50; Edgard Roquette-Pinto, *Rondônia* (São Paulo: Companhia Editora Nacional, 1950).

70. Paiva, *A grande aventura,* 20.

71. Ribeiro, *O Indigenista,* 56.

72. Ibid., 52–68.

73. Paiva, *A grande aventura,* 20.

74. Rondon, *Lectures Delivered by Colonel Cândido Mariano da Silva Rondon,* trans. R. G. Reidy and Ed. Murray (New York: Greenwood, [1916] 1969), 12.

75. Roosevelt, *Through the Brazilian Wilderness,* 30, 43.

76. Rondon, *Lectures,* 37; Roosevelt, *Through the Brazilian Wilderness,* 30.

77. Roosevelt, *Through the Brazilian Wilderness,* 43.

78. Morris, *Colonel Roosevelt,* 14.

79. Rondon, *Lectures,* 28–29.

80. Diacon, *Stringing Together a Nation,* 116–25; Lima, *Um grande cerco.*

81. Richard J. Perry, *From Time Immemorial: Indigenous Peoples and State Systems* (Austin: University of Texas Press, 1996), chap. 8.

82. Jim Igoe, *Conservation and Globalization: A Study of National Parks and Indigenous Communities from East Africa to South Dakota* (Toronto: Wadsworth/ Thomson Learning, 2004), 143.

83. Garfield, "Nationalist Environment," 139.

84. As translated and quoted in ibid., 156.

85. Orlando Villas Bôas and Claudio Villas Bôas, "Saving Brazil's Stone Age Tribes from Extinction," *National Geographic* 134, no. 3 (1968): 424–44, quotation on 427.

86. Quote from Igoe, *Conservation and Globalization,* 145.

87. Beth A. Conklin, "Body Paint, Feather, and VCRs: Aesthetics and Authenticity in Amazonian Activism," *American Ethnologist* 24, no. 4 (1997): 711–37.

88. Garfield, "Nationalist Environment," 152.

89. Mark Dowie, *Conservation Refugees: The Hundred-Year Conflict between Global Conservation and Native Peoples* (Cambridge, Mass.: MIT Press, 2009), 266.

90. Kent Redford, "The Ecologically Noble Savage," *Cultural Survival Quarterly* 25, no. 1 (1991): 46–50; Manuela Carneiro da Cunha and Mauro W. B. de Almeida, "Indigenous People, Traditional People, and Conservation in the Amazon," *Daedalus* 129, no. 2 (2000): 315–38.

91. Rondon and Roosevelt could only converse in French, which neither spoke well, or by using Roosevelt's son Kermit as translator. Moreno and Roosevelt could converse in English since Moreno learned to speak the language as a young child from his mother.

92. Ian Tyrell, "America's National Parks: The Transnational Creation of National Space in the Progressive Era," *Journal of American Studies* 46, no. 1 (February 2012): 1–21.

93. "Para parque nacional: Donación de Dr. F. P. Moreno," *La Nación* (Buenos Aires), November 8, 1903.

94. Dan Brockington, Rosallee Duffy, and Jim Igoe, *Nature Unbound: Conservation, Capitalism, and the Future of Protected Areas* (London: Earthscan, 2008), 6; Dowie, *Conservation Refugees,* xi.

95. For an overview of these "lost generations," see Franco and Drummond, "Wilderness and the Brazilian Mind (I)."

96. Maria Buchinger, "Conservation in Latin America," *BioScience* 15, no. 1 (1965): 32–37; Exequiel Bustillo, *Huellas de un largo quehacer. Discursos, conferencias, artículos y publicaciones diversas* (Buenos Aires: Ediciones Depalma, 1972).

97. Sachs, "Ultimate 'Other,'" 132.

98. Millard, *River of Doubt*, 80.

99. Hosne, *Moreno,* 8.

100. Thomas Lovejoy, "Glimpses of Conservation Biology, Act II," *Conservation Biology* 20, no. 3 (2006): 711–12.

FIVE

Nature's Laboratories?

Exploring the Intersection of Science and National Parks

PATRICK KUPPER

IN THE LATE 1950S AND EARLY 1960S, international conservation gained momentum. The process of decolonization had reached a critical mark and the conservation establishment was feverishly preparing for a postcolonial world. In this period of transformation the International Union for the Conservation of Nature, founded in 1948, became a key actor. The IUCN compiled information on the global state of conservation and facilitated the international exchange of conservation experts. While the organization in its founding years was dominated by Europeans, the influence of U.S. citizens rapidly increased. At the turn of the 1960s U.S. conservationists had already assumed important positions, and it was the Americans who took the responsibility to organize the First World Conference on National Parks, held in Seattle in 1962.[1] It was in this context that the American Committee for International Wild Life Protection, whose chairman, Harold J. Coolidge, also chaired a joint IUCN and United Nations committee on national parks, invited a number of conservationists from around the world to express their views on the need and uses of national parks. The result was published in the wake of the Seattle conference, in spring of 1962, in a small brochure titled "National Parks—A World Need."[2] How the issue of science is treated throughout this high-profile brochure is most instructive for our topic.

In charge of "National Parks—A World Need" was Victor H. Cahalane, vice chairman of the American Committee, assistant director of New York's State Museum, and former chief biologist of the U.S. National Park Service. In his introduction to the volume Cahalane remarked with delight that "in less than a century, the national park idea has spread around the world. Every continent now has national parks." At the same time, Cahalane noted that under "different nationalities and cultures the term 'national parks' has acquired some diversity of meaning."[3] On the ground of their main purpose Cahalane distinguished four types of national parks. Most African parks were "primarily wildlife sanctuaries." In North and Central America "superlative scenic quality" was at the forefront, giving rise to a "wilderness type of park," in which "permanent human residence is terminated" while "visitation by large numbers of people is encouraged." A third kind of national parks Cahalane saw realized in Britain and Japan, which both "sought to reconcile modern man and Nature in *joint* occupation of outstanding scenic areas." Parkland was in private ownership and agricultural use continued while public access was secured and visitation encouraged. Cahalane's forth type, which actually topped his list, contained parks in Switzerland, the Congo Republic, and French West Africa. These parks "were founded primarily as field laboratories for research on natural resources." Their recreational use was "definitely secondary or even non-existent."[4]

After this categorization Cahalane went on to defend national parks against the charge of being limited to a single use by portraying them as multiple-use areas. By preserving vegetation they would reduce soil erosion and safeguard watersheds to the benefit not only of the parkland but also its environs. A second service fulfilled by parks beyond their borders was the restocking of surrounding regions with game. As a third field Cahalane mentioned the parks' use as "extensive outdoor laboratories," where soil, plants, animals, and their interactions could be studied under natural conditions providing, among other things, insights about the best methods of land management. This kind of use was championed in the research-driven type of national parks. As prime examples Cahalane referred to Switzerland and the Soviet Union, which were both represented in the collection with a contribution.[5] As a fourth use of national parks Cahalane discussed tourism and its economic stimulus, which would especially profit less developed regions. Finally, he stated that there was still another use, which could not be expressed in money but was nevertheless real and demonstrable: "the aesthetic and spiritual value of wild places, which are best preserved in national parks."[6]

Science figures prominently in Cahalane's introduction. It goes without saying that scientific research was at the core of national parks' use as outdoor laboratories. Furthermore, Cahalane attributed to scientists a crucial role in the management of wildlife populations as well as in the proper designation of sanctuaries. Thus, national parks could serve as sites of scientific research while scientific research could contribute to the rational conceptualization and the proper management of parks. Himself a biologist, Cahalane was predisposed to highlight the mutual benefits performed by national parks and scientific research. A nonscientist may have put less emphasis on science and more on aesthetic, ethical, and recreational components of national parks. In this regard a close reading of the whole volume is revealing. Science is a central feature in only two of the thirteen contributions, in Lev Konstantinovich Shaposhnikov's account on "Nature Reserves in the Soviet Union" and in Jean G. Baer's "The

Swiss National Park." In the other eleven contributions, which cover East Africa and French West Africa, Australia, India, Japan, Europe, Canada, the United States, and Brazil, science is hardly mentioned, and if it was, then mostly as a residual yet neglected field of national parks.[7]

Taken as a whole, Cahalane's compilation leaves us with an ambiguous picture. Science is acknowledged as a force within national parks. Its importance, however, seems to greatly vary between different locations, and in many instances its involvement in national parks appears as more a plea than a fact. This leads us to the main goal of this chapter, namely, to reconstruct the evolution of the intersection of science and national parks in global context. In doing so, I am mainly interested in two aspects: the use of national parks as research sites for science on the one hand, and the use of science in guiding park management on the other. In order to be able to show both how this history unfolded in different places and how and to what effect such places were linked, I will have to combine and connect investigations on different scales. I will try both to draw the big picture of the park-science intersection and to give evidence to it by providing some close-ups of instances in which we can trace the intersection's unfolding in situ. Four research questions are guiding my investigation: I first ask what contexts favored an interconnection of science and national parks. Second, I am eager to establish how scientists networked national parks beyond national borders. Third, I am interested in what tensions the involvement of science caused in the parks, and, finally, I want to explain to what degree scientific knowledge and practices transformed national parks.

Involvement

When the United States shaped its national park system in the late nineteenth and early twentieth centuries, the preservation of scenic beauty was the main purpose. National parks (and other units of the expanding park system) were seen as national icons to be preserved from private speculation and become a public good. Scientific arguments played only a peripheral role. The Antiquities Act of 1906, which gave the president executive authority to establish national monuments, made reference to "scientific interest" but did not instill scientific reasoning into the park system in a durable way.[8] Ten years later, when the U.S. National Park Service (NPS) was created as a federal agency to administer all current and future national parks and monuments, the founding act did not mention any scientific interest in those reservations. The stated purpose was "to conserve the scenery and the natural and historic objects and the wild life therein and to provide for the enjoyment of the same in such manner and by such means as will leave them unimpaired for the enjoyment of future generations." In order to facilitate the public enjoyment the act of 1916 permitted the service to "grant privileges, leases, and permits for the use of land for the accommodation of visitors."[9] This prioritization had an enduring impact: over the following fifty years and more, NPS history became dominated by the service's efforts to expand, develop and manage its park system for outdoor recreation. Landscape architects and not biologists or other natural scientists became the dominating expert community within the U.S. National Park Service.[10]

The national park idea almost immediately went beyond the United States. Shortly after Yellowstone's establishment in 1872, various British settler societies adopted the idea and adjusted it to their own needs. National identification again ranked high, while science was not a motive.[11] The latter only related to national parks when such parks were created in Europe after the century's turn. For three decades national parks had been regarded as a typical New World endeavor and thus had not been imitated by any European nation. This perception fundamentally changed around 1900 with the rapid spread of what can be called the first global nature conservation movement. The same decades witnessed a remarkable increase in ecological sciences and thinking. All over Europe, scientists became key actors in the nascent nature conservation movement. Scientific reasoning not only bolstered their aims and claims to preserve nature from spoliation but also provided their proposals on how to safeguard nature with both guidance and rationale. In this context, national parks were turned into a promising tool for both nature conservation and the advancement of science.[12]

In the first decade of the twentieth century, plans for national parks concurrently developed in several European countries, although some countries gave preference to other names: *zapovednik* in Russia or *Naturschutzpark* in Germany and Austria.[13] In most countries park plans were backed by prominent scientists as well as scientific institutions and organizations. In Sweden and Switzerland, which in 1910 created the first national parks in Europe, the national academies of sciences were deeply involved in park formation. While in Sweden scientists became more and more sidelined by tourist interests,[14] in Switzerland they maintained a dominant role in the conceptualization and realization of the national park. In its first paragraph the Swiss park decree of 1914 mandated the "scientific observation" of the national park.[15] Similar primacy to research was given in the *zapovedniki*.[16]

In contrast to Russia, where the *zapovedniki* survived war and revolution, conservation initiatives in other parts of Europe were severely hampered by the Great War.[17] All the Alpine countries (France, Italy, Germany, Austria, and Habsburg Slovenia) had initiated park projects modeled after the Swiss park, and France had even designated a small area as its first national park. The French government intended to enlarge the park, but as in most other European countries, support for nature conservation and the creation of national parks and other protected areas rapidly faltered after 1914 and did not recover after the war. The same happened to an initiative by the Swiss scientist, explorer, and conservationist Paul Sarasin to build an international organization for nature protection (*Weltnaturschutz*). It had a promising start in 1913 when delegates of almost all major powers assembled in Switzerland's capital, Bern, and agreed on the establishment of an Advisory Commission for the International Protection of Nature, but it never managed to operate properly, despite repeated vigorous efforts by its initiator.[18]

An exception to the rule was the creation of Italy's Gran Paradiso National Park, which was prompted by Italian king Victor Emanuel III's decision to donate his royal hunting reserve to the state. It was made a national park in 1922, and an independent scientific commission was installed that focused on the ibex, the area's iconic animal.[19] An initiative similar in many ways unfolded at the same time in Belgium, although its object of study was located in central Africa. Belgium's King Albert passionately supported the creation of Africa's first national park, Albert National Park in the Belgian Congo, in 1925. The park focused on the protection of the mountain gorilla, the rare charismatic primate highly threatened by extinction. Scientific research was declared the park's main goal, while tourism and other uses were openly discouraged to the extent of barring anyone without special permission from accessing the area. An international advisory body was appointed and scientific explorations were organized from the 1930s onward.[20]

In those years Africa also moved into the center of French conservation efforts. The approaches taken varied significantly between the different regions of French Africa. The establishment of national parks in the northern territories was strongly influenced by French settler communities, while in the tropical West African colonies, foresters were the dominant force in creating forest reserves aimed at giving sustainable yield. In Madagascar, Paris's renowned Muséum national d'Histoire naturelle took charge and in 1927 established ten *reserves naturelles intégrales* dedicated to the sole purpose of scientific research. Later on, such reserves were also created in other colonies and in France itself.[21] These initiatives furthermore triggered the founding of an international office of nature protection by Belgian and Dutch conservationists in Brussels in 1928, the Office International de Documentation et de Cooperation pour la Protection de la Nature.[22]

The preservation of African fauna had already been addressed in 1900 by an international agreement passed in London.[23] In the 1930s British conservationists and colonial administrators sought to reestablish British leadership in this matter. Game reserves, which had been set up only on a temporary basis, were supposed to be turned into permanent national parks and supplemented by new parks across the continent. The role model was South Africa's Kruger National Park, established in 1926.[24] In 1933 London hosted another conference on African conservation, resulting in a new accord. The signing parties agreed that the constitution of protected areas built the cornerstone for the preservation of African flora and fauna. For the purpose of the convention the term "national park" was defined as follows:

> The expression "national park" shall denote an area (a) placed under public control, the boundaries of which shall not be altered or any portion be capable of alienation except by the competent legislative authority, (b)

set aside for the propagation, protection and preservation of wild animal life and wild vegetation and for the preservation of objects of aesthetic, *geological, prehistoric, historic, historical, archaeological or other scientific interest* for the benefit, advantage and enjoyment of the general public, (c) in which the hunting, killing, or capturing of fauna and the destruction or collection of flora is prohibited except by or under the direction or control of the park authorities. In accordance with the above provisions, facilities shall so far as possible be given to the general public for observing the fauna and flora in national parks.[25]

On French intervention a second category was introduced, the "strict natural reserve." Areas in this category had to undertake rigorous preservation measures. Access was subjected to written permission and "scientific investigations" had to be authorized.

The convention became highly influential beyond Africa and the signatory parties. It influenced park creation in Latin America, for example, in Brazil, and its legacy is clearly visible in IUCN's international categorization work. By emphasizing state control and public enjoyment, the definition of national parks at the London convention in general stuck to the meaning that was given to national parks in the United States but it broadened the park idea by portraying scientific interest as a legitimate reason of preservation. The new category of "strict natural reserve," which stressed science as the area's raison d'être, eventually mutated into Category Ia of IUCN's protected area category system, the strict nature reserve.[26]

In the United States, the interest of scientists in national parks was awakening in the interwar years as well.[27] The University of California at Berkeley, the Carnegie Institute in Washington, D.C., and the Ecological Society of America became early academic strongholds encouraging research in national parks. President Calvin Coolidge's proclamation of Glacier Bay National Monument in 1925, for example, went back to a proposal by the Ecological Society and was one of the rare instances in which scientific values were explicitly mentioned.[28] By the late 1920s scientists with an interest in national parks were networked on the national level and thus ready to start a concerted effort to lobby the National Park Service. Among its fruits were the creation of a research unit within the service, the designation of research reserves in national parks, the first system-wide survey of park wildlife (the landmark "Fauna Report No. 1"), and a reformulation of predator control policy. The honeymoon of academic scientists and national park authorities was short-lived, however. Already in the New Deal era, the scope for scientific activities dramatically decreased again. In the following decades the status of both scientific research and natural resource management within the park service remained precarious. National Park Service scientists and natural resource managers were time and again frustrated by the service's reluctance to both listen to their advice and integrate their findings into park management. Internal and external reports from the 1960s to the 1990s continuously blamed the service in strong terms for the inappropriateness of its scientific structures and programs.[29]

The lack of a coherent science policy neither prevented the realization of research projects within park units nor encouraged the adoption of scientific management practices by the park service, but it prohibited the national parks from developing a strong scientific identity of their own. This made their scientific policy prone to administrative reorganizations, changing political priorities, or shifting public opinion, which in turn undermined their credibility as cooperating partners in scientific projects. As long as the U.S. National Park Service did not display a scientific attitude and did not promote an internal research culture, university-based scientists hesitated to collaborate with the service. Questions such as who was in charge of the research areas, who directed the studies, and who owned the results loomed large and hindered the building of effective long-term collaborations. In such a context individual scientists and scientific institutions tended to opt for research areas outside national parks.[30] From this it also follows that the appeal of national parks for science cannot be compared in absolute terms but must be seen in context of the alternatives for field research available at a given place and time. In other words, one has to take into account the different opportunity costs of choosing a national park for research.

Networks

Scientific research in parks was conducted beyond regional and national borders: international cooperation occurred with publication of results, review of publications, compilation of material on foreign parks, and informal sharing of information. Scientists and park staff also traveled abroad and visited other nations' parks and protected areas. Transnational expert communities evolved, and scientists as well as diplomats and park managers gathered at international congresses and workshops, as national NGOs and governmental bodies became internationally active and international organizations were founded. The degree to which decisions and actions in a park were influenced by such transfers is often difficult to assess. The issue is complex and time-consuming to research as one needs to reconstruct in detail not only the networks of exchange and the formation of opinions but also how network and opinion building interfered. A few well-researched examples indicate that transnational networking and exchange were of considerable importance.[31]

In the late 1920s and early 1930s, John F. Merriam, president of the Carnegie Institution, became a main actor in pushing U.S. National Park Service to both support scientific research in its parks and build up its own scientific capacities. Before Merriam started his campaign in the United States, he was engaged by the Belgian government as an expert in the conceptualization of

Albert National Park in the colonial Congo. Referring to this experience, Merriam suggested creating "complete reservations" within U.S. national parks.[32] Looking for further inspiration, Merriam sent his collaborator, Harvey M. Hall, to Europe to collect information on European parks. Hall, a botanist and expert on Californian flora, knew Carl Schröter, the eminent Swiss botanist and cofounder of the Swiss National Park. The two men had first met in Yosemite National Park in 1913, when the International Phytogeographic Excursion traveled the United States and stopped in the park.[33] Hall visited the Swiss park with Schröter in 1928 and was deeply impressed by the scientific approach taken there. He shared his insights with Merriam in many letters in which he dissected how both natural conditions and institutional arrangements in the United States differed from those in Switzerland. Hall scrutinized what lessons the United States could learn from the Swiss. He considered action to be urgent because the unspoiled areas that were of most interest for ecological research were disappearing fast. Furthermore, he asked the scientific community to get involved much more actively and to strive for responsibility and control over research areas and programs. Within the service, chief naturalist Ansel Hall, who had toured the Italian national parks in the mid-1920s, supported the creation of research areas.[34]

While scientific research was no priority to the park service directors of the time, Stephen Mather and Horace M. Albright, both were well informed about the international developments in this respect. A steady flow of papers passed over their desks, among them letters and publications from the Swiss National Park and Albert National Park that emphasized scientific topics.[35] In 1933 Albright published an article on "Research in the National Parks" in the *Scientific Monthly,* delineating the service's attitude toward research at that moment. Albright opened his text with the following sentences: "Being equipped by nature with the most complete and magnificent laboratories imaginable, it was inevitable that scientific research should become an important and popular activity of the National Park Service. Nevertheless, it is one of the newest developments in national park work, which is primarily of a human welfare nature." In the remaining text Albright pondered the ability of science to inform park management regarding problems with wildlife, fire, insect infestations, and tree diseases. As to the value of parks as research laboratories in their own right, he only touched on this topic toward the end.[36]

While the U.S. National Park Service was late in integrating science into the national parks and did so only hesitantly and erratically, U.S. wildlife biologists became global leaders in game management. In 1933 Aldo Leopold, a forester by training, published the first textbook on this subject, and in the same year he was appointed the first professorship of game management by the University of Wisconsin–Madison.[37] In Switzerland, where universities did not teach wildlife biology, research on ungulates was neglected for decades. Furthermore, wildlife research, which required all-season observation in the field, was time-consuming and expensive. In the Swiss National Park, serious investigations only started in the 1950s when, on the one hand, national structures for research funding were created. On the other hand, the park came under political pressure as soaring numbers of red deer destroyed agricultural land adjacent to the park and large numbers of deer were found dead from exhaustion at every few winters' end. The research program inaugurated at the end of the 1950s involved the trapping, anesthetizing, and tagging of animals in order to investigate their life cycles and migratory patterns, using a scientific approach and technology adapted from the United States.[38]

In the U.S. national parks, wildlife research was pioneered by George M. Wright in the 1930s. After 1945 U.S. biologists not only resumed research activities but expanded them to national parks in other countries. Etienne Benson has investigated in detail how former students of the Craighead brothers, Frank and John, who pioneered radio-tracking technologies in their research on grizzly bears in Yellowstone National Park, became involved in tiger research in Nepal's Chitwan National Park and how Western research methods, local knowledge, and traditional hunting practices entangled.[39] Such scientific enterprises were multi-actor ventures and could easily become enmeshed with international politics. The Cold War affected scientific exchange, as in, for example, the removal of American scientists from Project Tiger in India in the 1970s.[40]

The most important postwar platform for international scientific exchange on park matters was the International Union for the Conservation of Nature. Under the presidencies of Roger Heim of France (1954–58) and Jean G. Baer of Switzerland (1958–63), both of whom were highly regarded scientists actively involved in national park research, IUCN became a science-based organization.[41] Its international network of experts connected scientists from all over the world, distributed information, set global standards, and evaluated conservation schemes on a scientific basis. For example, all of the World Wildlife Fund's international projects were scientifically approved by IUCN before being funded.[42] Over recent decades IUCN has played a key role in shifting national park administrations worldwide toward a more scientifically informed management.[43] In addition, by devising and promulgating a categorization system of protected areas, IUCN fostered an international understanding of what national parks actually were (and were not) about. Ecosystem preservation came first, permanence as guaranteed by the highest state authority second, and visitation and other uses (including research) third. At the occasion of the Second World Parks Conference in Yellowstone in 1972, then–IUCN president Harold Coolidge still hailed the Swiss National Park as a "splendid example of a scientific national park."[44] In the 1990s, however, parks with an emphasis on research were eventually recategorized. Thus, Category II National Parks in the World Database on Protected Areas (WDPA) includes neither the Swiss National Park nor the *zapovedniki* of the Russian Federation and other former Soviet Republics. Instead they are Category Ia, Strict Nature Reserves.[45]

Tensions

Science's involvement in national park affairs did not proceed without tensions; indeed, sometimes it provoked outright and long-standing controversies. Most scientists looked at science as a universal endeavor (though practice could differ, of course). National parks, in contrast, bore the national viewpoint in their name. In some cases the two standpoints could be reconciled quite easily. The conservation of local species, for example, was argued to be of scientific as well as national importance. "In this regard (the inventory of the local fauna) our science, which does otherwise not know national borders, has to fulfill also a national task," as one of the founders of the Swiss National Park and university professor of zoology wrote in a report to the Swiss authorities in 1911.[46] Throughout the twentieth century "species-based conservation, and especially captive breeding, had a particular appeal to governments," as Bill Adams notes in his global conservation history.[47] Endangered species provided a clear target for government-sponsored action. The success and failure of conservation measures could be assessed by establishing national inventories and monitoring species numbers. Furthermore, ascribing national identity to rare species evoked patriotic affection and helped bolster species conservation with both elite and popular support.

In other instances, scientific interests and national representativeness associated with national parks did not coincide as smoothly. The Swiss case also provides a nice example of how scientific approaches could become mixed up in national politics. In their original plans the Swiss scientist-conservationists envisioned several national parks, one in each of the country's main biogeographical units. Within a few years this ecological concept was replaced by one based on federal linguistic politics, Switzerland's single most important policy field in fostering national cohesion. The new plan involved one park in each of the nation's linguistic units, but it too was never realized. As support for additional national parks crumbled after 1914, what was originally intended to be Switzerland's first national park soon became *the* Swiss National Park.[48]

Public access to national parks was another source of tension. Scientific research in parks often demanded that access to the area be controlled and limited. Visitors to the Swiss National Park were not allowed to leave the official trails; some colonial authorities even closed their parks entirely to the public, which was also the rule in French nature reserves. Such politics violated the basic principle of national parks as places for public enjoyment. In his "hundred-year appraisal" of national parks at the Second World Parks Conference in Yellowstone in 1972, Jean-Paul Harroy, chairman of IUCN's International Commission on National Parks and former director of the Institut des Parcs Nationaux du Congo Belge, remarked that one of the parks established under Belgian administration in Africa (Garamba) "is still closed to tourists, which really should exclude it from the national park category." The same applied to the scientific reserves in colonial Madagascar, which "did not permit visits and, therefore, could not be considered national parks."[49]

Furthermore, scientific practices could conflict with the park experience of visitors (and more so with the park staff designing and arranging these experiences). Installations for scientific experiments, such as fences, destroyed the illusion of untouched nature, and animals with collars, tassels, and transmitters did not match visitors' expectation of wilderness. "A tiger with a radio hanging around its neck is no longer an attraction," as the research director of a tourist lodge in Nepalese Chitwan National Park said in 1973, expressing his concerns regarding a Smithsonian Institution tiger research project involving extensive radio tracking of animals.[50] For others, it was not a question of tourism but park ethics. For them, even invisible scientific intrusion corrupted the very value of national parks. They worshiped national parks as sanctuaries where any human intervention was destroying the magic of wilderness and therefore was principally improper. Moreover, national parks were seen as refuges from modern life. Science, however, not only typified modernity but also exposed parks to the most advanced technologies. In the Swiss park, despite its strong research orientation, park scientists and the park board struggled for decades over the degree of intervention that was tolerable in the park, slowly moving toward a more liberal understanding that allowed a shift from pure observation to experimental research.[51] Much of the resistance John and Frank Craighead encountered with their research on Yellowstone's grizzly bears from the late 1950s to the early 1970s was also rooted in a hands-off national park ethics.[52]

This points to another fundamental tension pervading national park history: the one between philosophies of intervention and nonintervention. James Pritchard discusses the history of nature resource management in Yellowstone National Park as a dispute between "two profoundly different world views, the first advocating human intervention as essential to the proper management of wildlife within the national parks, [and] the second view suggesting that humans are not required to intervene in nature, and that we can learn something from watching nature at work."[53] This dispute went far beyond the scientific community and the park administration and at times involved both national politics and popular voices. Over the course of the twentieth century, however, the importance of scientific opinions in those debates tended to increase. At a time when science acquired a higher value than ever before; when politics, economics, and social life became increasingly science-based; when scientific management entered public administrations, private firms, and even the governing of the self; and when scientific research and thinking deeply affected the ways in which nature was perceived, it was unsurprising that advocators of both human intervention and nonintervention were eager to scientifically bolster their world views. Can ungulate populations be left to nature's regulation, or must management intervene? Must fires or diseases be suppressed and combated, or can nature deal with them on its own? And what about nonnative species? Expert opinion, however, was seldom unanimous and subject to change. Furthermore, scientific disciplines cultivated individual doctrines. For example, foresters, botanists, and wildlife biologists analyzed the interaction between ungulate populations and forests and grasslands from different angles and consequently expressed divergent opinions about an area's carrying capacity. Such discussions not only unfolded in similar ways in different world regions but also became increasingly interconnected through scientific exchange.[54]

Transformations

The discussion up to this point has demonstrated that most parks undoubtedly had at least to reckon with science as an interfering force in the twentieth century, even if the degree to which science was embraced by national parks differed widely from place to place. In what ways did scientific research transform national parks at large? As this section will argue, science's involvement had the potential to deeply alter parks: the perceptions and goals of parks, their social meaning, and, last but not least, their physical appearance. Transformations occurred in various ways. First, the criteria for selecting park areas shifted. When scientist-conservationists chose the region for the establishment of a Swiss national park in the early 1900s, scenery played virtually no role. As the park was intended to serve as a large field laboratory where scientists could study natural processes in real settings, the founders sought an area where past and present human interference with nature was as low as possible. A second important criterion was biological diversity, and a third was the ability to protect vast connected areas in order to preserve the habitats of the park fauna. At the same time, ecological criteria could be and were used to argue against the creation of national parks. Hugo Conwentz, the most influential German conservationist of the time, deemed national parks to be unsuitable for Central Europe as he regarded its landscape as thoroughly altered by human action. Conwentz propagated *Naturdenkmäler* (nature monuments) as the appropriate means of preserving Europe's natural heritage. He believed that protecting biological diversity in a huge variety of small places instead of a few large spaces was both more efficient and more rewarding. By balancing the pros and cons of parks and monuments, this discussion foreshadowed the big debate in twentieth-century conservation biology about how sizes and numbers of habitats affected biodiversity, the so-called SLOSS debate ("single large or several small").[55]

Second, the inclusion of science confronted national parks not only with a new set of rationales but also with new actors and interests. University-trained scientists entered national parks as researchers, experts, advisers, and managers. Scientific institutions became engaged as research partners and as external reviewers of park performance, either through permanent commissions or ad-hoc investigations; these efforts could lead park administrations to either invest in building their own scientific capacities or rely on outside expertise. In either case, national parks in most of the world had to respond to increasing public and political pressure to become science-based, a pressure that since the 1960s was well orchestrated by internationally organized scientists and conservationists. In return, scientific intrusions into national parks depended on political power.

Third, scientific approaches changed the way national parks were managed and conceptualized. In this respect it makes sense to distinguish two basic applications of science. On the one hand, national parks were used as "nature's laboratories," as research sites for the scientific study of natural processes. To transform a park into a workable and scientifically sound outdoor laboratory, special conditions had to be created and maintained. Classical labs set the standards: inventories of the site had to be created and repeated over time, the site's conditions had to be surveyed and controlled, external disturbances had to be ruled out or narrowed down, and access had to be limited to scientific personnel. On the other hand, scientific research was dictated by the needs of park management. Scientific knowledge and authority was used to inform as well as legitimize practices of park management, and scientific results could not only trigger profound changes in management but even lead to territorial adjustments. The attitudes toward and management of predators and large mammals, of vermin and plagues, of fire and exotic species all changed considerably under the influence of scientific investigations. This process was not a linear one arising from irrational prejudices toward scientific enlightenment but an often extremely complicated means of gaining (and sometimes also losing) different kinds of knowledge. Neither the natural nor the social environments in which national park managements were operating were stable. Older management practices did not simply disappear but were often sustained for quite some time and occasionally became amalgamated with new approaches.

After 1945 ecosystem theory and wildlife biology had a huge impact on old and new parks. Ever more refined methods to track animals were developed and used in research projects and programs. Individually colored collars and ear tassels were gradually replaced by electronic transmitters, beginning in the United States in the 1960s. This allowed the radio tracking of animals from a research station and made the laborious detection and observation of animals in the field unnecessary.[56] Radio tracking was later complemented by geographic information and global positioning system technologies (GIS and GPS). Research results and technologies were transferred into park management and used, for example, to individually handle large predators and collectively manage population dynamics of ungulates. They also served to define park boundaries along the habitat range of species designated to protect (or trying to adjust the boundaries accordingly). Bernhard Grzimek and his son's study of the Serengeti's migratory animals in the 1950s or the reconceptualization of Yellowstone as the "Greater Yellowstone Ecosystem" are the best-known cases in point.[57]

As an invasive practice, on both symbolical and material grounds, scientific research and applications were contested to various degrees from site to site. Scientific research often entered parks to solve management problems and gained acceptance when doing so. In contrast to applied science, pure research remained at the fringes in many parks. National parks as nature's laboratories was a fascinating idea, but to practice pure research in national parks often proved difficult. In a handbook article published in 1924, Carl Schröter predicted a bright future for national park research within the biological sciences. He expected national parks to complement the herbarium, the botanical garden, and the laboratory as archetypical research sites and to support the transformation of ecology into a "genuine" field science.[58] His model was the Swiss National Park, whose scientific research commission Schröter led for its first decade and a half, from 1916 to 1929. Scientific practice in the Swiss National Park, however, had to cope with various obstacles. As with many other parks, the Swiss National Park was located far from research institutions and for many decades lacked basic research infrastructure. To monitor and control the conditions of research plots was demanding, and tensions sometimes complicated the essential collaboration with the park authorities. Furthermore, the space for experimenting was narrowly defined, and to wait for nature to perform on its own could be extremely frustrating.[59] Thus, neither the Swiss National Park nor national parks in general assumed the crucial role for science as imagined by Carl Schröter. In the last few years, however, in the context of global warming and biodiversity loss, their importance has increased.[60] Long-term data and records combined with relatively stable natural (so human-induced) conditions made them interesting sites for the study of changing environments.

The context of creation seems to have a long-standing impact on the development of a park.[61] In sites such as the Swiss National Park, whose establishment was centered on science, scientific research continued to rank high in importance, while recreation-oriented parks were usually slow in adding scientific practices to their assets. The reason for this can be seen in a general averseness of park authorities toward change. The national park as a category proved to be extremely malleable and was adapted to a whole range of different needs and wants. Once a national park was given a certain form, however, this form was disposed to endure. This conservatism was probably infused into the park idea with the very act of designating such areas for *permanent* preservation. Thus, on the one hand, historical legacies matter as do national regimes, which explains why the extent of scientific influence on parks has varied widely both in place and time. On the other hand, there can be no doubt that scientific reasoning profoundly altered the national park idea at large and initiated changes that were global in reach. Neither expert networks nor scientific exchange stopped at national borders. In the course of the twentieth century, national parks all over the world became objects of scientific management and, to a lesser degree, sites of scientific research.

Notes

1. Cf. Bernhard Gissibl, Sabine Höhler, and Patrick Kupper, eds., *Civilizing Nature: National Parks in Global Historical Perspective* (New York: Berghahn, 2012).

2. Victor H. Cahalane, ed., *National Parks—A World Need* (New York: American Committee for International Wildlife Protection, 1962). On Coolidge and the American Committee, see Mark V. Barrow, *Nature's Ghosts: Confronting Extinction from the Age of Jefferson to the Age of Ecology* (Chicago: University of Chicago Press, 2009), chap. 5.

3. Cahalane, *National Parks*, 6.

4. Ibid., 6–7.

5. Jean G. Baer, "The Swiss National Park," in Cahalane, *National Parks;* L. K. Shaposhnikov, "Nature Reserves in the Soviet Union," in Cahalane, *National Parks.*

6. Cahalane, *National Parks*, 6–7.

7. Ibid., 15, 30.

8. American Antiquities Act of 1906, 16 USC 431–433, approved June 8, 1906. Cf. Hal Rothman, *Preserving Different Pasts: The American National Monuments* (Urbana: University of Illinois Press, 1989). During the Theodore Roosevelt administration the U.S. government initiated a central resource management policy based on expert knowledge, which served as a model for the building of national administrative power. See Bruce J. Schulman, "Governing Nature, Nurturing Government: Resource Management and the Development of the American State, 1900–1912," *Journal of Policy History* 17 (2005): 375–403.

9. National Park Service Organic Act, approved August 25, 1916. See also Richard W. Sellars, *Preserving Nature in the National Parks: A History* (New Haven, Conn.: Yale University Press, 1997).

10. See, among the vast literature, Ethan Carr, *Wilderness by Design: Landscape Architecture and the National Park Service* (Lincoln: University of Nebraska Press, 1998); Alfred Runte, *National Parks: The American Experience,* 4th ed. (Lanham, Md.: Taylor, 2010).

11. Melissa Harper and Richard White, "How National Were the First National Parks? Comparative Perspectives from the British Settler Societies," in Gissibl et al., *Civilizing Nature,* 50–67. See also the chapters by MacEachern, Catton, and McGrath in this volume.

12. Patrick Kupper, "Translating Yellowstone: Early European National Parks, *Weltnaturschutz* and the Swiss Model," in Gissibl et al., *Civilizing Nature,* 123–39. Bowler is wrong in believing that "there was no immediate link" between ecological sciences and the conservation movement. Peter John Bowler, *The Fontana History of the Environmental Sciences* (London: Fontana, 1992), 309. His misjudgment probably derives from his one-sided focus on the Anglo-American world where the science-conservation link was indeed weaker than in continental Europe.

13. There is no decent translation of *zapovednik,* but its meaning is a strict nature reserve. *Naturschutz* means "nature protection." On the German-Austrian

movement for *Naturschutzparke,* see Patrick Kupper and Anna-Katharina Wöbse, *Geschichte des Nationalparks Hohe Tauern* (Innsbruck, Austria: Tyrolia, 2013).

14. See Tom Mels, *Wild Landscapes: The Cultural Nature of Swedish National Parks* (Lund, Sweden: Lund University Press, 1999); Sandra Wall Reinius, "A Ticket to National Parks? Tourism, Railways, and the Establishment of National Parks in Sweden," in *Tourism and National Parks: International Perspectives on Development, Histories, and Change,* ed. Warwick Frost and C. Michael Hall (New York: Routledge, 2009), 184–96.

15. "Der Nationalpark wird der wissenschaftlichen Beobachtung unterstellt." Bundesbeschluss betreffend die Errichtung eines schweizerischen Nationalparkes im Unter-Engadin, vom 3. April 1914, Art. 1. See http://www.amtsdruckschriften.bar.admin.ch/viewOrigDoc.do?id=10025346 (accessed August 29, 2015).

16. Douglas R. Weiner, *Models of Nature: Ecology, Conservation, and Cultural Revolution in Soviet Russia* (Bloomington: Indiana University Press, 1988).

17. An exception is Spain, where two tourism-oriented national parks were established during World War I. See Andreas Voth, "National Parks and Rural Development in Spain," in *Protected Areas and Regional Development in Europe: Towards a New Model for the 21st Century,* ed. Ingo Mose (Aldershot, U.K.: Ashgate, 2007), 141–60.

18. Kupper, "Translating Yellowstone."

19. Wilko Graf von Hardenberg, "Fascist Nature: Environmental Policies and Conflicts in Italy, 1922–1945" (PhD thesis, Cambridge University, 2007).

20. Jean-Paul Harroy, "Contribution à l'histoire jusque 1934 de la création de l'institut des parcs nationaux du Congo belge," *Civilisations* 41 (1993): 427–42; Marc Languy and Emmanuel De Merode, eds., *Virunga: The Survival of Africa's First National Park* (Tielt, Belgium: Lannoo, 2009).

21. Adel Selmi, "L'émergence de l'idée de parc national en France: De la protection des paysages à l'expérimentation coloniale," in *Histoire des parcs nationaux: Comment prendre soin de la nature?* ed. Raphaël Larrère et al. (Versailles: Editions Quae, 2009), 43–58; Caroline Ford, "Imperial Preservation and Landscape Reclamation: National Parks and Natural Reserves in French Colonial Africa," in Gissibl et al., *Civilizing Nature,* 68–83. In West Africa a single strict natural reserve (Nimba) was established in 1944. Theodore Monod, "Reserves and National Parks in French West Africa," in Cahalane, *National Parks,* 20.

22. Martin W. Holdgate, *The Green Web: A Union for World Conservation* (London: Earthscan, 1999), 12–13. On the Dutch conservation movement, see Henny van der Windt, "Parks without Wilderness, Wilderness without Parks? Assigning National Park Status to Dutch Man-Made Landscapes and Colonial Game Reserves," in Gissibl et al., *Civilizing Nature,* 206–23.

23. Bernhard Gissibl, "German Colonialism and the Beginnings of International Wildlife Preservation in Africa," *GHI Bulletin* Supplement 3 (2006): 121–43.

24. See John M. MacKenzie, *The Empire of Nature: Hunting Conservation and British Imperialism,* 2nd ed. (Manchester: Manchester University Press, 1997), 200–224. On Kruger National Park, see Jane Carruthers, *The Kruger National Park: A Social and Political History* (Pietermaritzburg, South Africa: University of Natal Press, 1995).

25. Convention Relative to the Preservation of Fauna and Flora in the Nature State 1933, Art. 2.1 (italics added). See Mark Cioc, *The Game of Conservation: International Treaties to Protect the World's Migratory Animals* (Athens: Ohio University Press, 2009), 14–57.

26. On the evolution of IUCN-protected area categorization, see Adrian Phillips, "The History of the International System of Protected Area Management Categories," *Parks* 14, no. 3 (2004): 4–14; Bernhard Gissibl, Sabine Höhler, and Patrick Kupper, "Towards a Global History of National Parks," in Gissibl et al., *Civilizing Nature,* 13–16. On Brazil: H. E. Strang, "National Parks in Brazil," in Cahalane, *National Parks,* 96; José Luiz de Andrade Franco and José Augusto Drummond, "Wilderness and the Brazilian Mind (I): Nation and Nature in Brazil from the 1920s to the 1940s," *Environmental History* 13, no. 4 (2009): 737. See also the essay by Drummond in this volume.

27. Patrick Kupper, "Science and the National Parks: A Transatlantic Perspective on the Interwar Years," *Environmental History* 14, no. 1 (2009): 58–81.

28. Theodore Catton, *Land Reborn: A History of Administration and Visitor Use in Glacier Bay National Park and Preserve* (Anchorage: National Park Service, 1995), chap. 3.

29. Sellars, *Preserving Nature.*

30. Victor Shelford's attitude toward national parks nicely illustrates this point. See Kupper, "Science and the National Parks."

31. The chapter by Wakild in this volume nicely speaks to this argument.

32. Library of Congress, Papers of John C. Merriam, Box 132, Letter of Merriam to the Members of the Committee on Study of Educational Problems in National Parks, December 26, 1929. Merriam chaired this committee appointed by the secretary of the interior in 1929.

33. The International Phytogeographic Excursion, an institution founded in Geneva in 1910, gathered the leading ecological botanists of the time for periodical excursions lasting for several weeks. See Kaat Schulte Fischedick and Terry Shinn, "The International Phytogeographical Excursions, 1911–1923: Intellectual Convergence in Vegetation Science," in *Denationalizing Science: The Contexts of International Scientific Practice,* ed. Elisabeth Crawford, Terry Shinn, and Sverker Sörlin (Dordrecht, Netherlands: Kluwer, 1993), 107–31.

34. Kupper, "Science and the National Parks."

35. Record Group 79, 3.1, Entry 7, 0–30 Foreign Parks, National Archives, Washington, D.C. On the Belgian Congo and Switzerland, see ibid., boxes 629 and 632. See also Kupper, "Science and the National Parks."

36. *Scientific Monthly* 36 (1933): 483–501, reprinted in Lary M. Dilsaver, ed., *America's National Park System: The Critical Documents* (Lanham, Md.: Rowman & Littlefield, 1994), chap. 3.

37. Aldo Leopold, *Game Management* (New York: Scribner's, 1933).

38. Patrick Kupper, *Creating Wilderness: A Transnational History of the Swiss National Park* (New York: Berghahn, 2014), 162–63.

39. Etienne Benson, *Wired Wilderness: Technologies of Tracking and the Making of Modern Wildlife, Animals, History, Culture* (Baltimore: Johns Hopkins University Press, 2010).

40. Michael Lewis, "Globalizing Nature: National Parks, Tiger Reserves, and Biosphere Reserves in Independent India," in Gissibl et al., *Civilizing Nature,* 232.

41. Holdgate, *Green Web.* See also Anna-Katharina Wöbse, "Framing the Heritage of Humankind: National Parks on the International Agenda," in Gissibl et al., *Civilizing Nature,* 140–56.

42. See Alexis Schwarzenbach, *Saving the World's Wildlife: WWF—The First 50 Years* (London: Profile Books, 2011).

43. See also the chapter by Rodriguez in this volume.

44. Harold J. Coolidge, "Evolution of the Concept, Role and Early History of National Parks," in *World National Parks: Progress and Opportunities,* ed. Jean-Paul Harroy (Brussels, Belgium: Hayez, 1972), 33. See also Jean-Paul Harroy, "National Parks: A 100-Year Appraisal," in *World National Parks,* 16.

45. http://www.protectedplanet.net (accessed January 25, 2012). WDPA was established in 1981 as a joint project between the United Nations Environment Programme and IUCN.

46. "Unsere Wissenschaft, die sonst keine Landesgrenzen kennt, hat also in diesem Fall auch eine nationale Aufgabe zu erfüllen." Report Friedrich Zschokke, in Schweizerische Naturschutzkommission, *Sixth Annual Report 1911/12* (Basel, Switzerland: Schweizerischer Bund für Naturschutz, 1912) 95–98.

47. William M. Adams, *Against Extinction: The Story of Conservation* (London: Earthscan, 2004), 145.

48. Kupper, *Creating Wilderness,* 49–50, 222–24.

49. Harroy, "National Parks," 18. On his role in Belgian Congo, see Harroy, "Contribution à l'histoire."

50. Charles McDougal to Secretary of Forests, His Majesty's Government of Nepal, September 8, 1973, as cited in Etienne Benson, "Demarcating Wilderness and Disciplining Wildlife: Radiotracking Large Carnivores in Yellowstone and Chitwan National Parks," in Gissibl et al., *Civilizing Nature,* 182.

51. Kupper, *Creating Wilderness,* chap. 5.

52. Benson, *Wired Wilderness,* chap. 2. For bear management policy and practices in U.S. and Canadian national parks, see the chapter by Routledge in this volume.

53. James A. Pritchard, *Preserving Yellowstone's Natural Conditions: Science and the Perception of Nature* (Lincoln: University of Nebraska Press, 1999), xiv.

54. For example, on the United States, see R. Gerald Wright, *Wildlife Research and Management in the National Parks* (Urbana: University of Illinois Press, 1992), and on Yellowstone in particular, see Frederic H. Wagner, *Yellowstone's Destabilized Ecosystem: Elk Effects, Science, and Policy Conflict* (Oxford: Oxford University Press, 2006). On the Swiss National Park, see Kupper, *Creating Wilderness.* Another source of tensions, pure versus applied science, is treated in the following section.

55. See Michael L. Lewis, "Wilderness and Conservation Science," in *American Wilderness: A New History,* ed. Michael L. Lewis (New York: Oxford University Press, 2007), 205–22. In the new century the debate has shifted to questions of fragmentation and connectivity.

56. Benson, *Wired Wilderness.*

57. Michael and Bernhard Grzimek's research and conservation efforts became most famous through their Oscar-winning film *Serengeti Shall Not Die* of 1959. Robert B. Keiter and Mark S. Boyce, The Greater Yellowstone Ecosystem: Redefining America's Wilderness Heritage (New Haven, Conn.: Yale University Press, 1991).

58. Carl Schröter, "Die Aufgaben der Wissenschaftlichen Erforschung in Nationalparken," in *Handbuch der biologischen Arbeitsmethoden, Abt. 11: Allgemeine Methoden zur Untersuchung des Pflanzenorganismus,* ed. Emil Abderhalden (Berlin: Urban & Schwarzenberg, 1924), 387–94.

59. For an in-depth analysis of pure research in the Swiss National Park over a century, see Kupper, *Creating Wilderness,* chap. 5.

60. For the nexus of national parks and climate change, see chapter 12, by Carey, in this volume.

61. On this argument see Carruthers, *Kruger National Park,* 1–6, and chapter 6 in this volume.

PART II | Ideas

| # "Why Celebrate a Controversy?"

South Africa, the United States, and National Parks

JANE CARRUTHERS

SOUTH AFRICA HOSTED THE FIFTH WORLD PARKS CONGRESS in Durban in 2003, a conference with the theme "Benefits beyond Boundaries." To commemorate the first of these decennial gatherings to be held in Africa, South African National Parks (SANParks)[1] published a lavish book titled *South African National Parks: A Celebration.*[2] In the preface, Hector Magome, director of conservation services, asked a critical question: "Why celebrate a controversy?"[3]

From the time of their inception and continuing into the present, national parks and other protected areas in South Africa have been extremely controversial. In this respect, national parks appear to have generated responses in South Africa very different from those evoked in the general public of the United States, where, apparently—at least in the popular imagination—they "have become a source of patriotic pride, highlighting not only the grandeur and diversity of the national landscape, but also the benevolent farsightedness of the Federal Government in preserving the 'American' environment."[4] While recent scholarship has critiqued many aspects of U.S. national parks, the seeming importance of that country in the global history of national parks continues to drive what appears to be an enduring aspiration to understand the dynamic of U.S. primacy, significance, and reach in the international national park movement. Because the very term "national park" conveys to most U.S. citizens an aura of public benefit, it is instructive to dissect national parks in places where the "nation" itself is highly contested, and so thus are institutions that have a national label. Even in those cases where the nation-state is very different from the United States, it may be useful to analyze whether the North American influence on national parks is discernible, whether that influence has been reciprocal, and how it may have changed over time.

This chapter argues that, despite references to the U.S. experience, the evolution and development of South Africa's national parks have primarily been the outcome of its internal history, and their course has mirrored that past. With one exception that soon lost traction, at first embedded in a settler colonial (and British imperial) ethos of game protection, game reserves first became national parks with the aim of providing recreational facilities at a time when black South Africans were increasingly marginalized, disempowered, and disadvantaged by racially discriminatory organs of state. National parks were solely for whites, and in numerous instances African interests—particularly relating to access to land—were compromised by the establishment of national parks. Scientific research and management that were later conducted in national parks ignored and denigrated indigenous knowledge and valorized a Western approach to biodiversity. Since 1994 and the advent of democracy, national parks have had to reposition themselves by exploring aspects of modern national parks that resonate with a nation that is overwhelmingly black, poor, and young, and in a country in which the creation of employment is the paramount national goal. Consequently, national parks remain playgrounds for whites, but what ensures their survival is that they are deemed to be net creators of wealth through their economic profitability. Supporting arguments for the continuance of these places as national parks relate, for example, to issues around environmental health, such as the study of ecosystem services, research into clean water, and other projects that are conducted in protected areas.

In terms of the U.S. influence on this history, there is no doubt that the rhetoric around the "good" of national parks that emanated from the United States was repeatedly invoked to support the genesis of many national parks in South Africa in the early years of the twentieth century. There is also evidence of sporadic mutual communication on issues relating to wildlife management science, tourism, and bureaucratic organization until well into the 1960s. The present mission of SANParks is to transform the image of South Africa's national parks from the apartheid era, and indeed from the global era in which national parks had no rivals in the international protected area estate as there are today (e.g., World Heritage Sites, biosphere reserves), to meet the needs and aspirations of the "new" nation. In particular, these transformations relate to including black South Africans in structures of management, to encouraging previously disadvantaged population groups to visit national parks, to generating the largest possible income from international and local ecotourism, to meeting obligations in terms of international treaties around biodiversity, and also to delivering on a national priority to utilize resources sustainably. The transformation mission is clear: "South African National Parks (SANParks) is striving to transfer power and control of resources from the minority that had been appointed and privileged by an undemocratic system, to the majority that participates in the new democratic process. It is also directing the benefits of its activities to providing for all South Africans, rather than only to the more wealthy and privileged sections of society."[5]

As countries, the United States and South Africa have little in common. In contrast to the United States, which is geographically vast and is a long-standing and economically well-developed democracy with a generally well-educated population, South Africa (while ecologically diverse) is relatively small and agriculturally poor. It is a new democracy (since 1994), with a high level of violence and prevalence of HIV/AIDS. In South Africa 76.4 percent of the population (total 51 million) is black, 9.1 percent white, 8.9 percent colored (mixed race), and the remainder of Indian or Asian extraction. Urbanization is high and growing, and towns and cities are surrounded by informal settlements in which services such as housing, education, electricity, fresh water, sewage, and other infrastructure are woefully inadequate. In terms of world per capita income, South Africa comes in about eightieth on the list, while the United States is among the top ten. The literacy level is extremely low; the unemployment rate, although officially 25 percent, is unofficially closer to 40 percent. South Africa is a union (not a federation) with nine provinces, three national capitals, and eleven official languages. IsiZulu is the most commonly spoken home language. The legal system is a hybrid of Roman Dutch law, British common law, and African customary law.[6]

In 1899 South Africa did not exist as a state: there were two British colonies (the Cape and Natal) and two Boer Republics (the Transvaal and Orange Free State). After the South African (Anglo-Boer) War of 1899–1902, the four polities became separate British colonies and formed the "nation" of the Union of South Africa in 1910. In 1961 South Africa became an Afrikaner-dominated republic outside the British Commonwealth. Until 1994 it was ruled by the white minority, first under a policy of racial segregation, and after 1948 under apartheid. In 1994, after a negotiated settlement, South Africa became a democracy with universal franchise and a new constitution. The government is a tripartite alliance of the African National Congress, the South African Communist Party, and the Congress of South African Trade Unions. This context is directly relevant to national parks because governance frames the national milieu in which such parks come into being and continue to exist as a form of land use.

Although the historiography is relatively rich in national park biographies, there are few comparative studies.[7] Moreover, little detail has been unearthed about the real influence of U.S. national parks on parks elsewhere, although allusions are often made. Sometimes these are misleading and give more credit to the United States than may be factually correct.[8] This chapter shows that although Yellowstone was invoked as an example, doing so was of a rhetorical, or exhortatory, nature. On the other hand, during the early 1930s, with financial support from the Carnegie Corporation Visitors' Grant Committee, a thorough study over three months was made by R. H. Compton, professor of botany at the University of Cape Town and director of the National Botanic Gardens of South Africa (Kirstenbosch, in Cape Town), of a large number of national parks and national monuments in the United States with a view to transposing their philosophies, administration, education, financial, and other dimensions into South Africa as a practical model.[9] It might be argued also that a major contribution that the U.S. National Park Service (NPS) made to South Africa's national parks was in continuing to welcome South Africa as a member of the "national park community" at scientific and other meetings and conferences during the Cold War, thus supporting the park authorities during apartheid and in this way legitimizing an otherwise pariah government in an international forum.

How Does One Recognize a "National Park"?

The question of what constitutes a national park has been thoroughly debated.[10] Although the term indicates that the involvement of a nation-state is required, this is not invariably the case, and constituent provinces (or states) can own and manage a "national park," in partnership with local communities or the private sector.

The term and entity "national park" existed for almost a century before there was an attempt to define it. The current categories of the International Union for Conservation of Nature (IUCN), first formulated only in 1969, are based on management objectives. Category II (one of six), National Park, is particularly vague. First, it is a "protected area," separately defined.[11] As a subset a national park is a "natural area of land and/or sea, designated to (a) protect the ecological integrity of one or more ecosystems for present and future generations, (b) exclude exploitation or occupation inimical to the purposes of designation of the area and (c) provide a foundation for spiritual, scientific, educational, recreational and visitor opportunities, all of which must be environmentally and culturally compatible." South Africa has a number of national parks that do not fit this definition and, more correctly, should be placed in other IUCN categories that are more appropriate.[12]

That Category II leaves much to be desired was discussed during the 2003 Durban congress. It was agreed that improvements were required on definitions, criteria, and principles; the relationship of national parks to ecological networks and regional planning; design; management objectives; private national parks; links to appropriate structures of governance; and clarification of the categorization process.[13] This long list indicates that the defining characteristics and principles of national parks include a great deal that is extremely woolly and unclear. Nonetheless, because it would be extremely contentious unilaterally to recategorize present inappropriate "national parks" around the world, it was decided to leave matters as they stand. Essentially, "national park" is an international brand, but it is not a product with clear reproducible standards and components, as is Coca-Cola, or a franchise such as McDonald's. It is only a name. However, an important point also to bear in mind, as expressed in a preparation discussion paper in advance of the 2014 World Parks Congress held in Sydney, Australia, is that "in Durban (and Caracas) a deliberate move was to use the term World Parks Congress in favour of the official title World Congress on Protected Areas. The former is a more brandable event in terms of the public, however, many countries prefer the term Protected Area to park. On the flip side is the broadening understanding of the word park which goes beyond the IUCN Categories. A more fundamental issue is the messages that the term 'protected' implies."[14]

Thus, national parks are open to extremely wide interpretation and are almost endlessly adaptable. However, it is perhaps this very ambiguity, flexibility, and imprecision that has contributed to their longevity and popularity. National parks serve a purpose that may be political, social, or economic, but the purpose is clearly not the same for every polity in which national parks exist. Moreover, purposes change over time, and as national objectives change, so do national parks. The titles of the world congresses are suggestive of their changing roles. In 1972 they were "A Heritage for a Better World"; in 1982, "Parks for Development"; in 1992, "Parks for Life"; and in 2003, "Benefits beyond Boundaries."

On March 1, 1872, Yellowstone was established by the U.S. federal government as a "public park." In justifying why the term "national park" was not used of Yellowstone in its founding legislation, Roderick Nash alluded to the fact that, at the time, any mention of "nationalism" had unsavory socialist implications for "a society dedicated to unrestrained private enterprise."[15] In view of the power and resonance in later years of the word "national" in "national parks," this is ironic.

In order to analyze any influence of U.S. national parks on South Africa, what Yellowstone *is* requires unpacking. Yellowstone can be described as a large, remote, and scenically and topographically attractive tract of land, owned by the central (federal) government, in which wildlife and other natural assets of the landscape are not "wantonly" to be destroyed and in which extractive industries are prohibited through explicit rules and regulations. Its purpose is to provide the "people" with ground to use for their "benefit and enjoyment" and that the visiting public be provided with recreational accommodation.[16] As will be explained, some of these features were mentioned in South Africa at various times in order to bolster the formation of national parks. The founding act of Yellowstone makes no mention of alienating the park from indigenous inhabitants,[17] but dispossessing black Africans and setting up "fortress conservation"[18] as recreational sites for the white settler elite was, and remains, the burning and controversial issue in South Africa.

The Origins of South Africa's Early National Parks and Their Comparison with the United States

A NATIONAL PARK IN NATAL

There is a contrast between South Africa's first national park that, as a mountain landscape with interesting and valuable indigenous flora and fauna and outdoor recreational potential, was similar to those in the United States, and the later

Kruger National Park and others that were based on a "game reserve" philosophy and thus very different from any "Yellowstone model." "Game reserves" connote the overriding focus on protecting a single aspect of the biota (mammals), while the "Yellowstone model" includes appreciation of scenery and landscape, geology, flora, and fauna, as well as promoting a recreational outdoor experience, such as cannot be contemplated in an area replete with many dangerous large animals. The term "national park" in southern Africa was first officially used in 1901, when Natal Legislative Assembly member Maurice Evans raised the question of "preserving areas of natural beauty in the colony as national parks."[19] The first protected area and national park (conforming to the modern IUCN definition of both) was situated in the Drakensberg, a mountain landscape straddling the boundaries of the present provinces of KwaZulu-Natal and Free State and the country of Lesotho. Some elements here compare well with Yellowstone. The northern Drakensberg is a large, scenically and geologically splendid area with high peaks and dramatic waterfalls, relatively far from (but accessible to) urban centers and suitable for recreational use such as hiking, mountaineering, and fishing. The mountains are fossil-rich and contain many interesting (and endemic) plants, as well as rare wildlife such as eland, but no dangerous animals such as elephants or lions. The Drakensberg is also culturally valuable, because the San (Bushman) hunter-gatherers who inhabited the mountains for millennia, and who perished with the intrusion of African pastoralists and white settlers, left behind what is probably the richest store of rock art in the world.[20]

On July 1, 1908, four state farms in the more northerly area of the Drakensberg were "reserved for the purposes of forming a National Park" in the colony of Natal.[21] In 1916 the Executive Council of the Union (central) government reaffirmed the reservation together with a large adjoining area as a formally named "national park."[22] In 1919 the promoters of the park had "in view . . . the preservation and maintenance for ever [*sic*] of the area as a public park and pleasure ground for the benefit, advantage, and enjoyment of the people of the Union of South Africa."[23] Certainly, these words echo the founding legislation of Yellowstone almost exactly, although the word "advantage" has been added.

In 1926 the Natal provincial secretary visited the United States and was impressed by the high regard that the public there had for its special landscapes, and he returned to South Africa appreciative of what his own country had to offer in this regard.[24] By 1929 the Natal National Park had better roads, a passable bridge, visitor accommodation, and plans for a golf course, tennis court, and swimming pool.[25] This was surely a "public park and pleasuring ground" for the public, indeed, a proper "national park."

There was, however, a particularly South African obstacle confronting the Natal National Park. Yellowstone was a very large area, while the Natal National Park was relatively small. Government eyes turned to the adjoining Upper Tugela Native Reserve. The conservator of forests argued that this reserve contained some of the finest scenery in the national park locality and advocated alienating

2,740 acres from the African community that lived there. He argued that doing so "will not represent a serious loss . . . especially as the land is required for a public and National purpose."[26] Although contested by an official in the Native Affairs Department, who believed that it was unjust to deprive rural Africans of their land merely "for the sake of satisfying pleasurable sentiment,"[27] any dissent was overruled and the authorities placed part of the Upper Tugela location under the administration of the National Park Committee, although it was not actually excised from the location.[28]

Until 1920, references in South Africa to the U.S. national parks had not been based on any official communication but rather on knowledge gleaned from personal networks and U.S. publicity. The first formal overture came from the U.S. National Park Service when an approach was made in June 1920, via the Colonial Office in London to the governor-general of South Africa, as follows: "Desire of US National Park Service to obtain information as to the measures adopted in other countries for the development of their natural attractions." The request was relayed to each of South Africa's four provinces, and a report was compiled and submitted in December 1920. The covering letter explained that control of "natural attractions" in South Africa was vested in each province, but that in the Transvaal and Natal there were areas that "may be regarded as coming under the heading of national parks."[29] Unlike the U.S. national parks, however, none of these areas was adequately developed for tourists, but it was stated that the long-term goal was to do so.

THE KRUGER NATIONAL PARK

The origins of the Kruger National Park differ from those of the Natal National Park and have been fully explained elsewhere.[30] However, the influence of U.S. national parks on Kruger has not been recorded in detail. The park was established in 1926 by combining two provincial game reserves in the tropical (and then remote and disease-ridden) eastern Transvaal (now Mpumalanga and Limpopo provinces) on the Mozambique boundary: the Sabi, between the Crocodile and Sabie Rivers (1898), and the Singwitsi, between the Olifants and the Pafuri Rivers (1903). They were closed to the public, and their founding object had been to recover the numbers of wildlife that had been decimated during the South African War (1899–1902).

The most pressing wildlife issue in the post–South African War Transvaal Colony was to protect what was left of large mammal life by curtailing or preventing market hunting. On July 24, 1907, there was a debate in the Transvaal Legislative Assembly over tightening the game laws. The U.S. example was invoked as an argument in favor of restricting the market hunter. Their operations, it was alleged, "will have the same result as those of 'Buffalo Bill' and his followers in America. We are aware in America, where the buffalo once was in millions, there is now one small herd in the Yellowstone Park, kept up at national expense, and guarded by two troops of cavalry."[31] Another member of the assembly stated that he had "many discussions with American sportsmen, and they have always regretted that [Yellowstone] was not set aside one hundred years earlier; if that had been done, they would have had a far larger number of animals than they have to-day. I think it is the duty of the Government to take this matter in hand, and make these preserves something in the nature of a national institution."[32] The question was being clarified in South Africa: which was more important to a national park—saving endangered wildlife or protecting scenic landscapes as recreational "pleasuring grounds"?[33]

Nothing, however, came of the "national institution" suggestion while the Transvaal was a colony. As the First World War drew to a close, the Transvaal Provincial Council moved to abolish part of the Sabi Game Reserve, which precipitated a fact-finding commission of inquiry. So impressed were the commissioners that they recommended the Sabi and Singwitsi be combined and be transformed into "a great national park where the natural and prehistoric conditions of our country can be preserved for all time." The commission's report listed the four main attributes it considered essential for a "national park": uninhabited state (not private) land; established "for all time" under a legislative mandate from national government; managed in such a way as to give recreational visitors a glimpse of the South African landscape before the "advance of civilisation"; and a training ground for scholarly research and education.[34]

After the commission had completed its work and a national park had been accepted in principle, years of activity lay ahead in expropriating the private land that lay within the proposed area, demarcating appropriate boundaries and mounting a publicity campaign. In this regard, James Stevenson-Hamilton, the warden of these two reserves who was determined to have them protected against mining and agricultural development, mentioned the benefits of nationalization and the pride of having the first national park in the British Empire, "along the lines of the American national parks, to popularize it with the public by making roads and rest-camps and advertising tours and picnics in suitable places."[35]

Some delay was occasioned by a change of government in 1924, when the Afrikaner-dominated National Party won the general election, but this may have been an advantage because it brought English and Afrikaans speakers, divided by language, culture, history, and war, closer together in founding the Kruger National Park. The Union of South Africa had been an uneasy amalgamation of the "two white races," and the Afrikaans-speaking community yearned for a republican form of government outside the British Empire. Specific Afrikaner historical and cultural symbols were employed to this end, but at the same time, there was a broader platform of white nationalism as many English speakers came to appreciate that the time had arrived to set aside old hostilities. In these circumstances it is not surprising that the "national park" was among the reconciliation symbols that were accepted as politically viable, particularly as it was situated in the Transvaal. Legislation for the national park passed unanimously. It did not define a South African national park. Instead, as does the modern

IUCN definition, it delineated management objectives. In 1926 these were "the propagation, protection and preservation therein of wild animal life, wild vegetation and objects of geological, historical or other scientific interest for the benefit, advantage and enjoyment of the inhabitants of the Union."[36] Appreciation for, or conservation of, grand and magnificent scenery and landscape so important in Natal or the United States was not mentioned in Parliament, although the existence of the Natal National Park was noted, but only in terms of having "no faunal wealth worth mentioning."[37] This remark indicates that in the opinion of some people, wildlife preservation on a large scale and of a large number of rare or charismatic mammal species is an essential element of an African "national park" rather than landscape or scenic beauty. Divergence between the U.S. national park experience and that of South Africa was widening.

The United States, however, was used as leverage in the parliamentary debate on founding the Kruger Park. The minister of lands, Piet Grobler, said: "In America there are eighteen national parks, of which the largest and best known is the Yellowstone Park, which occupies an area of 2,142,000 acres or 3,348 square miles. It is the largest national park in America and I think the second largest in the world. Our park will be slightly larger, 2,400,000 morgen."[38] Grobler acknowledged that Yellowstone was of "greater importance" with rare components such as geysers and fossils, "but as regards variety of game it cannot compare with our park here."[39]

Amid all the enthusiastic accolades for the idea, there was an objection to the phrase "national park" as not being appropriate in Afrikaans. The legislature agreed, and the suggestion of having the Afrikaans official name as the "Nasionale Kruger-wildtuin" (the National Kruger Wildlife Reserve) was accepted. Thus, anomalously, in Afrikaans Kruger is not named a "national park" even today.[40] Moreover, Stevenson-Hamilton, South Africa's leading wildlife conservator and warden of the KNP area from 1902 to 1946, preferred "National Faunal Sanctuary" or "National Wild Life Sanctuary." Despite his public rhetoric regarding the first national park in the British Empire, in fact, he rejected "national park" because he wanted to encourage a philosophy of wilderness, of "roughing it" in the bush in order to experience nature. A national park was a mere public "pleasuring ground," not a place in which to experience "nature" and wildlife. The tension between the philosophy of "wilderness preservation" and a "national park" endures into the present.

There was a significant difference between the United States and South Africa in terms of administration and bureaucratic status. South African national park legislation inaugurated an independent parastatal management structure under a National Parks Board of Trustees. South Africa's legislation did not create a division or department within the civil service like the National Parks Service in the United States, nor did it provide detail about responsibilities in terms of education, research, or even visitor amenities and finance.[41]

In publicizing the park after its establishment, comparisons with the United States were made. A prominent politician was quoted in a newspaper report of

August 1927: "Just as the Yellowstone Park is famous in America, we aim to make the Kruger National Park a world-wide attraction. Of course, we don't want to kill off all the lions, so we will have to make provision for the protection of the public who visit there,"[42] thus raising a problem that did not confront Yellowstone, or indeed the Natal National Park.

In the late 1920s and into the 1930s the Carnegie Corporation financed a number of programs in South Africa. Most famously was Carnegie's assistance in the 1930s, during the Great Depression, with a commission into "Poor Whites." In May 1929 the corporation established a Visitors' Grants Committee in order to facilitate visits by South Africans to the United States in order to study various aspects of U.S. society of relevance to South Africa.[43] In 1929, as mentioned above, prominent Cape Town academic botanist R. H. Compton received such a grant and, together with his wife, visited some thirty-five parks, monuments, gardens, and other related sites, traveling some 8,000 miles by car and traversing eighteen states between April 2 and June 29, 1930. Compton's report, the aim of which was to give South African readers a general account of "what the American National Park system is . . . and to suggest in what ways a similar system might be usefully established in South Africa," begins with a section titled "The National Park Problem in South Africa" in which he includes praise for the new Kruger National Park in becoming so quickly "one of the chief attractions of the country, ranking with the Victoria Falls as one of the world's most marvelous spectacles," and expresses his appreciation of the "enormous pride and advantage" relating to preserving "the nature features of a country." However, Compton strongly critiques South Africa's prioritizing of large mammal protection, attributing myopia in this regard to the extermination of these fauna on account of various factors, to the detriment of other aspects of "nature," particularly mountain scenery and indigenous flora. He also notes that the example of Mont aux Sources National Park (Natal), and that of the U.S. national parks with which it resonates, should be followed in South Africa, particularly in the Cape Province where mountain areas in the Cederberg and elsewhere should become designated national parks. The Natural and Historical Monuments Commission had made a strong recommendation to the appropriate cabinet minister regarding "suitable areas for reservation," noting that "there for the moment it rests." Nothing came of such ideas until many decades later, and one presumes that it was not politically expedient to proclaim them at that time.[44] It is interesting to note that in the minutes of the meetings of the Carnegie Corporation Visitors' Grants Committee on April 29 and July 12, 1933, there is reference to Compton having been requested to "tone down" his "wholesale condemnation in the strongest possible terms of what has been done and omitted in South Africa" expressed in his draft report because his sentiments "may give rise to contentions which may not serve the interests he has at heart." Presumably he did so, but the critique is nonetheless evident. The National Parks Board was asked to assist with publication costs of this report; it did not do so, but it did undertake to assist with the sale of printed copies.[45] The number of sales is not on record, but Compton's report was not widely circulated and is extremely difficult to find.

In all discussions, of course, it was assumed that Africans would have no part in national parks, although it was they who paid the price for these parks, not only—in some places—in terms of alienation from land but also because their government taxes—at least in the Kruger National Park—ensured financial viability, which paying recreational visitors did not accomplish.[46] Africans were not permitted to visit the park for recreation, and in the 1940s and 1950s, in a push to employ Afrikaners even in menial positions, many Africans lost the few job opportunities that national parks could provide.

The United States and South Africa: National Parks 1930s to 1950s

During the 1930s South Africa established more national parks—officially named as such, and administered and managed by the National Parks Board, which by then had a small paid staff in Pretoria. They were not founded with any system in mind—which is what Compton had suggested[47]—but came under the aegis of the board on an ad hoc basis via donations of private land, or in order to preserve a population of endangered mammals. Only one, the Kalahari Gemsbok National Park (1931), a large arid area bordering on Botswana and Namibia, met the modern criteria of "national park," though for many years it had no visitor facilities.

If Compton wished to advocate the adoption of the U.S. system for South Africa, the United States was equally curious, it seems, about South Africa's national parks at the very same time. Ralph J. Totten, the first U.S. ambassador to South Africa, was asked by the NPS to provide information on the KNP, which he did in July 1931. In particular Totten was struck by the presence of Africans who assisted visitors with their barbeque fires in the Kruger Park (there were no large communal campfires as there were in U.S. national parks). Another difference was that no hiking was permitted in the KNP. Because of the danger of predators such as lions, all traveling had to be done by motor car on roads that "are nothing more or less than tracks."[48] Further official correspondence between the two countries in the 1930s relates to an exchange of annual reports between the National Parks Board and the NPS, without comment on either.[49]

In 1932 D. J. Esselen, a South African agriculturalist studying in the United States, was offended by a remark he heard made by a ranger in Yellowstone that it was the "world's biggest game reserve." Esselen wrote at length to Yellowstone's superintendent, emphasizing the impressive mammal fauna that visitors could enjoy in the Kruger Park that was spectacular in comparison with Yellowstone.[50] In 1938 there was an exchange between Stephan du Toit, the clerk of the South African Senate, and Victor Cahalane, head of the NPS Wildlife Division.

(Because of the importance of wildlife conservation and management to South Africa's national parks, communication was generally with Cahalane rather than any other official in the NPS.) From the NPS and Cahalane, du Toit requested information that would assist in revising the South African National Parks Act. Together with other changes, du Toit believed that the South African legislation needed to include an injunction to preserve "scenic beauty," an aspect that had been omitted in the law of 1926[51] but which was, of course, seminal to the establishment of Yellowstone.

During the Second World War, when it became obvious that long-standing warden Stevenson-Hamilton would retire, and that a change of conservation philosophy would occur, there was sporadic correspondence between H. S. van Graan, secretary of the National Parks Board, and the NPS, as well as between Rudolph Bigalke (director of the National Zoological Gardens in Pretoria) and the NPS. The chairman of the National Parks Board, Afrikaner nationalist J. F. Ludorf, wrote to the director of the NPS in 1944, asking for "guidance and assistance" in handling the influx of tourists and problems with diseases. In response, Newton B. Drury asked for more information about management and legislation, and he emphasized the bureaucracy that governed the NPS. The following year, Cahalane recommended to Drury that "the Director and as many members of the staff as possible should visit South Africa and inspect conditions in these exceedingly interesting national parks," a suggestion that may have led to his own visit a few years later. In May 1946 Cahalane gave Drury a comparative list of the problems besetting U.S. and South African national parks, identifying what they had in common and where the differences lay.[52] It may be suggested that as the British Empire had collapsed after World War II, South Africa was aligning itself more closely with the United States—an alignment that lasted after the rise of apartheid in 1948 until almost the end of the Cold War.

However, the late 1940s saw South Africa's parks administration in disarray, which may well have been the catalyst that sparked the desire to discover how the U.S. national parks operated and to emulate it. The National Parks Act of 1926 had not defined organizational responsibilities, and lines of authority were unclear.[53] In 1952 the strains that had been built up over many years of lackadaisical administration erupted before a (closed) commission of inquiry during which graphic testimony was presented about mismanagement; unqualified, indolent, drunken, and corrupt staff; interpersonal enmity; poor financial record keeping; and the absence of any accountability.[54] The problems that had arisen in South Africa's parks had been avoided by the United States, where the NPS consisted of three separate divisions: "Ranger," "Naturalist," and an educational division that was, in fact, the "birthplace and nursery of ecological ideas within the National Park Service."[55] Biologists such as Bigalke who advocated similar structures for South Africa were familiar with the U.S. national parks, as the scientific head, Victor Cahalane, had visited South Africa as a guest of the National Parks Board in 1951—an inopportune moment in terms of the internal dissension that was evident to all and the commission of inquiry that was being contemplated.[56] The result of the commission was the dismissal of a number of staff members and a complete revision of the organization, although not along U.S. lines. Three separate departments were established—biology, management, and park development and tourism—and it was recommended that the chairman of the board be a biologist (in the event, a politician took the helm) and that there should be a formalized management position called "director" of National Parks to whom other department heads would report.[57]

In the arena of South Africa's postwar protected areas, two processes were at work. The first brought national parks firmly within the Afrikaner cultural ambit, which was relatively quick and very successful after 1948. By the mid-1960s the observation could be made that South Africa's national parks were part and parcel of the *volkshuishouding*,[58] a complex word suggesting that they had been integrated into the ethos of the Afrikaner *volk* (people), who were, by then, the ruling elite. Thus, it can be argued, South African national park philosophy had been wrested from a British "game reserve" philosophy to fit better a romanticized definition of Afrikaners as frontiersmen who defined themselves, and valorized a rural settler past, through the national parks, hitherto dominated by English-speaking South African visitors, wardens, and game rangers.

National parks were called upon to champion government policy in a manner that did not occur to the same extent in the United States. In many respects they were political tools of the apartheid era. Many Africans who had used or lived in what were to become national parks had been evicted, and those who remained were tolerated only as laborers and rent-payers. At this time Africans were restricted from entering national parks as visitors, and many of those who were employed within the park lost their jobs and were replaced by white Afrikaners. In the 1950s and 1960s all South African national parks became hemmed in by development and, more importantly, by the growing numbers of Africans who were forced into "tribal homelands" by the policies of apartheid. Some homelands bordered on national parks, and it was clear to those living on these boundaries that the well-being of animals and the recreational demands of whites far outweighed concerns over the basic quality of life of millions of rural disenfranchised Africans.

The second process was the introduction of a wildlife management strategy based on a specific scientific paradigm, an area in which South Africa attempted to learn from the United States. As early as 1936 Bigalke had asked the NPS for educational material and advice,[59] and he intended to emulate the scientific framework and educational outreach of the U.S. national parks. Bigalke was partially thwarted in this endeavor by the vicious internal politics that the Hoek Commission of 1952 uncovered and that led to his resignation from the National Parks Board. Bigalke's interest in the scientific value of national parks did not, however, abate, leading him to visit a number of U.S. parks in the

1950s. Bigalke's arguments that formal academic research should be conducted in the national parks so that they had a serious purpose and were not merely "pleasuring grounds" for visitors[60] may have been instrumental in including a provision for the "study" of wildlife, which had not been included in the 1926 National Parks Act (as it was not in the case of Yellowstone, either), in an altered National Parks Act of 1962.[61] By far the majority of the scientists appointed then and later by the National Parks Board (most of them in KNP) were Afrikaners trained at the Afrikaans-medium University of Pretoria; there was little engagement with English speakers (many of whom joined the Natal Parks Board) or with international scientists.

The United States and South Africa: National Parks, 1960 to 2003

Although the National Parks Board and the provincial departments of nature conservation devoted the 1950s and 1960s to molding national parks into an authoritarian and Afrikaner-dominated society, the country remained part of the international structures of nature conservation and became a member of the International Union for the Conservation of Nature (IUCN, 1948) and the World Wildlife Fund (WWF, 1961). These connections were to prove useful platforms for parading South African achievements at a time when the country was vilified in many other international forums.

There is evidence of considerable communication between the NPS and the National Parks Board in the 1960s. In 1966, at the height of apartheid, Rocco Knobel, then director of the National Parks Board, and C. Gordon Fredine, chief of the Office of International Affairs at the NPS, were on good terms.[62] Arrangements were made for R. J. Labuschagne, head of the Nature Conservation Division in the reorganized National Parks Board, to attend the "Short Course in Administration of National Parks and Equivalent Reserves" in the United States and to meet prominent conservationists in Washington, D.C. Thereafter he was treated to an extensive tour of a number of U.S. national parks.[63] A "US–South African Leadership Exchange Program" for conservation education and experience also existed throughout the 1960s.[64] However, there was no direct effect of these exchanges either on the wildlife management or the administration of South Africa's national parks, and the OIA website does not include South Africa among the national parks plans that the OIA wrote or contributed to in the 1960s and beyond.

Because of its apartheid policies, South Africa was banned from the international Olympic movement and had been expelled from the International Labor Organization and UNESCO. It was only thanks to a veto from the United States that South Africa survived expulsion from the United Nations in 1962. The United States supported and protected the South African government for many years, and the national parks movement provides further evidence of this close relationship. Nash makes a similar point in connection with the 1970s and the United States, in that after the debacle of the Vietnam War, the Yellowstone congress of 1972 had as its aim to show that the United States was a leader in something as "wholesome as the national park movement."[65]

Post-apartheid, the Fifth World Parks Congress and "Benefits beyond Boundaries"

As negotiations toward a "new" South Africa took shape in the early 1990s, there were calls to abolish national parks—the Kruger National Park in particular. As expressed in the *Baltimore Sun* in May 1995:

> To the tens of thousands of people who enter it each year, Kruger National Park offers the chance to mingle with lions, elephants and the other wild beasts of Africa. But for the impoverished millions of black people who live on the park's border, it represents an anachronistic bastion of white privilege. For generations, the people on the outside of the park's electrified fence have been like street urchins with their noses pressed up against the window of a showplace. In South Africa's new democracy, those people are now demanding to be allowed inside, to benefit from the potential riches there.[66]

In view of this history, changing the name of the park was also an option considered at this time. Many cities, towns, streets, and large dams were being renamed in South Africa in a process of reconciliation and redress, and for a while it seemed that Kruger would be no exception.[67]

The protected area estate in South Africa has altered greatly since 1994 and is coming into line with the national priorities of the country at large. Once relegated to menial positions in national parks, black South Africans (including women) now occupy top positions as managers and scientists in the renamed SANParks. Given the pressure on the country's national parks not to drain the treasury—with priorities being to provide housing, education, health care, and other services—they are expected not only to pay for themselves but to make a profit. In other words, protected areas have to be a productive form of land use that outcompetes agriculture. The pressure on the tourist industry to generate employment and revenue is critical to stability and progress, given the high rate of unemployment mentioned earlier. Some areas within the national parks have been given out as concessions so that SANParks receives royalties on a par with the luxury private ecotourism market. Sustainability through appropriate commercialization in all national parks is firm policy.[68]

South Africa's national parks are legitimate targets of land restitution, and successful claims may well result in considerable adjustments to park boundaries and ownership. Transfers of national park land to the originally dispossessed communities have meant that not only state land comprises national parks, which is the Yellowstone model. Contract parks and community ownership are already realities, cases in point being the Makuleke who have regained ownership of the northern portion of the Kruger Park, the Nama of the Richtersveld National Park, and the Khomani San of the Kalahari Gemsbok National Park.[69]

Since its readmission to the international community, South Africa has been able to establish World Heritage Sites and biosphere reserves. While in the early years of the twentieth century national parks and game reserves were the sole means by which to accomplish nature protection, in the twenty-first century it may be argued that the international significance of national parks has been eclipsed by these newer international structures and the global focus on sustainability and resilience rather than "conservation" and recreation. In comparison with national parks, these forms of protected areas have been uncontroversial, not only because they postdate apartheid but also because they result from consultation and partnership that was absent in an earlier era of national parks. A sense of community has generally been developed in the process of establishment, and there is a greater sense of national ownership. In addition, greater emphasis has been placed on transfrontier national parks, some of which are indeed controversial but which are designed to improve relations between neighboring governments that were also detrimentally affected by apartheid.[70] It can be argued that national parks are more divisive than other forms of protected areas and biodiversity conservation because of their history.

In his opening message to *South African National Parks: A Celebration,* former president Nelson Mandela alluded to some of the reasons for the contested nature of South Africa's national parks: "Many of our protected areas have their origins in the colonial past and there is a legacy of alienation from the local people who were excluded from enjoying them or benefiting from them in any way." Mandela believed that, because the title and mission of the Durban gathering was "Benefits beyond Boundaries," it "provides us with the opportunity to break with this history"[71] by utilizing national parks to provide a "better life for all." In the United States there is no similar mission for national parks to provide benefits and upliftment for communities that lie outside their boundaries.

In this discussion, however, it needs to be recalled that SANParks does not have a monopoly on South African "national parks." In preparation for the Durban World Parks Congress, South Africa drafted new legislation to provide for its protected areas under an updated and comprehensive law. This entrenches the power of SANParks but does not oblige other areas to give up their named "national parks," any more than SANParks is obliged to give up its national parks where they do not meet IUCN criteria. SANParks owns just over half of South Africa's total protected area, but most of this comprises two enormous national parks, the KNP (1,948,528 ha) and the Kalahari Gemsbok National Park (960,029 ha). There are no SANParks-managed national parks in KwaZulu-Natal.

Conclusion

Although elements of the U.S. national parks have been invoked from time to time, and there has been sporadic communication between the national parks organizations of the United States and South Africa, there has not been a U.S. pattern to which South Africa has adhered. Moreover, the country has such a wide variety of national parks that it cannot even be said that South Africa itself has anything like a general template. South Africa's many national parks vary greatly in size, in conservation value, in scientific and administrative management objectives, in land regime, and in partnership and consortium arrangements.

Over the years, the changing vision and mission of South Africa's protected areas have been the result of the prevailing government priorities. Just as the apartheid state had a political mission for national parks, so too has the current government. "The focus for SANParks in the first decade of democracy has been to make national parks more accessible to tourists in order to ensure conservation remains a viable contributor to social and economic development in rural areas." Accordingly, SANParks's mission is "to develop and manage a system of national parks that represents the biodiversity, landscapes, and associated heritage assets of South Africa for the sustainable use and benefit of all."[72] Note the absence of the word "enjoyment." The current benefit with which national parks can provide South Africans is economic. Ecotourism is extremely beneficial to South Africa, being a sustainable industry and creating employment and building capacity in the service sector, while ecotourism in the remote rural parts of the country brings skills development and money into impoverished areas.[73]

As the decades have passed, South Africa's national parks have increasingly diverged from anything that might be termed the "Yellowstone model." It is for their potential to contribute to the treasury that national parks have some political support in a country in which the majority of citizens are extremely poor and would not, themselves, have the opportunity to visit a national park, nor would they wish to do so, as recreational visits to "pleasuring grounds" (or wildlife viewing) are not a cultural norm. Certainly, the numbers of black South Africans visiting national parks is growing, but it is far from being a national pastime. For most people, the costs of transport, entry fees, and overnight accommodation are prohibitive. Most black South Africans who visit do so for the day (not overnight), and the statistics include organized visits for school learners and neighboring communities free of charge. Because of the past history, wealthy black South Africans have no tradition of family holidays in national parks and would prefer to travel abroad or to purchase luxury goods that they have previously not been able to acquire. SANParks visitor figures tell the story. During the year April 2010 to March 2011, visitor numbers for all national parks totaled 2,073,582, of whom 78.1 percent were South African residents, 0.9 percent were from the southern African region, and 21 percent were international. Of the 1,620,291 visiting South African residents, 1,219,794 were whites, out of a total population of 4.9 million. There were 397,618 "blacks" (around 25 percent of South African visitors), a category that includes colored, Indian, and Asiatic South Africans, from a total population of 45,700,000.[74] As indicated above, most black visitors come for a day, either as part of a school or community outing, particularly to national parks close to densely populated semirural areas. This cohort forms almost 25 percent of South African day visitors, but only 7.5 percent of the total South African overnight guests. The number of black day visitors increases each year with special efforts made in this regard by SANParks.

In terms of the exportation of "America's Best Idea," South Africa's national parks differ from those in the United States. In terms of national importance and "benefit" they are motors of economic growth and prosperity, employers and capacity builders. Moreover, they have generated interesting models of private/public partnerships, and particularly community partnerships that have also shifted the emphasis in national parks toward overt socioeconomic goals. Parks thus serve local interests and, in this way, are integrated into national ideology and agendas as has been the case in the past.

Notes

1. The South African parastatal body tasked with managing certain protected areas expressly named in English as "national parks." The official website for SANParks is www.sanparks.org (accessed July 24, 2015).
2. Anthony Hall-Martin and Jane Carruthers, eds., *South African National Parks: A Celebration* (Johannesburg: Horst Klemm, 2003).
3. Hector Magome, "Preface," in Hall-Martin and Carruthers, *South African National Parks,* viii.
4. Project notes for "National Parks beyond the Nation" conference, arranged by the Public Lands History Center and the Department of History at Colorado State University in September 2011. See https://nationalparksbeyondthenation.wordpress.com/about-2/ (accessed July 24, 2015).
5. See http://www.sanparks.org/about/transformation.php (accessed December 26, 2011).
6. See http://www.southafrica.info/about/facts.htm (accessed June 24, 2011).
7. For comparison, see William Beinart and Peter Coates, *Environment and History: The Taming of Nature in the U.S.A. and South Africa* (London: Routledge, 1995); Warwick Frost and C. Michael Hall, eds., *Tourism and National Parks: International Perspectives on Development, Histories and Change* (Abingdon, U.K.: Routledge, 2009). See also Karen R. Jones and John Wills, *The Invention of the Park: From the Garden of Eden to Disney's Magic Kingdom* (Cambridge: Polity Press, 2005).
8. George H. Stankey, Vance G. Martin, and Roderick Nash, "International Concepts of Wilderness Preservation and Management," in *Wilderness Management,* ed. John C. Hendee, George H. Stankey, and Robert C. Lucas (Washington, D.C.: U.S. Department of Agriculture, Forest Service, 1978), 50.

9. R. H. Compton, *The National Parks of the United States of America: A Report Presented to the Visitors' Grants Committee for South Africa of the Carnegie Corporation of New York* (Pretoria: Carnegie Corporation Visitors' Grants Committee, 1934).

10. As examples, see Roderick Nash, "The Confusing Birth of National Parks," *Michigan Quarterly Review* 16 (1977): 216–26; Roderick F. Nash, "The American Invention of National Parks," *American Quarterly* 22 (1970): 726–35; W. C. Everhart, *The National Park Service,* 2nd ed. (Boulder, Colo.: Westview, 1983); Ann MacEwan and Malcolm MacEwan, *National Parks: Conservation or Cosmetics?* (London: George Allen & Unwin, 1982).

11. https://www.iucn.org/about/work/programmes/gpap_home/pas_gpap/ (accessed July 23, 2015). A protected area is "a clearly defined geographical space, recognized, dedicated and managed, through legal or other effective means, to achieve the long term conservation of nature with associated ecosystem services and cultural values."

12. Kevin Bishop, Nigel Dudley, Adrian Phillips, and Sue Stolton, *Speaking a Common Language: The Uses and Performance of the IUCN System of Management Categories for Protected Areas* (Cardiff University: IUCN and UNEP, 2004); Adrian Phillips, "Turning Ideas on their Head: A New Paradigm for Protected Areas," *George Wright Forum* 20 (2003): 8–32.

13. IUCN, *Guide to the WPC Recommendations Procedures Fifth World Parks Congress* (Gland, Switzerland: IUCN, 2003), 50–51.

14. Zoe Wilkinson (supported by Trevor Sandwith), "Preparing for the 6th IUCN World Parks Congress 2014: A Draft Background Discussion Paper," http://cmsdata.iucn.org/downloads/iucn_wpc_2014_draftdiscusionpaper_4_1_11 .pdf (accessed December 26, 2011), 20.

15. Nash, "Confusing Birth," 224.

16. An Act to set apart a certain Tract of Land lying near the Head-waters of the Yellowstone River as a public Park. Forty-Second Congress, Session II, Ch. 21–24, 1872, Library of Congress Web Guides, Primary Documents in American History, Act Establishing Yellowstone National Park, http://www.loc .gov/rr/program/bib/ourdocs/yellowstone.html (accessed July 25, 2015). See also Roderick Nash, *Wilderness and the American Mind* (New Haven, Conn.: Yale University Press, 2001).

17. See, for example, Mark D. Spence, *Dispossessing the Wilderness: Indian Removal and the Making of the National Parks* (Oxford: Oxford University Press, 1999).

18. A phrase popularized by Daniel Brockington in *Fortress Conservation: The Preservation of the Mkomazi Game Reserve, Tanzania* (Oxford: James Currey, 2002).

19. Natal Archives Depot (hereafter NAB) MJPW 85 LW 3215–1901.

20. David Lewis-Williams, *Images of Mystery: Rock Art of the Drakensberg* (Cape Town: Double Storey, 2003). The uKhahlamba-Drakensberg, inscribed in 2000, is a mixed site, recognized for both natural and cultural criteria. It is perhaps instructional of South African national park history that the Drakensberg has never been managed by SANParks, although today almost the entire range is a World Heritage Site (the uKhahlamba-Drakensberg), a Ramsar Wetland, and a Transfrontier Conservation Area.

21. South African National Archives, Central Government (hereafter SAB) LDE-N 11 29, Acting Surveyor General Natal to Magistrate Bergville, May 17, 1915;

SAB NTS 9489 76/400 J.S. Henkel, Conservator of Forests, Natal, Preliminary report on the National Park, division of Bergville, October 31, 1917.

22. SAB URU 289 2288.

23. S. H. Haughton, "The Natal National Park," *Annual of the Mountain Club of South Africa* 22 (1919): 32.

24. SAB NTS 9489 76/400. Provincial Secretary to Sir William Hoy, February 4, 1926.

25. Alfred H. T. Perry, *National and Other Parks* (n.p.: n.p. [1929]), 10.

26. SAB NTS 9489 76/400. J. S. Henkel, Conservator of Forests, Natal, Preliminary report on the National Park, division of Bergville, October 31, 1917.

27. SAB NTS 9489 76/400. Dower to Provincial Secretary, May 1, 1918.

28. SAB NTS 9489 76/400. Note after discussion between Provincial Secretary and Chief Native Commissioner, September 13, 1918.

29. SAB GG 163 & 158 3/31. Prime Minister Minute 1280, December 10, 1920.

30. Jane Carruthers, *Game Protection in the Transvaal 1846 to 1926* (Pretoria: Archives Year Book for South African History, 1995); Jane Carruthers, *The Kruger National Park: A Social and Political History* (Pietermaritzburg, South Africa: University of Natal Press, 1995).

31. Union of South Africa, House of Assembly Debates, 3rd Session, 5th Parliament, January 22 to June 8, 1926. See May 31 cols. 1421–1422.

32. Ibid., May 31 col. 1425.

33. See John F. Reiger, *American Sportsmen and the Origins of Conservation,* 3rd ed. (Corvallis: Oregon State University Press, 2001); Thomas R. Dunlap, *Saving America's Wildlife: Ecology and the American Mind, 1850–1990* (Princeton, N.J.: Princeton University Press, 1988).

34. Transvaal Province, *Report of the Game Reserves Commission, TP 5-'18* (Pretoria: Transvaal Province, 1918), 10–11.

35. SAB GG 2186 76/7. Note from Stevenson-Hamilton, November 17, 1924.

36. National Parks Act No. 56 of 1926 section 1. SAB, GG, Vol. 393, 7/4013, 1926.

37. Union of South Africa, House of Assembly Debates (n. 31). See May 31 cols. 4366–4380 and col. 4372.

38. Ibid., cols. 4367–4368. A morgen is a South African land measurement of 2.10 acres or 0.850 ha. The Kruger National Park is substantially larger than Yellowstone. Currently Yellowstone is 3,468.4 sq mi or 8,983 km^2 or 2,219,791 acres or 898,318 ha. Kruger is 7,332 sq mi or 18,989 km2 or 6.2 million acres or 2.5 million ha.

39. Union of South Africa, *The Senate of South Africa: Debates,* Vol. 7, June 3, 1926, cols. 1080–1081.

40. Union of South Africa, *House of Assembly Debates*, Vol. 7, May 31, 1926, cols. 4377–4378.

41. In his report, Compton detailed the various practical aspects of U.S. national parks in terms of organization, the concession system for visitor amenities, wildlife preservation policies, protection from fire, education, finance, and publications, indicating the many areas in which South Africa might emulate the United States. See Compton, *National Parks of the United States of America,* 27–54.

42. *Rand Daily Mail,* August 18, 1927, "Motor Roads for Game Reserve."

43. Another grant holder was the ornithologist Austin Roberts, whose report was published as *Museums, Higher Vertebrate Zoology and Their Relationship to Human Affairs* (Pretoria: Carnegie Visitors' Grants Committee, 1935).

44. Compton, *National Parks of the United States of America,* 1–8, 57–58.

45. University of South Africa Archives and Special Collections (hereafter UNISA), Minute book: Carnegie Corporation Visitors' Grants Committee, 1929–1948, minutes of meetings held on April 29 and July 12, 1933.

46. SAB LDE 571 [545] 7748/44 contains the correspondence in this regard.

47. Compton, *National Parks of the United States of America,* 11.

48. Archives of the National Park Service (hereafter NPS), Selected documents File 0–30. Foreign Parks, South Africa, Totten to Secretary of State, July 6, 1931.

49. Ibid., Potgieter to Director NPS, June 21, 1933; Cahalane to du Toit, January 5, 1938; Cammerer to de Wet, July 14, 1938; Bryant to du Toit, August 29, 1938.

50. Ibid., Esselen to Superintendent, Yellowstone, November 15, 1932. The ranger may, of course, have been confusing KNP with Parc Albert in the Congo.

51. Ibid., du Toit to Cahalane, October 21, 1938.

52. Ibid., Ludorf to Drury, November 11, 1944; Drury to Ludorf, February 19, 1945; Calahane to Drury, August 1, 1945; Ludorf to Drury, September 18, 1945; Calahane to Drury, May 8, 1946.

53. South African National Parks (hereafter SANParks) Minutes, November 13, 1950.

54. P. W. Hoek, "Verslag oor 'n Ondersoek van die Bestuur van die Verskeie Nasionale Parke en die Nasionale Parkeraad se Administrasie," unpublished report, 1952, Archives of the National Parks Board of Trustees, Pretoria, 21–25. The report detailed the staff complement: Head office, 8 whites, 1 black; Addo, 1 white, 8 blacks; Bontebok, 1 black; Kalahari, 1 white, 4 blacks; Mountain Zebra, 1 white, 1 black; Kruger National Park, 18 whites, 400 blacks, and about 150 more in the tourist industry. Of the total of 29 permanent white staff, three were called "scientists and educationalists."

55. James A. Pritchard, *Preserving Yellowstone's Natural Conditions: Science and the Perception of Nature* (Lincoln: University of Nebraska Press, 1999), 48–52.

56. See Richard W. Sellars, *Preserving Nature in the National Parks: A History* (New Haven, Conn.: Yale University Press, 1997), 165, 169; John Ise, *Our National Park Policy: A Critical History* (Baltimore: Johns Hopkins University Press, 1961), 699; Hoek, "Verslag," 21, 52.

57. Hoek, "Verslag," 39–41.

58. SANParks, Minutes, March 11, 1968.

59. NPS, Selected documents File 0–30. Foreign Parks, South Africa, Bigalke to Director NPS, August 5, 1936.

60. Rudolph Bigalke, "Science and the Conservation of Wild Life in South Africa," *Journal of the South African Veterinary Medical Association* 21 (1950): 166–72; Rudolph Bigalke, *National Parks and Their Functions, with Special Reference to South Africa,* Pamphlet no. 10, South African Biological Society (Pretoria: Biological Society, 1939).

61. Union of South Africa, House of Assembly Debates, vol. 108, May 1–26, 1961, cols. 7291–7315. The final wording of Section 4 in Act No. 42 of 1962 was: "The object of the constitution of a park is the preservation and study therein of wild

animal and plant life and of objects of geological, archaeological, historical, ethnological and other scientific interest and the benefit and enjoyment of visitors to the park."

62. NPS, Record group 79, Administrative files, 1949–1971, Box 2182, "Foreign Parks and Historic Sites, South Africa, 1966–1967," Fredine to Knobel, January 4, 1966. At the early high point of U.S. prominence in protected area public relations (1962, the First World Conference on National Parks, Seattle) the NPS founded its Office of International Affairs (OIA).

63. Ibid., Fredine to Knobel, January 7, 1966; Fredine to Labuschagne, January 19, 1966; Fredine to Labuschagne, February 3, 1966; Fredine to Labuschagne, March 8, 1966.

64. Ibid., Fredine to Petrides, January 10, 1966.

65. Nash, "Confusing Birth," 216.

66. Michael Hill, "Fenced-out Villagers Await South African Park Reforms," *Baltimore Sun,* May 24, 1995, http://articles.baltimoresun.com/1995-05-24/news/1995144104_1_kruger-national-park-south-africa-wildlife-parks (accessed July 25, 2015).

67. African National Congress, Daily news briefing, November 29/30, 1994, South African Press Association, Pretoria.

68. See, in Hall-Martin and Carruthers, *South African National Parks,* the following essays: Peter Fearnhead and David Mabunda, "Towards Sustainability," 84–99; Mohammed Valli Moosa and Murphy Morobe, "The Future," 246–52.

69. Jane Carruthers, "South Africa: A World in One Country," *Conservation and Society* 5, no. 3 (2007): 292–306.

70. M. Pabst, *Transfrontier Peace Parks in Southern Africa* (Stuttgart: SAFRI, 2002). See also www.peaceparks.org.

71. Nelson Mandela, "Message from Nelson Mandela," in Hall-Martin and Carruthers, *South African National Parks,* vi.

72. www.sanparks.org/about (accessed June 30, 2015).

73. Anna Spenceley and H. Goodwin, "Nature-based Tourism and Poverty in South Africa," *Current Issues in Tourism* 10 (2007): 255–77.

74. The most visited national park is the KNP, with a total of 1,386,287 visitors for the period of April 1, 2010 to March 31, 2011. The least visited is Tankwa Karoo with 4,328. I am grateful to Glenn Phillips for these statistics. See SANParks Annual Reports, in particular for these figures, the annual report for 2010, http://www.sanparks.co.za/assets/docs/general/annual-report-2010.pdf (accessed August 27, 2015).

| # Dragons or Volcanoes?

National Parks and Nature Loving in New Order Indonesia, c. 1980–1998

STEVEN RODRIGUEZ

DURING THE AUTHORITARIAN NEW ORDER REGIME of General Suharto (1966–98), the Indonesian government established thirty-four national parks, all of them between 1980 and 1998. Yet by the end of Suharto's reign, it had become evident that Indonesia's celebrated system of national parks had not succeeded in achieving its purported biodiversity conservation and development objectives; the many parks created during his rule failed to protect biodiversity, contribute to the development of the local economy, provide recreation for the population, or provide meaningful locations for national identity.

This chapter focuses on the development of Indonesian national parks in the 1980s and 1990s. It examines why Suharto and other Indonesian elites promoted the implementation of a system of national parks and attempts to explain why Indonesia's national parks program proved "unsuccessful." In particular, this chapter argues that the national park model adopted by Suharto's New Order regime failed in large measure because—for a variety of reasons—it did not conform to the actual desires, interests, and tastes of Indonesians. By pointing to the disjuncture between official motives and popular interests, this chapter not only explains the puzzling failure of the New Order's parks initiatives but also highlights a more general problem with the establishment of national parks in countries controlled by authoritarian regimes.

In 1997 Indonesia's celebrated rain forest and orangutan sanctuary, the Kutai National Park in eastern Borneo, burned for three months consuming 95 percent of its remaining lowland forest and killing one-third of its orangutan population. The fires in East Kalimantan soon became an international affair as the unhealthy air pollution "haze" generated from the burning forest spread westward from Indonesia to Singapore, Malaysia, and Brunei. The Indonesian government blamed the El Niño effect, but conservation organizations and the neighboring countries affected placed the blame on Indonesia. International investigators discovered that careless and illegal logging procedures in and around the park had created the conditions for the catastrophic fires.[1]

The fire-induced "haze" was not the first international crisis involving Indonesia's national parks. The previous year, the OPM (Organisasi Papua Merdeka or Papuan Freedom Organization) had taken a group of European conservation workers hostage. These conservation professionals—representatives of the United Nations Educational, Scientific, and Cultural Organization (UNESCO) and World Wildlife Fund (WWF)—were investigating the 1.2 million ha of territory slated to become Lorentz National Park. The leaders of the OPM wanted compensation for the loss of the tribal lands to be included in the transaction, and failing to receive any recognition of their rights and any compensation, they decided to take direct action hoping to attract media attention to their cause. After several months of lengthy negotiations, including intervention by the pope and the International Red Cross, their strategic protest ended with the Indonesian military moving in and slaughtering the Papuan rebels (all the Westerners survived). Lorentz National Park was designated the following year.[2]

The Papuan hostage episode and the latest haze crisis were only the first major signs that Indonesia's national park system was malfunctioning. The Asian financial crisis and the subsequent demise of the Suharto regime in 1998 exposed the true magnitude of the problems with Indonesia's national parks. Decentralization of government control of natural resources was among the first acts initiated by the post-Suharto Reformasi (Reformation) government. Forests and forest revenue—including national parks—were put under the jurisdiction of the provincial governor, resulting in the acceleration of illegal logging as hundreds of small-time operators sought to take advantage of the situation and reap maximum profit as soon as possible. Their status as national parks provided little protection for these areas. Worse, the association of national parks with Suharto and his oppressive regime made these forest locations targets for exploitation.[3]

The deterioration of Indonesia's national parks program at the end of the twentieth century was a startling reversal of fortune for global flora and fauna conservation. During the 1980s and 1990s, conservation and protected-area development became significant features of Indonesian domestic policy. In 1980 the Indonesian government declared its first five national parks, and by 1998—the year the Suharto regime collapsed—Indonesia had developed a system of thirty-four national parks.[4] Starting in 1982, the Indonesian government committed US$12 million a year to enhancing its national park programs, and in the 1990s the annual investment for protected areas reached US$22–33 million a year. In 1977 the Indonesian nature conservation directorate, previously a paper bureaucracy, had been reconstituted with a new leader, and a school for nature conservation was created at Ciawi. Moreover, a conservation bureaucracy was established at the ministry level—the highest level of government—and this new Ministry of Population and the Environment (Menteri Kependudukan dan Lingkungan Hidup) was placed under the control of the economist Emil Salim, one of Suharto's most competent advisers.[5] The Indonesian government introduced a series of environmental laws related to the national parks and management of the living environment, in particular the Basic Environmental Law of 1982 and the Conservation of Living Natural Resources and Their Ecosystems Act of 1990. In this era, Indonesia also became a party to many global conservation initiatives, including the World Heritage Convention and the UNESCO Man and the Biosphere program.[6]

Although Indonesia's first national parks were not officially declared until the 1980s, the location and boundaries of these parks were based on an earlier system of "reserves," "nature monuments," and "wildlife sanctuaries" that had been established by the Dutch during the colonial era. Each of these protected-area categories represented different conservation objectives. "Reserves" were designed for the protection of wildlife or the preservation of natural areas for scientific research. "Natural monuments" were smaller areas managed for the protection of specific tourist sites such as volcanic craters, mountain lakes, boiling mud pools, and spectacular views. "Wildlife sanctuaries" were the largest protected areas, established to protect large animals and their habitats. Between 1932 and 1940, seventeen wildlife sanctuaries were created—eight in Sumatra, two in Java, two in Kalimantan, one each in Bali and Lombok, and three covering individual islands in the Komodo group.[7]

Although the Dutch colonial network of protected areas supplied the basis for Indonesia's new system of national parks, the Dutch never formally designated areas "national parks," nor did they provide any legal definition for such a category. It was not until the Conservation of Living Natural Resources and Their Ecosystems Act of 1990 that the Indonesian government established a formal definition for a *taman nasional* ("national park"). According to the 1990 act, "national park" was defined as "a nature conservation area which possesses natural ecosystems, and which is managed through a zoning system for research, science, education, supporting cultivation, recreation and tourism purposes."[8]

Indonesian national parks policy in the 1980s emphasized the protection of flora and fauna. In the initial management plans, devised with the assistance of the Food and Agriculture Organization of the United Nations (FAO) and the United Nations Development Program (UNDP), the purpose of Indonesia's system of national parks was to "preserve representative viable samples of all major ecosystem types and living-communities within Indonesia."[9] National parks were conceived as part of a larger network of 196 protected areas designed to conserve the entire range of Indonesia's flora and fauna.[10] In addition, the promoters of Indonesia's national parks also focused on the contribution of these areas to local development. According to Effendy Sumardja, Harsono, and John Mackinnon, three of the conservationists who contributed to formulating Indonesia's national parks program: "National Parks must be clearly seen to be in the regional interest so that their establishment will constitute a benefit, rather than an added hardship, to the rural people living around them. Such benefits can include: preservation of high quality living environment; protection of water sources; establishment, where necessary, of buffer zones; job opportunities (working the park or created by local tourism industry); [and] special developments around parks, e.g. schools, road improvements, irrigation improvements."[11] Their stress on what would increasingly be referred to as "sustainable development" reflected the dominant agenda for international conservation in the 1980s. Designed to replace earlier "fortress conservation" models that encouraged setting apart and locking up resources, the new modus operandi for national parks emphasized that these areas must contribute to local and national economic development processes. Thus envisioned, Indonesia's national parks program would be a flagship example of how national parks could be advantageously integrated into the development programs of the emerging nations of the global South.

There was a striking incongruity between the promotion of national parks by the New Order government and the regime's record for environmental destruction. Suharto was infamously corrupt, and during his reign the Indonesian government had been notorious for its policies of reckless deforestation. In the 1970s Indonesia became the world's biggest exporter of tropical timber; gross foreign exchange earnings from timber rose from US$6 million in 1966 to US$2.1 billion in 1979. Suharto's military dictatorship thrived on the exploitation of the forests, and these resources were carefully controlled by a cartel that included Suharto and his personal allies.[12]

Consequently, many scholars dismissed Indonesia's national parks programs as merely another scheme that enabled Suharto to monopolize control over the timber industry. National parks were created in all regions of Indonesia not because of environmental concerns, but because this justified the creation of a compliant national bureaucracy. Placing all of the provinces' parks under the national government facilitated Suharto's command of the nation's forests and prevented governors and local elites from exploiting these resources. National parks also legitimized the central government's use of force to protect forests "for the nation." Far from a method for helping local people with sustainable development, national parks were intended to control their access to forest products and to justify removing people entirely from an area.[13]

Scholars also pointed out that Suharto coveted the financial and technological assistance provided by the international conservation and development organizations that supported Indonesia's national parks program. In the 1980s and 1990s, the UNDP, FAO, World Bank, WWF, Smithsonian Institution, (U.S.) Nature Conservancy, and a range of other international and foreign-national bodies funded Indonesia's national parks programs.[14] These groups contributed to the establishment of research stations, provided training for Indonesian scientists, and supplied valuable maps, geological surveys, and biological assessments to Indonesian scientific institutes. The international conservation organizations contributed to creating an infrastructure for the maintenance of Indonesia's conservation bureaucracy and thus helped facilitate the reach of the central government.[15]

Participating in the global national parks movement also provided important prestige and propaganda benefits for Suharto and his cronies. In the 1980s Indonesian elites aspired to become leaders of the developing world, and Suharto pressed for Indonesia to acquire all of the trappings of modern nations, from national airlines to national parks.[16] Indonesia's impressive national parks program garnered international accolades and elevated its status as a leader in global conservation, inspiring IUCN, FAO, and UNDP experts to convene the prestigious World Parks Congress in Bali in 1982.[17] This international recognition helped shelter Suharto from criticism for his otherwise environmentally destructive policies. Since the 1970s widespread corruption and incompetence in the Indonesian Ministry of Forestry had resulted in the wasteful ruin of the forests of Sumatra and Kalimantan. The Indonesian government used national parks as a method for placating international critics as well as the growing domestic environmental movement that was emerging among middle-class Indonesians.[18]

Despite the various ulterior motives, to dismiss the New Order regime's national parks program as merely a scheme to centralize control of natural resources, a method for gaining prestige benefits and cash from conservation organizations, or a fig leaf covering environmental destruction would be inaccurate. In particular, tourism was promoted alongside conservation as an important motivation for the development of national parks. According to one of the architects of Indonesia's park system, these areas would "attract tourists and foreign revenue and increase the opportunities for education and development of national pride in Indonesia's natural heritage."[19] Tourism became an important feature of Indonesian economic development in the 1970s, and by the 1980s it was a booming industry. Export earnings from tourism increased from 2.8 percent in 1985 to 5.8 percent in 1989. By 1987 inbound visitation hit the 1 million mark and continued to rise. During the period 1984–88, US$ 918.6 million in foreign money and 2.15 billion rupiahs from local interests funded a total of 264 investments, including 183 hotels and 40 recreation facilities.[20]

Nature tourism was a major component of Indonesia's tourism portfolio, and the government was keen to cultivate national parks for this purpose. As a representative of the Ministry of Education and Culture pointed out at the time: "The capacity of the tropical forests in Southeast Asia in fulfilling the growing needs of recreation and tourism is largely undeveloped, partly because of failure to recognize their scenic attributes and the potential interest in wildlife they contain."[21] The NGOs that pressed for the creation of national parks also promoted the tourism benefits of national parks. Conservationists and NGOs did a very good public relations job, leading the Indonesian elite to believe they could make money from national parks through tourism.[22]

However, the government's efforts to develop national parks as a source of tourism revenue were not particularly successful; throughout the 1980s and 1990s, few of Indonesia's national parks received more than 10,000 tourists a year. Domestic tourism was particularly meager, with only a handful of Indonesian parks achieving more than 5,000 Indonesian visitors a year. Komodo and Ujung Kulon, two of Indonesia's most prestigious national parks and both World Heritage Sites, provide striking examples of the lack of national park tourism in this era. In the 1990s Komodo National Park, home of the legendary Komodo "dragon," received approximately 30,000 tourists a year, but fewer than 10 percent of these visitors were Indonesian.[23] Ujung Kulon National Park in western Java, the sanctuary for the precariously surviving endangered Javan rhino, had very few visitors at all. In the 1990s roughly 4,000 people a year visited this park, and only an estimated 2,000 were Indonesian.[24]

The lack of Indonesian tourism at the national parks was often attributed to socioeconomic and cultural factors. The notion of a transcendent nature, separated from culture, was considered a unique feature of the West, while the anxieties over species extinction and the nostalgia for "wilderness" were concerns suited to a modern, Darwinian, secular culture. Indonesian economic and cultural factors prevented them from possessing the aesthetic appreciation of nature necessary for a national parks project to succeed in either its environmental or domestic tourism objectives.[25]

However, it is unclear whether the framers of Indonesia's national parks policies ever actually investigated Indonesian attitudes toward nature tourism. Often conservation reports attributed to Indonesians the same wilderness values popular amongst North American and European audiences. In one glaring example, the Western authors of the 1982 FAO report *National Conservation Strategy for Indonesia* presumed that the desire to walk in "wild unpeopled places" was a sentiment shared by Indonesians.[26] In fact, many Indonesians—irrespective of ethnicity or religion—are extremely wary of the forests. Throughout Indonesia, the forest is widely considered to be *hantu* ("haunted")—a frightening and dangerous place where a person would certainly never wander alone.[27]

The employment of the phrase *taman nasional* also reveals the problems that develop when experts transplant conservation concepts. The formulators of Indonesia's national parks policy translated "national park" as *taman nasional,* a literal translation of "national park" but a phrase that does not coincide with the connotations of "national park" within the international conservation community. The Indonesian term *taman* is generally employed in the sense of "amusement park" or "public park" for recreation. Hence, the formulation *taman nasional* conjured up the idea of playgrounds and rides, and generally gave associations of artificiality rather than wildness, while the majority of Indonesia's national parks were developed in remote locations, difficult to get to, and undeveloped.[28] The failure of most national parks to provide popular locations for public recreation did not necessarily reflect an Indonesian inability to appreciate nature, but rather that the national parks implemented in Indonesia were not suitable to the tastes of the citizens of that nation. What required greater examination was how Indonesians experienced pleasure in nature, and the type of nature areas that they would be interested in visiting.

The popularity of mountain climbing in the Suharto era is an example of a prominent feature of Indonesian culture underexploited by the promoters of national parks. Mountains, of course, have an acute objective reality in Indonesia, which is after all an archipelago of thousands of volcanic islands. Mountains and volcanoes, both active and extinct, are a shared experience for all residents of Indonesia and figure prominently in the art, mythology, and ceremonies of many of Indonesia's ethnic groups. Certain mountains have an especially strong cultural resonance for Indonesians and have provided important symbols for the nation. For example, Mount Merapi and Mount Semeru were among the popular motifs employed by the government and socially engaged artists of the 1950s for the unity and endurance of the new nation.[29]

Although many Indonesian ethnic groups have traditionally celebrated annual ceremonies in which they climbed or otherwise interacted with local mountains, it was in the 1960s that modern "alpinism" became associated with Indonesian nationalism. One of the first indications of the new status of mountain climbing was the mission to conquer Puncak Jaya, the peak in West Papua (on the island of New Guinea) that had been designated the highest in Indonesia but was also considered the highest in all of Southeast Asia. Shortly after the Republic of Indonesia took control of West Papua in 1964, President Soekarno renamed the remote peak Puntjak Soekarno (Mount Soekarno) and sent a team organized by the Indonesian military to climb this daunting mountain; the expedition planted the Indonesian flag at the summit on March 1, 1964. As the first time an Indonesian team had reached the peak, the event was considered a great triumph for the nation and was widely reported in the media. Soekarno's command to the team, "Advance . . . no retreat," became a popular slogan for Indonesian mountaineers and nationalists.[30]

When the celebrated journalist and youth leader Soe Hok Gie formed the student hiking group Mapala at the University of Indonesia in 1964, the link between mountain climbing and nationalism entered a new phase.[31] Responding to the tumultuous political violence of the era, Soe turned to mountain climbing as an activity that could organize students outside of political struggles. Within a few years, mountaineering groups had proliferated at other major Indonesian universities, and by the 1990s most universities and many high schools had "nature loving" clubs that regularly organized hiking expeditions.

The "nature loving" youth movement did not directly associate itself with social change–oriented politics, which allowed these groups to flourish during Suharto's authoritarian regime. After the implementation of the 1978 Campus Normalization Law and the subsequent prohibition of student organizations, mountaineering clubs were one of the few groups allowed to organize on university campuses.[32] Deprived of other social outlets, middle-class students embraced mountain climbing, which soon became a widespread feature of Indonesian youth identity.

These "nature loving" groups coalesced into a national constituency: Indonesian youth came together around Indonesian nature, and through the networks of university groups, student nature lovers across the country learned similar ways of talking about and being in nature. These student groups participated in a nationwide network—they organized, traveled, and climbed mountains together. Moreover, they learned to love a national nature, one that was made accessible through their national citizenship and schooling. Through the student nature clubs all of these impulses were brought together into a national movement.[33]

An important feature of these university mountaineering groups was the active participation of women, who were prominent members from the very first expedition organized by Soe Hok Gie. Hiking expeditions, often lasting for several days, were one of the few acceptable activities in which young women in Indonesia could function outside of adult supervision for extended periods of time. And hiking groups provided a rare opportunity for young men and women to travel together and socialize.

Mountain climbing also allowed for a close association between the students and the military. In the 1980s mountain climbing became an important component of the military culture of the New Order regime. Starting in 1980, Kopassus (the military's special forces) began to employ mountaineering as part of their training. For technical assistance, Kopassus turned to the Mapala groups, and from 1983 they trained together. Thus, mountaineering became a nationalist activity capable of combining and organizing the energies of the educated middle class and the military.[34]

The 1990 expedition to climb Puncak Jaya (no longer named Puntjak Soekarno), intended to commemorate the twenty-fifth anniversary of the first Indonesian ascent of the mountain in 1964, was a culmination of the Indonesian mountaineering movement in this era. Organized in conjunction with the military and Mapala groups, the 1990 expedition was a much more sophisticated affair than the 1964 mission. Facilitated by modern transport—buses, cable cars, and helicopters—the expedition was completed without a flaw, and the Kopassus and Mapala mountaineers reached the peak on April 17, 1990. According to the report in the national newspaper *Kompas,* the climbers "helped bring about our nation's national ideal."[35] Together the student nature lovers and the military made the scaling of mountains an important nationalist enterprise—to conquer a peak was to conquer it for the nation.

Considering the popularity of mountaineering, it is unsurprising that one of the Indonesian national parks that has enjoyed widespread appeal for domestic tourists has been Mount Bromo-Tengger-Semeru National Park, designed specifically to protect the most outstanding and monumental geological features of eastern Java. The national park contains a cluster of active and semi-active volcanos located in and around the enormous Tengger Caldera—8 km in diameter with almost sheer walls rising to 130 meters. The frequent eruptions of Mount Bromo have kept much of the region in the caldera free of vegetation and have created a field of ash and lava known as the Tengger Sand Sea.[36]

The tourism potential of the Bromo-Tengger-Semeru region was recognized by the Dutch, who during the 1920s and 1930s promoted the area as one of the main attractions in Java. The image on the cover of the colonial government's official tourism guide, *Come to Java,* was a photo of the magnificent Bromo-Tengger-Semeru landscape. The book noted that Mount Bromo was an excellent region for scenery, hiking, and recreation, and that it was also conveniently located close to the major transportation hubs and easily accessible.[37]

During the rise of the Indonesian tourism industry in the 1980s and 1990s, this region became a popular destination for both Indonesians and international tourists; notably, Bromo-Tengger-Semeru National Park was one of the few Indonesian national parks where more than 70 percent of the tourists were Indonesian. In the 1990s approximately 100,000 Indonesians a year journeyed to the location—a domestic visitation rate that far exceeded most of the country's other national parks during this time.[38] Interviews with the Indonesian tourists at Bromo-Tengger-Semeru National Park revealed a range of motivations for their visit. Many of the Indonesians interviewed visited the location seeking solitude and a transformational experience of nature, and they complained about the litter, commercialization, and crowding that they felt detracted from the experience. Some traveled with their families and complained about the lack of refreshments, air-conditioned restaurants, and other luxuries. Some were disappointed that there was not more information on the park or more trails, and others complained about the manure from the horses that tourists could ride around the volcanic region. Some groups were concerned with hiking, others with conservation, and some just "liked the view."[39]

Interviews conducted with members of Mapala groups in the 1990s also revealed a range of motivations for their hiking expeditions. Some of these students desired a transformative experience through contact with nature, and others sought wisdom and healing through an intense and spiritual experience. Many were hedonists and wanted to be rebellious, or they were seeking freedom and mobility. For many students mountain climbing was a serious, intellectual mission, an exercise in organizational leadership skills and personal discipline, while a large number felt responsible as personal guardians of nature and participated in the Mapala groups as a feature of their commitment to "environmentalism."[40]

What the responses by the tourists at Bromo-Tengger-Semuru and the members of Mapala reveal is that Indonesians have emotional responses to nature often considered uniquely "Western." Moreover, the failure of parks to achieve one of their principal objectives, recreation and tourism, cannot be attributed to an innate Indonesian insensitivity to nature tourism in general. Rather, it is a consequence of privileging a particular type of nature tourism based on the experience of wilderness and the observation of wildlife, instead of the sublime experience of monumental landscapes.

The question, then, is why didn't Suharto and other leaders of the New Order regime implement a national parks program that focused on the conservation of mountain landscapes and other regions of monumental scenery? As I have indicated above, implementing a national park system that privileged biodiversity conservation provided a range of benefits to the Indonesian government. It is also not impossible that the effects of environmental damage may have actually concerned Suharto and other Indonesian elites, who were swayed by the opinions of international conservation experts and their arguments for protecting biodiversity. Whatever the motivations, it would not be unreasonable to suggest that if the leaders of the New Order regime had focused more attention on developing a system of national parks that protected and promoted the magnificent mountain landscapes of Indonesia, they might have created parks of more enduring national significance. National parks that attracted Indonesians might have generated more substantial tourism benefits and also may have enhanced national pride and prestige, contributing to the patriotic sentiments that the state hoped to obtain from national parks. National parks founded on sublime landscapes may also have provided locations for satisfying the aesthetic yearnings for nature that conservationists have so often argued is crucial for the human spirit. Most important, if Indonesia's national parks actually provided these multiple benefits, it might have allowed them to weather the changes in administration and economy, as well as prevented the accelerated exploitation of the national parks after the fall of Suharto. Indeed, if the parks were a source of tourism, recreation, and nation-building benefits, local elites and the nation would have had a much stronger stake in protecting and preserving the areas.

Admittedly, the cultivation of national parks for the protection of celebrated mountains and volcanic landscapes would not have immediately addressed the biodiversity concerns of the international conservation organizations. However, a system of national parks based on outstanding geological features, on old-fashioned scenery, as national parks began in the United States, might have contributed to the creation of a widespread Indonesian sentiment toward nature in which biodiversity conservation became an important idea. As a middle-class youth movement, as a display of military power, and as a shared cultural experience, mountains and mountaineering were a part of the emerging fabric of the Indonesian nation. These were diverse paths, but each contributed to understandings of nature that led to its consideration as an object of reflection, discussion, or advocacy. Through the conservation of mountain environments, the Indonesian public might have developed a foundation for the broader sensitivity toward ecosystems and biodiversity conservation that the international organizations hoped to achieve.

Notes

1. James Cotton, "The Haze over Southeast Asia: Challenging the ASEAN Mode of Regional Engagement," *Pacific Affairs* 72 (1999): 331–35; S. Robert Aiken, "Runaway Fires, Smoke-Haze Pollution, and Unnatural Disasters in Indonesia," *Geographical Review* 94 (2004): 66–72.
2. Daniel Start, *The Open Cage: The Ordeal of the Irian Jaya Hostages* (New York: HarperCollins, 1997). The Indonesian western part of New Guinea has a complicated history of names. Irian Jaya had been used prior to Papua.
3. Christopher Barr et al., "Decentralization's Effects on Forest Concessions and Timber Production," in Christopher Barr et al., eds., *Decentralization of Forest Administration in Indonesia: Implications for Forest Sustainability, Economic Development and Community Livelihoods* (Bogor, Indonesia: Center for International Forestry Research, 2006), 87–93.
4. "National Parks in Indonesia," *Ministry of Environment and Forestry, Republic of Indonesia*, accessed August 14, 2015, http://www.dephut.go.id/uploads/INFORMASI/TN%20INDO-ENGLISH/tn_index_English.htm (accessed May 16, 2014).
5. Herman Haeruman, "Conservation in Indonesia," *Ambio* 17 (1988): 218–22; Robert Cribb, *The Politics of Environmental Protection in Indonesia* (Melbourne: Monash University, 1988), 12–14.
6. IUCN, *Protected Areas of the World: A Review of National Systems. Volume 1: Indomalaya, Oceania, Australia and Antarctic* (Cambridge: IUCN, 1992), 49.
7. Cribb, *Politics of Environmental Protection,* 3–6; J. H. Westermann, "Wild Life Conservation in the Netherlands Empire, Its National and International Aspects," in *Science and Scientists in the Netherland Indies,* ed. Pieter Honig and Frans Verdoorn (New York: Board for the Netherlands Indies, Surinam and Curacao, 1945), 418–20.
8. IUCN, *Protected Areas,* 54.
9. Effendy A. Sumardja, "First Five National Parks in Indonesia," *Parks* 6 (1981): 1–2.
10. FAO/UNDP, "A Second Five-Year Conservation Programme for Indonesia," *Unasylva* 35 (1983): 29.
11. Effendy A. Sumardja, Harsono, and John MacKinnon, "Indonesia's Network of Protected Areas," in *National Parks, Conservation, and Development: The Role of Protected Areas in Sustaining Society,* ed. Jeffrey McNeely and Kenton R. Miller (Washington, D.C.: Smithsonian Institution Press, 1984), 215.
12. Malcolm Gillis, "Indonesia: Public Policies, Resource Management, and the Tropical Forest," in *Public Policies and the Misuse of Forest Resources,* ed. Robert

C. Repetto and Malcolm Gillis (Cambridge: World Resources Institute, 1988), 51–59; Christopher Barr, "Bob Hasan, the Rise of Apkindo, and the Shifting Dynamics of Control in Indonesia's Timber Sector," *Indonesia* 65 (1998): 4–6.

13. Cribb, *Politics of Environmental Protection*; Nancy Peluso, *Rich Forests, Poor People: Resource Control and Resistance in Java* (Berkeley: University of California Press, 1992).

14. Haeruman, "Conservation," 219–20; Raisa Scriabine, "IUCN and Indonesia: Over 20 Years of Cooperation," *Ambio* 11 (1982): 318–20; IUCN, *Protected Areas,* 49–50.

15. The small Ministry of Population and the Environment (Kependudukan dan Lingkungan Hidup) was not one of Indonesia's more powerful departments and relied on the support of NGOs. See Emil Salim. "Recollections of My Career," *Bulletin of Indonesian Economic Studies* 33 (1997): 62.

16. Paul Jepson, "Biodiversity and Protected Area Policy: Why Is It Failing in Indonesia?" (PhD diss., University of Oxford, 2001).

17. According to Kathy MacKinnon, lead biodiversity specialist at the World Bank, "Indonesia was one of the first countries in Southeast Asia, and indeed the world, to use the best principles of conservation biology to plan a national protected-area system representing all habitats in seven biogeographic regions; many of these areas became national parks." Kathy MacKinnon, "Megadiversity in Crisis: Politics, Policies, and Governance in Indonesia's Forests," in *Emerging Threats to Tropical Forests,* ed. William Laurance and Carlos Peres (Chicago: University of Chicago Press, 2006), 293.

18. Philip Eldridge, *NGOs in Indonesia: Popular Movement or Arm of Government?* (Melbourne: Center for Southeast Asian Studies, Monash University, 1989); Joshua Gordon, "NGOs, the Environment, and Political Pluralism in New Order Indonesia," *Explorations in Southeast Asian Studies* 2 (1998): 14–32.

19. Sumardja, "First Five," 1.

20. Haruya Kagami, "Tourism and National Culture: Indonesian Policies on Cultural Heritage and Its Utilisation in Tourism," in *Tourism and Cultural Development in Asia and Oceania,* ed. Shinji Yamashita et al. (Bangi, Malaysia: Penerbit Universiti Kebangsaan, 1997), 64–65. In addition to increasing foreign exchange, tourism also was useful for promoting Indonesia's image as a "safe and peaceful" country. See Linda K. Richter, *The Politics of Tourism in Asia* (Honolulu: University of Hawaii Press, 1989).

21. Cited in Cribb, *Politics of Environmental Protection,* 13.

22. See Jepson, "Biodiversity."

23. Henning Borchers, *Jurassic Wilderness: Ecotourism as a Conservation Strategy in Komodo National Park, Indonesia* (Stuttgart: Ibidem, 2004), 56–57.

24. Donald E. Hawkins, "Sustainable Tourism Competitiveness Clusters: Application to World Heritage Sites Network Development in Indonesia," in *Cultural and Heritage Tourism in Asia and the Pacific,* ed. Bruce Prideaux (London: Routledge, 2008), 293–94.

25. In 1996 a conference at Ujung Kulon National Park addressed the issue of tourism in the region. The reports from the conference and editorials from newspapers focusing on the conference often disparaged the Indonesian public

for their lack of interest in the park. Ir. R. Sumarsono, "Pelestarian Keanekarag-aman Hayati dan Fungsi Taman Nasional," in *Pelestarian Keragaman Hayati dan Fungsi Taman Nasional, Jakarta dan Ujung Kulon, 9–12 Oktober 1996* (Jakarta: Konrad Adenauer Foundation and Perwakilan Jakarta, 1996), 45–52; "Is Ecotourism the Right Approach for Ujung Kulon?" *Jakarta Post,* October 27, 1996.

26. FAO, *National Conservation Plan for Indonesia, Introduction* (Bogor: UNDP/FAO National Park Development Project, 1982), 3.

27. Peter Boomgaard, "Sacred Trees and Haunted Forests in Indonesia: Particularly Java, Nineteenth and Twentieth Centuries," in *Asian Perceptions of Nature: A Critical Approach,* ed. O. Bruun and A. Kalland (Richmond, U.K.: Curzon Press, 1995), 48–62.

28. Janet Cochrane, "Indonesian National Parks: Understanding Leisure Users," *Annals of Tourism Research* 33 (2006): 989–91. For example, a *taman hiburan rakyat* is an amusement park and a *taman wisata* is a park for public recreation.

29. Lucas Sasongko Triyoga, *Manusia Jawa dan Gunung Merapi: Persepsi dan Kepercayaannya* (Yogyakarta: Gadja Mada University Press, 1991), 1–15; Astri Wright, *Soul, Spirit, and Mountain: Preoccupations of Contemporary Indonesian Painters* (Kuala Lumpur: Oxford University Press, 1994).

30. Lieut-Colonel Hadji Anwar Hamid et al., eds., *Madju Terus . . . Pantang Mun-dur!* (Jakarta: KTI, 1964).

31. "Mapala" is an abbreviation of Mahasiswa Pencinta Alam, literally "University Student Nature Lovers."

32. Edward Aspinall, "Indonesia: Moral Force Politics and the Struggle against Authoritarianism," in *Student Activism in Asia: Between Protest and Powerless-ness,* ed. Edward Aspinall and Meredith Weiss (Minneapolis: University of Minnesota Press, 2012), 160.

33. Anna Lowenhaupt Tsing, *Friction: An Ethnography of Global Connection* (Princeton, N.J.: Princeton University Press, 2005), 129–31.

34. Tsing, *Friction,* 133–35.

35. Norman Edwin, "Dua Puluh Enam Tahun, antara Perintis dan Penurus," *Kompas,* April 29, 1990, 1, 13.

36. Janet Cochrane, "Ecotourism, Conservation and Sustainability: A Case Study of Bromo Tengger Semeru National Park, Indonesia" (PhD diss., University of Hull, 2003), 169–72.

37. Official Tourist Bureau Weltevreden, *Come to Java* (Batavia: G. Kolff & Co., 1923), 77–80.

38. Cochrane, "Ecotourism," 184.

39. Cochrane, "Indonesian National Parks," 985–90.

40. Tsing, *Friction,* 148–53. Students' reflections on mountain climbing can be found on the website of Mapala UI and its Facebook, Twitter, and Instagram accounts. Many of the postings are in English: http://mapala.ui.ac.id/.

| # Nature Conservation in Africa's Great Rift Valley

A Study in Culture and History

CHRIS CONTE

HUMANITY EVOLVED IN AFRICA'S GREAT RIFT VALLEY. Our history with nature has therefore been playing out here for more than 100,000 years. In stark contrast, the history of parks and nature reserves in the African Rift dates back about a century. On maps, the parks appear oddly isolated from their surroundings, demonstrating that circumscription is foundational to the ethos of nature conservation across the East African Rift, as is forcible eviction. The process of enclosure essential to forming the parks has been largely ignored by hundreds of articles and photographs published in pseudoscientific publications such as *National Geographic*. While such literature has stimulated immense interest in Africa's wildlife, it also tends to portray the Rift Valley as a museum piece of natural history that predates humanity's fall from ecological grace. Nature conservation in the Rift Valley protects a myth.[1]

My own experience with the Rift Valley goes back to December 1981, when I arrived in what Kenyans called Maasailand to serve my stint as a Peace Corps teacher. Over the past thirty-three years, I have broadened my view of the Rift with trips to Tanzania and Mozambique. Based on that experience, I hope to draw a set of contrasts between the historical interpretations of the *National Geographic* accounting of the Rift, the one that buttresses the conservation ethic, and my own understanding of the environmental history of the people who reside in and move through the Rift's mountains, plains, and lakes.[2]

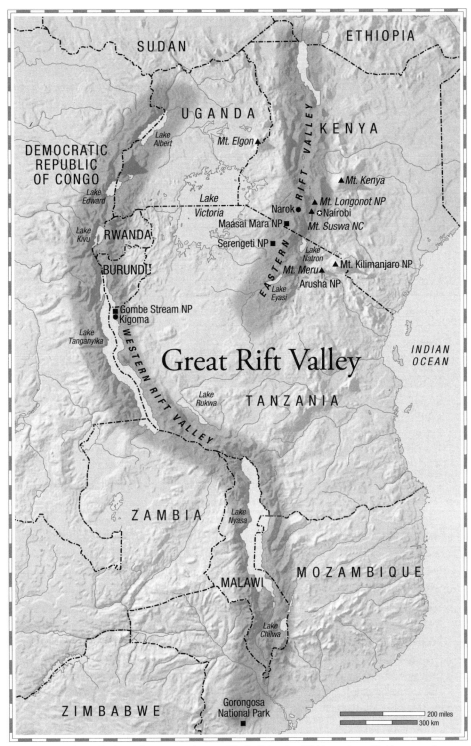

Great Rift Valley

SUDAN

ETHIOPIA

UGANDA

KENYA

Lake Albert

Mt. Elgon ▲

▲ Mt. Kenya

DEMOCRATIC REPUBLIC OF CONGO

Lake Edward

Lake Victoria

▲ Mt. Longonot NP

Narok ● ▲ ⊙ Nairobi

Lake Kivu

RWANDA

Maasai Mara NP ■

Mt. Suswa NC

BURUNDI

Serengeti NP ■

Lake Natron

▲ Mt. Kilimanjaro NP

Mt. Meru ▲

Arusha NP

Lake Eyasi

■ Gombe Stream NP
● Kigoma

INDIAN OCEAN

Lake Tanganyika

Lake Rukwa

T A N Z A N I A

Z A M B I A

Lake Nyasa

M O Z A M B I Q U E

MALAWI

Lake Chilwa

Z I M B A B W E

Gorongosa National Park
■

200 miles
300 km

Reading the Rift Valley Landscape:
A View from the Road through Kenya's Maasailand, 1982–83

I crossed the Rift Valley perhaps fifty times between January 1982 and December 1983. Sometimes I hitched rides with semitrucks, but mostly I rode on public transport between Narok, a town situated at about 6,200 feet above sea level along the foothills of the Mau Escarpment, and Nairobi, Kenya's capital city 182 km away. In the early 1980s, up-country public transportation in Kenya depended on large buses (the "country bus") and small Japanese pickups called *matatus* usually carrying twelve to fifteen people. The maintenance standards for these Narok-to-Nariobi matatus were fair to poor; road accidents occurred regularly between Narok and Nairobi. Despite the risk, people preferred these vehicles, which were fast and direct, over the country bus. The route led across the mountains of the Mau Escarpment on the Rift's western side, then down the escarpment face into the plains at the Rift Valley floor. Once across the valley, the matatu climbed up the eastern escarpment a thousand feet to Limuru Town, and then dropped down onto the plains at Nairobi.[3] An incident-free trip to downtown Nairobi took about three hours.

RETURN FROM NAIROBI

The first view of the Rift Valley appeared after an hour's climb out of Nairobi through Kiambu, an fertile upland area where the Maasai and Kikuyu ethnic groups had for several centuries intermingled for their mutual social and economic benefit. Earlier Rift inhabitants had likewise mixed herding and agriculture over several thousand years. In colonial times the state appropriated for British settlers these well-watered uplands, which became known as a piece of Kenya's White Highlands. The matatu's westward descent into the valley followed the "spaghetti road," so called because it had been built by Italian World War II prisoners who constructed the numerous tight turns and switchbacks. During my time in Kenya, this poorly maintained two-lane road was a major traffic artery that connected Nairobi with Lake Victoria. It carried heavy traffic, including large trucks hauling fuel and freight. Potholes, speed, and poor brakes led to road accidents, especially head-on collisions. Deadly accidents occurred with such regularity that junked cars and trucks covered the steep escarpment walls between the switchbacks of the spaghetti section. One's life was not in one's own hands over this stretch, so for sanity's sake, I liked to focus on the incredible crack in the earth that spread out to the west. Mount Longonot, a lone volcano with a ruggedly distinctive cone, came into view immediately, and then Mount Suswa, a big and heavily eroded volcano with concentric craters, would begin to rise in the background to the southwest. The characteristic openness of the Rift sky was on full display.

Once they reached the valley floor, the Narok matatus turned to the west at the crossroad hamlet of Mai Mahiu (which means "hot water" or "hot springs"), a small cluster of curio shops and eateries that served as a quick stop for tourist vehicles following the same road on their way to what was then known as Maasai Mara National Park (today, it is a national reserve). The road crossed the plains between the volcanoes and headed directly for the Rift's western wall. During the rainy seasons, wildlife and domestic stock together spread out across the valley bottom in search of sweetgrasses and water. Above these bottomlands, the Maasai homesteads (*manyattas*) with their characteristic enclosures of small mud huts and fenced cattle pens lay scattered across the rolling hillsides of the volcanoes' lower slopes. The settlement pattern was anything but haphazard, however. Families located these manyattas so that their animals could walk daily to water and grazing, then return before dark to the safety of the homestead corrals.

On the plains rain falls annually in two seasons, but precise timing, duration, and intensity are difficult to predict. Periodic drought is also a common phenomenon that herders must figure into their accounting of resource access and use. During the rains, I saw the kinds of landscapes that featured so prominently in *National Geographic*. Gazelles, wildebeest, giraffes, zebras, and other wild ungulates grazed in the open while the predators lurked nearby. I would also see grazing on the same pastures the herds of Maasai cattle, sheep, and goats, which were managed expertly by boys carrying spears and throwing sticks to protect their herds from thieves, lions, and leopards. During the driest parts of the year, the plains stood largely empty as the animals and people moved to wetter and higher pastures in a long-standing pattern of mobile pastoralism. Even as I began to develop a more informed view of the difficulties of Maasai life, Suswa always looked beautiful. The sheer Rift Valley walls, the open grasslands, and the dramatic volcanoes all suggested to my mind something essential about the origins of human life and the earth's natural history.

According to the *National Geographic* paradigm, the Maasai I saw at Suswa lived in nature. Their dress, adornment, rituals, and economy together constituted an anachronism awaiting modernity.[4] The historical record, however, tells a far more nuanced story of change and adaptation. Suswa actually formed part of the Maasai Reserve, created by the British colonial government in 1904 and 1912 in order to make room for British settlers. The colonial state herded Kenya's Maa speakers from the most productive areas of the Rift Valley onto a fraction of their former range.[5] The forced moves constituted one of the first acts of colonial eviction and land confiscation. Nor was Maasai occupation in Kenya a singular historical phenomenon at Suswa, or in the Rift Valley more generally. The Maasai represented only the most recent manifestation of a husbandry tradition that had existed in the Rift Valley for more than 6,000 years.[6]

CLIMBING TO NAIRAGIENGARE

When matatus reached the valley's western wall, they shifted down to second gear for the steep climb up the escarpment face to the Nairagiengare (in the Maa language, the name means "a place where water gathers") turnoff, an isolated intersection of tarmac and dirt where a few passengers usually alighted. After the 1982 coup attempt in Kenya, the place served as a police checkpoint. I saw people abused and detained arbitrarily at this spot. I visited Nairagiengare town a couple of times to look in on one of my Peace Corps buddies, a "water development worker." The hamlet looked down on green, gently rolling and heavily cultivated hills. True to its name, Nairagiengare, perched on the western wall of the Rift escarpment, had clearly acted as an important dry-season grazing site. However, by the early 1980s farmers from other parts of Kenya, especially Kikuyu from central Kenya, had been migrating to this well-watered place without government sanction since the 1920s. Additional Kikuyu migrants had come to the area in 1947 as part of an official resettlement program of landless Kikuyu families at a place called Olenguruone a few miles to the northwest. In the interests of project success, the colonial government excised about 33,000 acres from the Maasai Reserve.[7]

INTO TOWN

After a series of death-defying, high-speed plunges into valleys followed by slow climbs up the steep hills of the Mau, the matatus finally dropped down a long hill into Narok town, which sat in a depression along the Narok ("black") River. Towns have a relatively brief history in the Rift Valley, and there had never been one at Narok until the British colonial government established a district headquarters on the Narok River floodplain early in the twentieth century. The early site flooded regularly, and eventually all government offices had to be relocated to higher ground on the hillsides above the river, while the shops remained below. Colonial efforts in Narok District focused on the imposition of cattle quarantines designed to keep Maasai livestock off of the Kenyan colonial market. When I lived there, lower Narok town had evolved into a haphazard collection of hotel/bar/restaurants and small dry goods shops run mostly by Kenyans from other parts of the country. Maasai men entered only to sell their animals and to partake of town life for a short while.

Narok Hospital served the entire district, and many people came to town to visit the doctors. One of my first Peace Corps memories was of a line of people, mostly Maasai mothers and children, which began at the hospital and stretched for hundreds of yards down the road toward town. As we walked by, a fellow teacher informed me that these people were waiting to see the eye specialist, who came to the hospital only one day a week. It seemed like just about everybody in Narok town came from somewhere else. Most were civil servants who gathered nightly at the small bars and eateries. Many young women from other parts of Kenya also came to Narok to work as waitresses in those bars and to double as sex workers after hours in order to earn a living. They had few other options.

I lived on a hill two miles north of town in a stone and corrugated iron shack about a five-minute walk from Narok High School, where I worked. Like the town, our school had been dropped into Maasailand from the outside. And like just about everybody else in Narok, the students came from other parts of Kenya. My house had been hastily erected on the edge of a pasture bordering Narok District's Farmer's Training Center, where I drew my water. From my perch on the foothills above Narok town, I could look out southward over the Loita Hills and toward Maasai Mara National Park, the northern extension of Serengeti Plains. Narok could be a very cold place in July and August and brutally hot and dusty in December and January; the wind blew almost constantly in every season. The Mau hills all around us could be green, yellow, brown, or black depending on the seasonal applications of fire and rain. I lived with the almost constant smell and sounds of livestock as they moved past my window on their daily walk to the river and then back again to their corrals.

In the evenings I often sat on the veranda of the Mara Bar on Narok's main street. I'd see tourist vans fueling up in Narok town at the Shell station before heading off for another hour or so on their dusty journey to the national park. Maasai Mara was just a place I'd heard about until, as the Narok High School games master, I rode to the park in the high school's old Bedford lorry. We drove directly to the staff quarters' soccer pitch where our boys took on the park's African staff. I remember that we saw a lion on the way home.

Conservation areas have multiplied since the 1980s; current maps demarcate Mount Suswa as a nature conservancy. Mount Longonot has now been designated a national park. The Mau highlands north of Narok town are now called a national reserve, though the view from Google Earth shows numerous gardens on the western and southern sides of the reserve. The most recent migrants to the Mau have arrived over the past thirty years from the Kalenjin-speaking highlands to the north, and they did so at the encouragement of the Kenyan government under President Daniel Arap Moi. Like the Kikuyu, these landless migrants settled in dry season grazing zones above 1,800 meters. The migrants are now regularly blamed for massive deforestation and soil erosion.[8] I saw the beginnings of this particular land grab on the Mau hills thirty years ago. It began with commercial logging, which was followed by large-scale wheat production. In a scheme financed by the national agricultural bank, select investors gained access to land and interest-free government loans. They then contracted with agricultural machine operators to plant and harvest wheat, which grew well on the high cold mountains of the Mau. The rumor around Narok town was that these well-connected entrepreneurs were making piles of money.

That piece of the Kenya Rift Valley that I came to know as a singularly beautiful, evocative landscape now calls to mind as well a history of injustice, violence, corruption, and ecological degradation. Maasai Mara National Park was a piece of colonial space carved out of a colonial reservation originally created for Maasai refugees. By the early 1980s Kenyans in most other parts of the country considered Narok a backwater, a place for the landless and uncivilized—and the park was for white people.

Nature in Gorongosa and Gombe Stream National Parks

Over the past five years, I have twice visited two other Rift Valley Parks and hiked some of the farming areas that surround them with members of those communities.[9] Unlike the Maasai pastoral regions of southern Kenya, these highland sections of the Rift support small-scale arable agriculture. By the reckoning of *National Geographic* features on Gombe Stream National Park in Tanzania and Gorongosa National Park in Mozambique, the surrounding African populations constitute direct threats to the animal populations inside the parks. I spoke with farming people who live around the parks and found an alternate historical interpretation that is at once much broader spatially and temporally, and more locally grounded in ritual and story. The history related by the elders grows from a sense of place that is steeped in everyday life where the landscape is something intimately familiar and explicitly historical. In this cosmology, parks appear as no-go zones, now removed from historical experience.

The *National Geographic* histories I draw from for Gombe Stream and Gorongosa National Parks expertly employ film and photography to isolate nature from its larger geographical context. People producing a living in parks do not fit the iconic imagery, so farmers and fishermen have been removed from the frame. Indigenous people have been either evicted or relegated to reside in "buffer zones," where park managers can monitor and evaluate their behavior. At the same time, the parks aim to expand their territory into surrounding farmland.[10] In both cases, the expansionists argue that forests constitute the key ecological communities. At Gorongosa, the Mozambican government has recently annexed Mount Gorongosa, whose forested watershed is vital to both the ecological health of the park downstream and the regional agricultural economy. Talk of eviction on Mount Gorongosa has mixed with a larger national political crisis to create a particularly volatile situation. At Gombe Stream, the removal of agriculture from the surrounding hills would theoretically foster the necessary forest restoration to significantly enlarge chimpanzee habitat.

PARK ECOLOGY AND HISTORY AT GOMBE STREAM NATIONAL PARK

In July 1960 Jane Goodall brought her equipment and her mother to Kasekela, a fishing village on the shores of Lake Tanganyika, where she began her famous study of primate behavior. She recorded her early research experiences in the 1963 volume of *National Geographic*.[11] In a moving account written in a deeply sympathetic voice, Goodall provides *National Geographic*'s readers with a compelling vision of Africa. Her research site is apparently isolated from civilization in a dense forest and inhabited by a large number of wild chimpanzees.

Goodall's story is a study in persistence and passion. Early frustration, as the chimpanzee troops kept their distance, gave way to familiarity. She relates her futile attempts to approach a chimpanzee troop. Only after a dramatic and life-threatening display of aggressive behavior by a dominant male do the chimps accept her presence among them. Eventually, three males begin to show up at her camp and take bananas from the hands of Miss Goodall, who comes to know and name the entire troop.

The article's numerous anecdotes and illustrations draw on stereotypes that highlight African nature as primeval and Western presence as paternalistic and compassionate. In one story, Goodall conjures the image of a dramatic rain dance that she witnessed during an evening thunderstorm rolling in off of Lake Tanganyika. An artist's rendition appears in the article. It shows how in the midst of lightning strikes and high winds, mature male chimpanzees rushed headlong down the steep slopes screaming in demonstrations of sheer aggression. Goodall's prose describes a stunning and elemental response to nature's fury. Later in the essay, she tempers the image of physical violence with stories of chimpanzee compassion. Their aggressiveness appears as a behavioral anomaly spurred occasionally by nature, or by an intruder such as herself. The accompanying photographs suggest that the Africans at Kasekela come into camp out of curiosity or for medicines that Goodall and her mother have brought with them, but these local people remain mostly a prop in the larger story.

At no point does Goodall describe the landscape beyond the reserve except to stress the dangers of local farmers to the survival of her research subjects. Readers learn nothing of patterns of settlement and land use beyond Gombe Stream, an omission that allows the reader to enter an isolated forest that could be anywhere in Africa. The place is ancient, an ecological relic of earlier times when forests supposedly covered the entire area. Now, it seems, this last wild place is under dire threat from neighboring farmers eager to expand farming onto Goodall's research site. The lake's fishermen and upland farmers have been agglomerated into a single rapacious mass of resource seekers.[12] Leonard Carmichael, secretary of the Smithsonian Institution and chairman of the National Geographic Society's Committee for Research and Exploration, authoritatively echoes this looming threat in the essay's introduction: "Many experts fear that the African forest where these great, almost human animals with their good brains and strong hands now live may soon be taken over for agriculture. It is thus possible that the interesting account given here will stand for all time as a unique record in scientific zoological literature because the opportunity to make such field studies may soon be lost forever."[13]

Fifty years later, Gombe's steep mountainsides carry heavy forest, but there is little research on the human history within the park.[14] In fact, steepness may indeed be the reason the forest survived the regional deforestation events that occurred beginning in the eighteenth century. When Goodall arrived in 1960, there were fishing villages along the lakeshore inside what is now Gombe Stream National Park. In late May 2011 I counted three abandoned village sites within the park at the mouth of streams flowing into the lake; the sites were overgrown with mango groves, oil palms, and banana trees, clear signs of human habitation.[15] The overgrown hamlets suggest that fishing communities had built up a productive lakeshore economy long before GSNP existed. By the mid-nineteenth century, in fact, this section of the lake served a system of global scale exchange financed by Indian capitalists. Their agents, African coastal merchants, built their port city at Ujiji, less than ten miles south of what became GSNP. The trade moved slaves from central Africa to clove plantations owned by Omanis on the Indian Ocean islands of Zanzibar. Farmers on the western side of Lake Tanganyika produced crop surpluses to feed the hundreds of porters and visitors who arrived in the trade caravans, not to mention their captives. Local community leaders, in their turn, exacted heavy taxes on this trade. The emergence of fishing villages along the lake was linked to this episode of economic growth.[16] Nor is there any evidence that an extensive forest covered all the mountainsides on the lake's eastern shore.[17]

Yet what still strikes the visitor to GSNP is the sense of isolation in the dense forest, the very same place that Goodall traipsed through on her research tours. I twice took the 16 km trip boat trip that begins in Kigoma harbor. On both occasions, I accompanied a school group that had hired a fishing boat to take us to the park. A Tanzania Parks Authority boundary sign clearly marks the border of the park's land. Our boat ran easily in the clear blue waters close to the shore until we came ashore at Kasekela, just as Goodall had. After paying the entrance fees, we met our Tanzania National Parks guides who would escort us through the forest on our quest to see the chimpanzees. The guides were professional and well organized, and they knew the landscape. There are no roads in GSNP, so visitors walk up and down steep mountain paths that challenge even the fittest hikers. Our guides, in close communication with spotters, eventually located a chimp group. On our approach, they nonchalantly sauntered by us on the narrow trail. The hike was at once beautiful, strenuous, and exciting.[18]

THE OTHER SIDE OF THE RIDGE

In a compelling description of Kigoma Region's indigenous cosmology, Michele Wagner tells of a woman's ambivalence on a river crossing as she confronted the water spirits that inhabited the river. Wagner expands on what she refers to as the "conceptual ecology" that animates the landscape. Earth priests, diviners, and experts in human dysfunction created some cosmological order by mastering the knowledge of nature and culture. Wagner argues for the historical persistence of particular types of environmental knowledge tied to residence in particular ecological niches. "Nature in the mind," then, meant different things for people who lived along the lake or beside river courses, where water spirits animated life, than it did for the "lowlanders" or "hill people." In all cases, however, a distinct biophysical environment animated their own sense of the sacred.[19] When I visited the area in 2011, I was lucky enough to meet a man who still carried such conceptual maps in his memory.

Lake Tanganyika and its surrounding mountain and plateaus constitute a major escarpment of Africa's Great Rift Valley. The mountains immediately to the east of GSNP's border appear on Google Earth as a series of parallel north/south-running ridges, some of them deeply cut by stream valleys. The ridgelines run in elevation from 1,400 to 1,600 meters above sea level. Lakeside ridges rise the highest, then there is a steady dip in elevation as one moves eastward along the tilted and deeply incised plateau. The steepest westward-facing slopes lie immediately along the lake, where the elevation falls from 1,579 meters along the park's highest ridges to 780 meters on the shore at Kasekela. These rugged highlands run northward along the lake into Burundi, where they rise above 2,000 meters. As one moves upslope from the lakeshore, the heavily forested drainages give way to more open woodland and grassy parklands on the upper slopes. GSNP receives approximately 160 centimeters of annual rainfall, but rainy seasons can vary greatly in intensity. Because of their aspect, the mountainsides immediately facing the lake receive more rainfall than areas in the rain shadow to the east.

The Africans who live in the hills just to the east of GSNP occupy towns laid out on a recognizable grid pattern, a legacy of the resettlement policies that Tanzania began to institute in earnest a few years after independence. Julius Nyerere, Tanzania's first president, instituted a policy of rural collectivization called Ujamaa ("familyhood") in many parts of the country, including the regions around Lakes Victoria and Tanganyika. Throughout the 1970s cadres of government workers resettled thousands of Tanzanians into planned villages in order to build a communal, self-reliant, and uniquely African socialist society.[20] The grids in Kigoma Region, which includes GSNP, can easily be seen from Google Earth. From space, the park's deep green vegetation stands out from its surroundings. Outside the park, open fields and large groves of oil palms cover the gentler slopes and drainages.

I walked through this area in May 2011. My guides were two local African boys in their late teens who carried rucksacks and wore hiking boots, and Mzee Idi, an elderly gentleman in his seventies who accompanied us in flip-flops.[21] Mzee explained the meaning of important landscape features to all three of us, since the boys knew little about this world. It was late May, the beginning of the dry season, and we walked fairly quickly along the dry, well-worn paths leading up onto the ridge above GSNP. Nobody lived along these pathways. We saw no houses, churches, schools, or medical dispensaries, having left all of those things back at the grid. We nevertheless walked among hundreds of acres of gardens of mounded cassava plants and the large groves of oil palms that covered almost every hillside in the drainage. Mzee Idi pointed out the first oil palm that, according to his story, had been imported directly from Congo basin.[22] He continually described now unoccupied villages, markets, and homesteads, all of which he identified by the presence of specific trees. *Msufi* (kapok or cotton) trees stood on now abandoned market and village sites, including those on the secondary school campus where we had begun our day's journey. The kapoks in fact organized the spacing of several substantial settlements, while other specific trees, such as *mrumba* and *mgogo,* had been planted at the base of grave sites and remained places of prayer and sacrifice. Old homesteads now carried citrus and mango trees, some of them quite massive, in yet another signature of past habitation. Mzee named the landowners for us and told stories of some of these men, especially the healers whom people held in great esteem.

At one point, we passed the former estate of one Mzee Mazuru, who had been an *mganga* or healer. He had lived his entire life in the foothills above the Nkuruba River, which rose in nearby GSNP. People remember him for a number of amazing feats, the most famous of which was his refusal to move into the planned settlement during the Tanzanian Ujamaa era. According to the story, the government sent police, regular army, and even special forces to remove him from his farm. All efforts failed. Mzee had surrounded his house with trees of all sorts so that passersby would see only a forest. Those who came to consult his healing and magical powers could enter without fear, but those who wished him ill could never locate his house, or his livestock, among the trees. It was as if he and his farm had simply disappeared into a forest. Some of the government forces that entered his forest found a lake where his house should have been. Mzee Mazuru never did move and died at his home in Nkuruba in 1999. These days, huge mango trees mark his homestead.

THE STORY OF SOKO LA ULIMWENGU
There used to be an immense market near GSNP on the ridge that borders the territories of Kiganza and Mahemba. I suspect, but I don't know for sure, that this market, or similar markets, had sprung up in the nineteenth century as part of an exchange network tied to the growing trading lake port of Ujiji, which now sits on the lake just to the south of the town of Kigoma. Ivory and slaves brought from across the lake in Congo moved into and out of this bustling town, the slaves on their way to work the clove plantations on the Zanzibari islands of Uguja and Pemba. Ujiji is also where Stanley found Livingstone, but that's another story.

The Soko la Ulimwengu ("market of the universe") met four days a week instead of one, which was the norm for most area markets, and people came from many miles to trade for the copious produce available there. Merchants brought fish from the lake to trade for agricultural harvests from the hills. Protein bought carbohydrates and vice versa. Mzee Idi remembered that as a very small boy he often went there with his grandparents. The pathway we followed seemed to have been quite heavily used, as we found it still covered with large, flat paving stones. According to Mzee, sometime just after Tanzania's independence in 1961, a cholera outbreak induced the government to shut down the market, presumably to discourage large gatherings of people. But its popularity was such that people continued trading there at night, and Mzee Idi remembered that palm oil lamps lit the market stalls.

He said that people began to call the meeting place Soko la Ulimwengu to hide its identity from government officials. If asked where they were going, people simply said, "Soko la Ulimwengu," which simply meant to outsiders the "anywhere and everywhere" market. In 1974 the Ujamaa police closed the market permanently. What used to be a bustling commercial hub is now an empty field that the people of Mahemba pass through on their way to work in their fields.

Ujamaa forcibly altered settlement and trade patterns, but for Mzee Idi at least, the landscapes outside the park held onto their past. We passed through, for example, grove after grove of oil palms. Over many centuries, these trees had found their way to Kigoma Region from West Africa just as bananas had found their way across the continent from the east coast. At one point, we watched as an oil bootlegger boiled and pressed the palm seeds to produce a reddish oil that farmers can only legally sell in official markets.

After a difficult climb to the ridgeline above the lake, we illegally walked into Gombe Stream National Park. The younger men in our group did this with reluctance, fearing park guards would see us, but Mzee seemed unworried. He was on familiar ground, and we were several kilometers away from the regularly visited sections of the park. The hillocks that formed the ridge were covered in grassy glades and large, open groves of spreading migongo trees, also known to my companions as "the beautiful tree." We decided to have lunch under one of them on a site that looked down on GSNP and the Congo mountains to the west. Gardens, pathways, and oil palms covered the landscape to the east. A large group of baboons watched us from the trees a few hundred yards away, but they kept their distance. We were otherwise alone with our stories. The contrast between the inhabited hills and the park below us could not have been starker, both in terms of vegetation and the stories that various people told about them. Mzee did know that park as a cultural place and named for us the emptied villages that GSNP now enclosed: Bwavi, Mitumba, Kanyiru, Nyansanga, Mkenke, and Kakombe.[23]

Gombe Stream National Park, in cooperation with the Jane Goodall Institute, USAID, and the Nature Conservancy, has plans for a park expansion into the surrounding hill country to the north and east.[24] Towns fall outside of the planned park expansion, but as my own hikes demonstrate, the areas beyond that surround the towns are vital farming areas. The ambitious plan designates the entire region as the Greater Gombe Ecosystem, including all of the populated grids of the hill country. The GGE plans from the international aid and conservation organization are available freely online, though I have not been able to locate Swahili translations for them. Lillian Pintea, chief scientist for the JGI, told me that they had cooperatively developed conservation land-use plans for all of the area's villages, and as the map demonstrates, the JGI had collected and analyzed large amounts of satellite data in support of the expansion project.[25]

On our hike back to Kiganza town along the high ridges, Mzee pointed out a hillside just to the east of the park boundary. It had recently been taken out of cultivation in the interest of reforestation for the GGE. The expansion had indeed begun to chip away at the farmland margins. An agricultural extension agent with long experience working with village cooperative groups explained to me that many of the affected people are refugees from Burundi trying to make a living in Tanzania. They are not likely, he said, to protest the expansion for fear of raising the issue of their return to Burundi. It's hard to say what will happen when the expansion spreads onto lands farmed by the well-ensconced native Tanzanians.

THE SOUTHERN RIFT VALLEY AND GORONGOSA NATIONAL PARK, CENTRAL MOZAMBIQUE

Similar threats of evictions in the name of park expansion have spurred discontent in Mozambique's Gorongosa National Park. The simmering anger became clear in the spring of 2012 when two of my colleagues from Utah State University traveled to Mount Gorongosa in order to conduct an agricultural survey on the land that the Mozambican government had recently annexed to the park. On the morning of May 24, soon after the professors and their team arrived on the eastern side of the mountain, some villagers greeted them angrily. They claimed that under no circumstances would they leave the mountain as the government had mandated. My colleagues, an environmental engineer and an agricultural scientist, operated under the assumption that Gorongosa's ecological resiliency will depend on the seasonal recharge of the wetlands and therefore a healthy mountain watershed. They wanted to understand the nature of mountain production and the extent to which small-scale farming might be integrated into watershed function. Relatedly, they wanted to know the nature of deforestation on Mount Gorongosa, and they sought ultimately to suggest ways that watershed health could coexist with productive mountain agriculture. They had planned the survey for some time, but it seems that the government's annexation of the mountain and the implied threats of eviction had poisoned their work. Quite a brouhaha ensued as everybody tried to make sense of the whole affair.

In a chilling reminder of the violence that had enveloped Central Mozambique during that country's postindependence civil war (1978–92), RENAMO, the political party of the former rebel group, has resorted again to arms and set up camp on Mount Gorongosa. In response, the Mozambican government has sent its own troops to Gorongosa. Spasms of deadly violence have broken out in the form of ambushes along the main roads. Refugees have streamed into Gorongosa town.

In Nat Geo WILD's *Africa's Lost Eden* (2009) the war's history appears in grainy black-and-white snippets, an event in the past. The film also shows a small memorial and some bullet holes in a park building, but the war's human costs have not been chronicled.[26] The Nat Geo film's narrator focuses, rather, on how the violence decimated Gorongosa's renowned herds of ungulates, including elephants and hippos, and big cats, all of which were killed during the war. The narrator reads the census of dramatic declines in the large mammals. The decimation has furthermore upset the park's delicate ecological balance. With the large grazers gone, rapid vegetation growth has created fuel for wildfires, which the park staff fights almost continually during the dry season. The narrator explains that although some of the fires result from lightning strikes, poachers start a number of these damaging fires. Woven into the film's accounting of the decline in wildlife populations are the unprecedented restoration efforts now under way, aimed to return the park to its former glory. Gorongosa's chief game warden fondly recalls his own boyhood visit to the park in the colonial era, when the place was teeming with animals. To this professional, restoration means a return to that memory of an Eden, as the film's title suggests.

Park staff clearly has a difficult task. Nat Geo expertly films a risky episode during which park staff truck five male elephants over a thousand kilometers from South Africa's Kruger National Park. Not everything goes as planned, and when one forty-year-old bull elephant bolts from the park and invades the surrounding farm country, he must be recaptured and sedated. He dies tragically while on his way back to the sanctuary. Despite the heartbreak, the park's powers of recovery, of healing, are remarkable. A lion killing a wildebeest revives the viewer's hope for the park's recovery, as does a scene with Gregg Carr, an American philanthropist whose foundation has provided millions in the recovery effort. The Nat Geo WILD narrator introduces Carr's commentary by noting that his passion for park conservation stems from Carr's upbringing near Yellowstone, a truly iconic American park. Carr's vision is broad enough to couple Gorongosa's conservation to local economic development, though beyond a local fruit-drying factory, the broader character of development remains unclear in the film. Gorongosa is a long way from western Wyoming.

A final vague reference to African life bookends the film when the game warden argues that, ultimately, the fate of people around Gorongosa is tied to that of the park. If we lose the park, he explains, then the Africans will lose their soul. Nat Geo's narrator concludes hopefully with the observation that Gorongosa's restoration can become an engine of prosperity for an emerging nation and a signal to the world. "If this Eden can become again the place where Noah left his ark, there is hope for damaged ecosystems all over the world."

Redoubt or Historical Landscape?

Despite the fifty-year gap in *National Geographic* features on Gombe Stream and Gorongosa, the overriding conservation mentalities that animate these stories

have changed little. Parks, in this paradigm, should contain the sacred characteristics of nature—abundance, diversity, exoticism, and beauty. They should
remain accordingly uncontaminated from land use by African farmers, whose
apparent disregard for nature constitutes a direct threat to nature preservation.
The *National Geographic* media industry, furthermore, contrasts the land-hungry
Africans with benign conservation workers, funders, and researchers who protect
and nurture the land as something sacred.

The Gorongosa documentary makes a quick case for the role of violence
in ecological destruction, but the brief history lesson begins and ends with the
1979–92 war. The film's historical paradigm argues furthermore for parks as
places into which human life intrudes. Parks are therefore redoubts, imbued
with the ecological and spiritual value ascribed to them by conservation science.
Much scholarly literature argues, to the contrary, that these uninhabited conservation areas are historical landscapes covered in signs of past human land use.[27]
Africans moved over these places, mapping them, developing them for agropas-
toralism and fishing, and conserving them for their economic and cultural value.

I last visited Gorongosa in October 2012, when Muala Domingos and I hiked
on Gorongosa Mountain.[28] Domingos arranged for me to receive blessings from
the chief at Kanda, who prayed for my safety as I walked on the mountain. The
next day, off we went. We parked at the ranger camp and followed the road on
foot to an open valley. Maize was just beginning to show itself in the fields. We
followed the broad stream valley up through the farm country that covers the
mountain's lower slopes. As we passed between the gardens, we met by chance an
Mzee, a man Domingos called his teacher. Mzee accompanied us most of the rest
of the day. People greeted us reluctantly—not a good sign. After about an hour,
we reached a waterfall popular with tourists. We sat on the rocks and ate lunch.
Moving on, we then climbed a rocky, steep trail to get above the falls. On this
steep, uncultivable piece of ground, somebody had burned several acres of forest.

Above the burn we continued along the creek to a park-sponsored tree nursery, its seedlings earmarked for a riparian restoration project nearby. Most of the
young trees that park staff had planted streamside had recently been destroyed.
We heard that somebody had burned another of the park's nurseries. People had
begun to fight their eviction; they did not trust the Mozambican government
to relocate them to a place with similar natural endowments, not to mention
access to health care and education. Then things got worse last year when the
guns came out again.

Gorongosa's restoration aims to rebuild an ecological moment of wildlife
abundance and natural beauty in Africa's Great Rift Valley. That such a place
existed at all resulted from a decades-long ecological crisis of massive proportions
that hit the Rift Valley in the late nineteenth century. A historical conjuncture
of disease, drought, famine, and depopulation had created that very landscape
of the mythical past. Without their cattle, pastoralists had to adapt for survival.
When they left their pastures to the elements, woodlands quickly invaded the

abandoned spaces.[29] This new vegetation regime fostered an explosion of sleeping sickness. The Maasai, among other pastoralist groups, rebuilt their herds and brought some of the pastures back into what they viewed as ecological health. But by then the image of a wild savanna covered in mammals had been burned onto the colonial mind—and the pages of *National Geographic.*

Notes

1. See the introduction to Jonathan S. Adams and Thomas O. McShane, *The Myth of Wild Africa: Conservation without Illusion* (Berkeley: University of California Press, 1992). For a powerful description of conservation ethos in Tanzania, see Dan Brockington et al., "Preserving New Tanzania: Conservation and Land Use Change," *International Journal of African Historical Studies* 41, no. 3 (2008): 557–79.
2. For a few of the feature articles treating Rift Valley nature parks, see Theodore Roosevelt, "Wild Man and Wild Beast in Africa," *National Geographic* 22 (1911): 1–33; Georges-Marie Haardt, "Through the Deserts and Jungles of Africa by Motor: Caterpillar Cars Make 15,000-Mile Trip from Algeria to Madagascar in Nine Months," *National Geographic* 49 (1926): 651–720; W. Robert Moore, "Britain Tackles the East African Bush," *National Geographic* 97 (1950): 311–51; W. Robert Moore, "Roaming Africa's Unfenced Zoos," *National Geographic* 97 (1950): 353–80; Allan Fisher, "Kenya Says Harambee!" *National Geographic* 135 (1969): 151–206; David Quammen, "Jane: Fifty Years at Gombe," *National Geographic* 218, no. 4 (2010): 110–29.
3. For a nineteenth-century view of some of the same sections of the Rift Valley, see Joseph Thompson, "Through Maasailand to Victoria Nyanza," *Proceedings of the Royal Geographic Society* 16, no. 12 (1884): 690–712.
4. See especially Roosevelt, "Wild Man"; Moore, "Bush"; and Fisher, "*Harambee,*" for this kind of interpretation.
5. The Maasai are speakers of Maa and do not constitute a unified ethnicity. Rather, they are made up of a number of sections that often competed with each other over grazing resources from northern Kenya into central Tanzania. For the longer history of pastoralism in the Rift Valley, see John Sutton, *A Thousand Years of East Africa's History* (London: British Institute in Eastern Africa, 1990).
6. Richard Waller, "Ecology, Migration and Expansion in East Africa," *African Affairs* 84 (1985): 347–70.
7. For an excellent description of the project's genesis, see David W. Throup, *Economic and Social Origins of Mau Mau, 1945–53* (Athens: Ohio University Press, 1988), chap. 6, "Olengurone," 120–38; and Tabitha Kanogo, *Squatters and the Roots of Mau Mau* (Athens: Ohio University Press, 1987), chap. 4, "The Crisis: Decline in Squatter Welfare, 1938–48," 96–124.
8. In September 2009 the BBC World Service did an exposé on the Mau. For links to these articles, see "Kenya's Heart Stops Pumping," September 29, 2009, at http://news.bbc.co.uk/2/hi/africa/8023875.stm (accessed July 2, 2015).

9. For the National Geographic interpretation, see Curt Sager, "Africa's Great Rift," *National Geographic* 177, no. 5 (1990): 2–41. For an excellent map of the Rift Valley, see 11.

10. Quammen, "Fifty Years."

11. Jane Goodall, "My Life with the Wild Chimpanzees," *National Geographic* 124 (1963): 272–308.

12. For a more contemporary view, see Quammen, "Fifty Years," 115–21.

13. Goodall, "My Life," 274.

14. Most of the ecological work suggestive of human history in the area deals with a larger regional scale of the lake basin. For a regional view of vegetation history based on the history of deltaic deposits in the deltas of streams along the Lake Tanganyika lakeshore, see Andrew S. Cohen, Manuel R. Palacios-Fest, Emma S. Msaky, Simone R. Alin, Brent McKee, Catherine M. O'Reilly, David L. Dettman, Hudson Nkotagu, and Kiram E. Lezzar, "Paleolimnological Investigations of Anthropogenic Environmental Change in Lake Tanganyika: IX. Summary of Paleorecords of Environmental Change and Catchment Deforestation at Lake Tanganyika and Impacts on the Lake Tanganyika Ecosystem," *Journal of Paleolimnology* 34 (2005): 125–45.

15. According to Mzee Ramadhani, there were five settlements inside what today constitutes GSNP.

16. For a similar trend to the south along Lake Tanganyika, see Nobuyuki Hata, "Ndagaa Fishing and Settlement Formation along Lake Tanganyika," *Kyoto University African Studies* 2 (1968): 31–50.

17. Michele Wagner, "Environment, Community, and History: 'Nature in the Mind' in Nineteenth and Early Twentieth-Century Buha," in *Custodians of the Land: Environment and Famine in Tanzania,* ed. James Giblin, Gregory Maddox, and Isaria Kimambo (London: James Currey, 1995), 175–78.

18. I visited the park in late May 2011 while on a curriculum development grant from the Office of Global Engagement at Utah State University.

19. Wagner, "Nature in the Mind."

20. For a overview of Ujamaa policy, see James Scott, *Seeing Like a State: Why Certain Schemes to Improve the Human Condition Have Failed* (New Haven, Conn.: Yale University Press), chap. 7. For an interesting literary take on Ujamaa in western Tanzania, see Euphrates Kezilahabi, *Kichwa Maji* (Dar es Salaam: East African Publishing House, 1974).

21. In Kiswahili, "Mzee" means elder.

22. The area is called Mgazi Mmoja in Kiswahili, or "first palm."

23. For a fascinating discussion of sacred sites now within the Rift Valley's Serengeti National Park, see Jan Bender Shetler, *Imagining Serengeti: A History of Landscape Memory in Tanzania from Earliest Times to the Present* (Athens: Ohio University Press, 2007).

24. For expansion plans at Gombe Stream National Park, see Greater Gombe Ecosystem Conservation Action Plan, 2009–2039, version 1, circulated April 2009, sponsored by USAID, the Jane Goodall Institute, and the Nature Conservancy. For Gorongosa's expansion, see the park's website at http://www.janegoodall.org/media/news/GGE-CAP. For more about the controversy surrounding Mt. Gorongosa's inclusion in the national park, see by Carolien Jacobs, "Navigating

through a Landscape of Powers or Getting Lost on Mount Gorongosa," *Journal of Legal Pluralism* 42, no. 61 (2013): 81–108.

25. Personal communication, May 2013.

26. For an analysis of the war's psychological effects, see Victor Igreja, C. B. Dias-Lambranca, and A. Richters, "*Gamba* Spirits, Gender Relations, and Healing in Post–Civil War Gorongosa, Mozambique," *Journal of the Royal Anthropological Institute* 14, no. 2 (2008): 353–71.

27. See especially Shetler, *Imagining Serengeti*.

28. Many thanks to Domingos Muala, a Gorongosa National Park communications staffer who accompanied me on all my journeys both inside and outside the park. Thanks also to the staff of Gorongosa's Community Education Center for their warm hospitality.

29. Richard Waller, "'Emutai': Crisis and Response in Maasailand, 1883–1902," in *The Ecology of Survival: Case Studies from Northeast African History,* ed. Douglas H. Johnson and David M. Anderson (Boulder, Colo.: Westview, 1988), 73–114.

PART III | Boundaries

| # "100 Dangerous Animals Roaming Loose"

Grizzly Bear Management in Waterton-Glacier International Peace Park, 1932–2000

KAREN ROUTLEDGE

IN NOVEMBER 1971 A GRIZZLY BEAR broke into the Chief Mountain Port of Entry, a remote border crossing linking Glacier National Park in the United States to Waterton Lakes National Park in Canada. No one witnessed the rampage, but officials later discovered considerable damage. "Doors and windows were broken to gain entry and a general shambles made in at least six buildings," noted a report. "Interiors were badly damaged and decorative trimming bitten as though the animal was infuriated and in a blind rage."[1]

By 1971 grizzlies had gripped the attention of park staff on both the Canadian and American sides of the forty-ninth parallel. The bear population could not be contained: not within national parks, not within one nation, and not away from park visitors or infrastructure. Encounters between humans and bears sometimes proved fatal for both species, and they sparked hundreds of conflicting letters from concerned citizens. While some visitors mourned every time a bear was killed, others wondered, "Why do we have to put up with 100 dangerous animals roaming loose in a public recreation area?"[2]

The bear that tore up the border station had its own, presumably apolitical motivations. Nevertheless, the international boundary between Glacier and Waterton Lakes National Parks has long been traversed and affected by local forces. Grizzly bears, by refusing to respect human-created boundaries, forced Canadians and Americans to question the meaning of national parks and the place of animals within them. While national parks may serve as abstract symbols of "national" nature, grizzly bears remind us that they are also visceral, regional places that demand sensitive and varied management. In this history of bear management,

I look at how Canadian and American park staff have frequently shared services, information, and ideas across the international boundary line. Federal archives retain little evidence of this cross-border cooperation, but local records and knowledge show that informal collaboration has been both necessary and common. This raises broader questions about how federal records—likely in many other contexts as well—can fail to capture the importance and extent of local initiatives.[3]

"A Privilege to Be Associated with Them": The Creation of Adjacent National Parks

Waterton Lakes and Glacier National Parks have been linked by land, animals, and ideals since their creations. The parks lie adjacent to each other along the international border. The Canadian park was established in 1895, and the larger American one in 1910. Although the two national park systems have since become far more diverse, Waterton Lakes and Glacier were typical of early Canadian and American national parks. Both were relatively inaccessible and nonarable, with dramatic mountain scenery and large populations of game animals. The two parks reflected a growing conviction among middle-class Americans and Canadians that wild spaces such as these were limited, precious, and important to their nation's identity and the health of its citizens.

As Alan MacEachern discusses earlier in this volume, park promoters went to great lengths to attract international tourists, and considerable numbers of Americans and Canadians visited each other's parks. Animals also freely crossed over the border between Waterton Lakes and Glacier, shaping these two parks and their relationship to each other. In 1911 the Canadian government introduced legislation that gutted Waterton Lakes to an insignificant 13.5 square miles (35 km²), shifting its border northward and cutting it off from Glacier National Park. Complaints followed. The Waterton superintendent pleaded to Ottawa that "Goat, Deer, Sheep, and especially Bear are seen [in the park] daily," and he wondered how they could survive in such a small area.[4] James Harkin, who held the top job of commissioner in Canada's newly created Dominion Parks Branch, agreed. He argued that Waterton was useless for protecting game animals "since it is an easy matter for hunters to walk through the park and drive out the game into the surrounding country where it can be slaughtered at will." The slim parcel of land that now divided Glacier and Waterton, Harkin argued, "offers sportsmen a paradise of game shooting, thus nullifying to a large extent the usefulness of both parks."[5] In 1914 the Canadian government reenlarged Waterton to once again become contiguous with Glacier. It remained one of the smaller Canadian parks, and is today only one-eighth the size of its American neighbor.[6] The survival of Waterton's wide-ranging animal populations continues to depend on the protection afforded by Glacier National Park to the south.

In 1932 the two parks were formally united as the Waterton-Glacier International Peace Park, but this link had more to do with the peace movement that followed the Great War than with cooperative management. The Peace Park was the brainchild of Canadian and American Rotary Club members, who lobbied their respective governments to officially unite the two parks. As Canadian commissioner Harkin made clear, this was a symbolic union only, a "commemoration of the peace and good will that exists between Canada and the United States." "Undoubtedly," Harkin continued, "many objections would arise in either country if the proposal involved any joint administration."[7]

Park administrations may have remained separate, but Canadian and American park staff had little choice but to communicate extensively with each other. Americans and Canadians frequently attended each other's meetings and training programs. Roads and trails crisscrossed the border, hydroelectric dam proposals and fish stocking required international consultation, tourist bureaus benefitted from cross-promoting the parks, and forest fires and animal populations disregarded national boundaries. In 1935 a large forest fire started in Glacier National Park and quickly spread across the boundary line, where it was controlled by Canadian park staff. The following year, Waterton wardens crossed into Glacier to help fight a fire that threatened the Many Glacier Hotel.[8] Waterton staff sprinkled their reports with glowing reviews of their counterparts in Glacier National Park. "We continued to receive the utmost courtesy and co-operation from all members of Glacier National Park," commented the Waterton superintendent in 1941.[9] More than a decade later, another superintendent wrote, "We consider it a privilege to be so closely associated with them."[10]

The feeling was apparently mutual. In 1937 a Glacier National Park official wrote to the director of the U.S. National Park Service to protest a new regulation that would require him to seek permission every time he wanted to work with Canadians. He admitted that administrations of the two parks were technically separate, but he commented:

> In many ways our contact with the officials of the Canadian government are practically the same as with our own staff. . . . We have official government telephone lines which reach the office of the superintendent of Waterton and also maintain a short wave radio station of our Glacier Park system in the Canadian park. In the matters of fire protection and control, our relations are on the same cooperative basis as those with the US Forest Service and adjoining National Forests. You will see, therefore, that . . . [we] would be severely handicapped if we had to secure permission every time we had any business to transact with them. Much of this consists of handling important visitors; fire control; game management; fish planting; roads, trails, and telephone connections; and numerous other matters. I frequently make trips to Waterton Lakes.[11]

Most park communication with Washington or Ottawa never mentions these cross-border projects, some of which may have been deliberately conducted under the radar of national headquarters. Yet local collaboration between Waterton Lakes and Glacier employees was an everyday and important matter. Before the 1960s, however, it is unlikely that American and Canadian park staff spent much time discussing grizzly bears. While grizzly bears are now an iconic and controversial symbol of these two national parks, attitudes toward bears have changed radically since park establishment.

"Don't Blame the Bears": Early Bear Policy in Glacier and Waterton Lakes National Parks

In the early twentieth century, visitors to Glacier and Waterton Lakes rarely saw grizzly bears. Most visitors stayed close to infrastructure or traveled in groups on horseback; grizzly bears tended to remain at higher elevations and kept their distance from people. Grizzlies were therefore of almost no concern to park staff, except when the bears strayed into nearby provincial or state lands. Waterton records show a litany of grievances from Alberta ranchers, some of whom had grazing permits in the park. These cattlemen repeatedly sought permission to come in and shoot grizzlies. Their requests were denied, but a few were given honorary warden status to enter the park and kill wolves.[12] In 1930 rancher A. T. Cleaver presented Waterton with a bill for $140, for cattle he claimed bears had killed in the past year. "It is now time that the park kept the bear on their own property," Cleaver demanded.[13] Into the 1940s wardens generally shot grizzlies that killed cattle inside or outside the park. Their supervisors in Ottawa backed them up, commenting that while it was "desirable" to protect grizzly bears, "we are not unmindful of the difficulties to be encountered where there are grizzly bears in a small park such as Waterton."[14] Waterton and Glacier staff were presumably grateful for the border they shared, since grizzlies crossing between the two parks incited no public complaints.

The concerns of ranchers aside, grizzlies were seldom mentioned in park records before the 1960s. Waterton staff, at least, were initially confused over how to classify bears. In the early twentieth century, they divided animals into "game" and "predators"—two terms that were gradually subsumed into "wildlife" by roughly the late 1950s.[15] When grizzlies killed cattle, staff sometimes listed them as predators, but at other times they classified bears as game. Bears were sometimes not included in official animal reports at all, suggesting they posed no problems worth mentioning. And in early park reports, any discussion of "bears" generally referred to the smaller and more sociable black bears, which were the single wild animal most commonly seen by tourists. A former Canadian park warden had little to say about grizzly bears, but he said he "could go all night" talking about black bears. He recalled that when he would drive along the Banff-Windermere Highway in the Canadian Rockies in the late 1940s, he would see "maybe sixty, sixty-five [black] bears" begging for food on the 100-kilometer (60-mile) stretch between Radium Hot Springs and Castle Mountain.[16] The situation was similar in Waterton Lakes and Glacier. In 1931 a newspaper article claimed that "virtually every visitor" to Glacier came "into close contact with the lumbering [black] bears, that make a practice of holding up busses loaded with tourists to beg for candy or other dainties."[17] At first, Canadian and American park staff did little to discourage such behavior. Ursine antics were the highlight of many vacations. In 1924 the superintendent of Glacier National Park hinted at one reason bears were tolerated in this period, writing that mountain lions were "never seen by tourists and might be said, therefore, to have no public value."[18] If public value was defined at least in part by visibility, black bears certainly had it. Rangers and wardens destroyed aggressive bears or sent them to zoos, but there was never a systematic attempt to exterminate black or grizzly bears as there was with carnivores such as wolves, coyotes, and cougars.[19]

Although park staff recognized the appeal of bears, by the late 1930s they could not ignore a growing number of unwelcome incidents. Black bears and occasionally grizzlies ripped up cars and ruined property; they had scratched and maimed visitors in several of the most visited national parks in Canada and the United States. Canadian and American park officials communicated with each other on these issues and kept their policies closely aligned. The American secretary of the interior prohibited the feeding of bears in national parks in 1938. The same year, a Canadian federal parks official cited the new American regulations when instructing Waterton Lakes staff to erect signs telling visitors to stay away from bears.[20]

Ideas about controlling bear-human interactions flowed across the border in both directions. Furthermore, bear policy was never simply top-down. It was continually negotiated and developed between local, regional, and federal levels of the park service. In the early 1940s, when tourists persisted in approaching bears, the Canadian Park Service headquarters in Ottawa released a poster titled, "Don't Blame the Bears." It warned of potential dangers from black bears frequenting the highways and campgrounds, and from grizzly bears in the backcountry. It ended with the admonition, "Do not feed the bears or try to be intimate with them."[21] Although intimacy now carries different connotations, the slogan captured the extent to which mid-twentieth-century visitors saw black bears as familiar animals; they were exciting and potentially dangerous, but also charming and approachable. They were both wild and tame. Waterton Lakes parks staff freely handed "Don't Blame the Bears" posters out to visitors. In 1946 an American parks official from the regional office in Omaha obtained a copy while visiting Waterton Lakes. He sent the poster to Glacier and some other western American parks, suggesting that they might consider this approach.[22]

Glacier National Park was ahead of the Omaha official—they may even have inspired the Canadian brochure. Three years earlier, Glacier park staff had produced mimeographed notices warning visitors that "the feeding, touching, teasing, or molesting of bears is prohibited." A copy of this earlier leaflet does not appear to have survived, but Waterton Lakes park staff saved another Glacier promotional brochure from this decade that included a sketch resembling the illustration at the bottom of the Canadian "Don't Blame the Bears" poster. Park officials on both sides of the border shared their strategies and materials, and they were starting to take the bear issue seriously. By 1950 Glacier staff were passing out notices about bears to every visitor entering the park. Glacier also experimented with rhyming highway signs featuring a variety of whimsical slogans, such as "Highway bears are often rude; they eat fingers as well as food." A displeased grizzly bear reportedly clawed several of the signs and tore one of them to pieces.[23]

Simply asking tourists to stay away from bears was a weak deterrent, and there was not yet sufficient motivation, or perhaps staff capacity, to enforce regulations. In 1937 a Glacier ranger recommended several additional bear safety measures: incinerators; bear-proof garbage disposal; increased visitor information and enforcement of regulations; and mandatory education on bear safety for all concessionaires, contractors, and government employees. Most of his suggestions would not be broadly implemented until decades later. The same ranger admitted that begging bears offered an unforgettable experience, "something always remembered by anyone interested in wildlife and after all, who isn't interested."[24] As late as 1953, the Waterton superintendent recorded a similar hesitancy about reducing human-bear contact: "Bear . . . caused the usual amount of excitement and trouble. They are well worth while, however, as they provide the visitors to the area, with the type of excitement and interest that they are looking for. . . . Some tourists fed them, of course, and we had two people hurt, but generally the public took notice of the posters warning them against the feeding of these animals."[25] Both Waterton and Glacier opted to deal more strictly with bears than humans. Despite visitor education initiatives, improved garbage collection, and programs to live-trap and relocate bears, between 1949 and 1963 Glacier National Park rangers destroyed 103 black bears and 15 grizzlies. Waterton also shot black bears, but as a smaller park with much less backcountry habitat, they only very rarely had to shoot grizzlies. The culling was effective and does not seem to have incited broad public protest. By the early 1960s Glacier National Park was reporting far fewer "beggar bears" hanging around campgrounds and roadsides.[26]

In 1963 Glacier park rangers tried to find a less violent solution to deal with the remaining bears that solicited food: they tested a spray repellent called Halt on them. The experiment was a failure. Most of the sprayed bears kept strolling toward the staff members, hoping for a handout. "Halt" did not seem to incite its namesake reaction in other animals, either. A sprayed deer reacted by kicking a ranger so hard he had to be hospitalized. Canadian authorities had been eager to hear if "Halt" was worth trying, but the Glacier superintendent reported that in his opinion, "there is no effective substitute for prompt garbage collection, bear-proof garbage cans and the use of the culvert live trap."[27] On both sides of the border, no cheap, acceptable, and simple solution appeared for what had become known as the "bear problem"—but was in effect a human problem.[28]

Park officials became increasingly concerned about bears after the Second World War, when national parks became more popular vacation destinations than ever before. Visitation at Glacier tripled from 1945 to 1956, and Canadian and American parks as a whole experienced similar growth.[29] Still, the first flood of postwar visitors mostly stayed close to roads and infrastructure, where they were far more likely to encounter black bears than grizzlies. In 1955 the Waterton superintendent commented only vaguely: "[Grizzlies] have been reported as being seen at the higher levels, but I did not see one, and did not get a very authentic report of one being seen, although they were mentioned by our employees from time to time."[30] This would change as more visitors opted to hike and camp in the backcountry. In 1967 one terrible event in Glacier National Park—it became known as the "Night of the Grizzlies"—would push grizzly bear management to the center of park policy.

Night of the Grizzlies

On August 12, 1967, Julie Helgeson and Michele Koons went separately into the backcountry of Glacier National Park. They were both young students working summer jobs at park concessions. It was Julie's weekend off, and she and a companion walked into a small campground near the Granite Park Chalet. Michele hiked with some friends to a good fishing spot at Trout Lake. Both groups slept in the open, spreading their sleeping bags on the ground.

Just after midnight, a grizzly bear knocked Julie Helgeson and her friend out of their sleeping bags. Guests at the Granite Park Chalet heard their screams. Miles away at Trout Lake, Michele and her friends were repeatedly harassed by a different grizzly bear. The small group built a campfire, but the bear returned later that night. Everyone fled for the trees except Michele, who was unable to free herself from her sleeping bag. The bear hauled Michele out of sight. Both Julie and Michele died of their wounds, leaving behind devastated friends and family. Rangers shot four grizzlies in the days that followed, including the two that had killed the young women.[31]

Julie and Michele's deaths incited uproar. These were the first fatal bear attacks ever recorded in Glacier National Park. After the news broke, hundreds of impassioned and concerned letters poured into the park offices and to the Department of the Interior. Many visitors questioned whether national parks belonged to people or to bears. Citizens wrote fervent letters on both sides of the debate. Adrian Maas from New Jersey was upset that the rangers had been so quick to shoot four grizzlies, when there were only an estimated hundred left in the park. He pleaded with the parks service to save the bears. "When our universe is sterilized," he lamented, "we will be sterilized along with it."[32] Other letter writers disagreed, saying that the grizzlies left them with "a feeling of deep uneasiness" when they hiked in the park. Some tourists had avoided Glacier because of its bears.[33] Many of these citizens wanted rangers to keep shooting until there were no grizzlies left in Glacier National Park. "Their hides could be stuffed and placed in various chalets in the park . . . as mementoes of past wild life," suggested John Franklin Donahoo of Honolulu.[34] A former Glacier park ranger asserted in a local Montana newspaper, "There is only one thing to do with the grizzly bear in Glacier National Park and that is to get rid of him, and turn the park back to the people."[35]

A Canadian citizen, Carl Ellis, wrote an emotional response to this ex-ranger's letter, claiming that parks had room for people and grizzlies. As a frequent hiker in the International Peace Park, he said he took steps to protect himself from grizzly attacks and to warn grizzlies of his presence. The bears did not deter him or detract from his experience; rather, they were central to his enjoyment of the area. "Much of the thrill and excitement of trail hiking in Glacier and Waterton springs from the fact that wild animals including Grizzly, Cougar, are perhaps somewhere in the area," he argued.[36] Tony Hoyt from Ithaca concurred. He had spent weeks in the Glacier backcountry, where he reveled in the knowledge that grizzlies were in those mountains. Their presence renewed his "delight in the mysteries of earth, sea and sky" and preserved his "sense of wonder."[37] Ellis felt that if grizzlies were to be exterminated, "the National Park Service will have betrayed the people."[38]

North of the border in Waterton Lakes National Park, Canadian staff kept a close eye on the American reaction. The Superintendent of Waterton Lakes wrote that although his park had few problems with bears that summer, "the general public has become much more concerned over bears because of the two grizzly incidents."[39] He requested more information from the U.S. park service on the maulings, and he reported to his superiors that out of the first 121 letters received at Glacier National Park, only 14 percent favored the extirpation of grizzlies from the area.[40] Glacier was also careful to keep Waterton informed. When Glacier staff completed a report on the attacks the following spring, the superintendent of Waterton Lakes National Park was the first name listed as having received a copy.[41]

After the fatal maulings of Julie Helgeson and Michele Koons, both parks made intensive efforts to curtail human interactions with bears. It does not appear that anyone with authority seriously considered exterminating bears, and Canadian and American park policies remained aligned. Shortly after the incident, Glacier National Park released a somber twelve-page booklet in which the first sentence read, "Bears are wild, dangerous animals."[42] In 1968 the Waterton superintendent sent his Glacier counterpart a strikingly similar pamphlet titled "Bear Facts." He attached the note, "If the pamphlet seems vaguely familiar remember that while plagiarism may be a dirty word among authors, imitation is a compliment among Superintendents."[43]

Both pamphlets contained dire warnings for visitors, but also welcomed them. The first page of the American brochure ended with the statement, "We want you to enjoy the National Parks, too." The similar Canadian publication declared, "Obviously, man has a place in this wilderness too."[44] Both national parks chose the same path: to balance the interests of recreational users with those of bears. It was in keeping with both systems' mandates. The Canadian and American National Parks Acts contained, and still contain, the almost identically worded sentiment that protected areas "shall be maintained and made use of so as to leave them unimpaired for the enjoyment of future generations."[45] Still, keeping people and grizzlies in peaceful coexistence has never been an easy task. Neither bears nor people could ever be fully controlled.

In large part, the "new" bear policy consisted of better implementing existing regulations. A few years before the Night of the Grizzlies, after some nonfatal but gruesome attacks, and amid questions about the legal responsibility of national parks, Glacier National Park had created a bear management plan.[46] The plan included strategies for bear-proof garbage disposal and the education of concessionaire and hotel staff, as well as provisions for trail closures where bears were known to be behaving aggressively.[47] Yet these regulations were not universally followed. The grizzly that killed Michele Koons had frightened other campers in weeks preceding her death. One woman who wrote a letter to the National Park Service in August 1967 included a postcard of begging black bears that she claimed was sold at concessions throughout many western national parks. She also pointed out correctly that Julie Helgeson had died minutes from the Glacier Park Chalet, where there was a table-scrap pile frequented by six grizzlies.[48]

The 1967 attacks spurred Glacier into action. In May 1968 the superintendent of Waterton was greatly impressed by a public education event he attended in Montana. After the deaths, he said, the American park was "making a special effort to warn visitors of the dangers." He continued:

> The bear program [at Glacier] is based upon research, education and enforcement. They are going to be quite tough about laying charges for feeding animals, leaving dirty campsites, etc. They are going to close off trails or areas any time there seems to be justification because of bear incidents. I believe we should also be making a more positive effort. . . . While this effort is perhaps required in all our parks it is most important here because of our connection with Glacier.[49]

After the deaths of Julie Helgeson and Michele Koons, Glacier National Park cleaned up most outstanding garbage problems and closed trails when bears were present.[50] The new measures helped decrease aggressive encounters between bears and humans, but occasional maulings continued to occur. In 1976 a young woman was dragged from her tent in the Many Glacier Campground. The following year, a five-year-old girl was killed in Waterton Lakes. Both a black bear and a grizzly bear were destroyed following her death.[51] It was the small park's first and (as of 2015) only bear-induced fatality, although there have been other attacks that resulted in very serious injuries.

The parks also struggled to create better reporting and tracking systems for bear-human encounters. In 1984 Waterton Lakes issued a bear management plan that contained an organizational chart listing who should be notified of bear incidents and in what order. Oddly, Glacier National Park appeared nowhere in the chart. However, the same document later noted, "All bear observations and recording will be done in collaboration with Glacier National Park, USA. It is important that we maintain good liaisons with Glacier National Park officials as our bear populations do not recognize man-made territorial boundaries."[52] Canadian and American staff also kept records of bear sightings on a common computerized form that was in use across several American national parks. Effective management demanded cooperation, even if some of it happened outside the bounds of formal working agreements.

The gruesome nature of grizzly bear attacks understandably continued to hold a special terror for many potential visitors, most of whom did not encounter wild animals in their daily lives. In 1977 Glacier officials compiled a list of known deaths in their park to underscore the rarity of bear attacks. "You may be interested in a record of how grizzly fatalities compare to other fatal accidents in Glacier since 1913," wrote the chief park ranger to two concerned citizens. "Drownings 36; heart attack 19; vehicle 17, hiking falls 16, climbing falls 11, avalanches 7, falling rocks 6, exposure 4, grizzlies 3."[53] When humans and grizzlies encountered each other in the Peace Park, bears were in far more danger of dying. Indeed, the letters that had poured in after the 1967 bear attacks had raised two related issues: the need to keep people safe from bears, and the need to keep bears safe from people. As park administrators and legislators wrestled with how to protect the grizzly bear population from humans, they realized how little they actually knew about these bears, their numbers, and their habits. Grizzly bears became both a social and scientific issue.

Members at Large: Collaboration in the Era of the Endangered Species Act

At 6 A.M. on the morning of Sunday, August 13, 1967, a ranger knocked on Cliff Martinka's door. "We've got a bear problem," the ranger said. "Get your gun." Martinka was a wildlife biologist for Glacier National Park; it was only his second week on the job. In the weeks that followed the deaths of Julie Helgeson and Michele Koons, people from within and outside Glacier directed endless questions about grizzly bears to park staff. Martinka did not have the answers. He was the park's first dedicated research scientist, and as he later recalled, "There was amazingly little work going on with grizzly bears."[54]

Canadian wardens and American rangers, Native peoples, and others who lived in close proximity to grizzlies knew much about their habits, but this information was rarely communicated to or sought out by researchers. There had been a few previous scientific studies. The naturalist Adolph Murie spent twenty-five summers observing grizzlies and wolves in Denali. His brother, Olaus Murie, investigated grizzly bear habitat in Yellowstone in 1944. Most famously, the twins Frank and John Craighead researched grizzly bears in Yellowstone for more than a decade, beginning in 1959. They devised radio collars that enabled them and countless other scientists to track the animals' movements. Their results showed that even in a large park such as Yellowstone, grizzlies ranged far beyond the park borders.[55]

Such conclusions must have been troubling for Glacier and Waterton Lakes national park officials. Bears ranging back and forth between the two parks were generally a nonissue, but in order to ensure the grizzlies' survival, the parks would now have to work with the owners and administrators of land outside their other, equally porous, boundaries. The forty-ninth parallel is only one of many lines that matter here. These parks border on lands housing coal mines, railroads, national forests, First Nations reserves and Indian reservations, ranches, timber lands, oil and gas development leases, and sprawling residential developments.[56] The international border remains an issue today to the west of Waterton Lakes National Park, where Glacier's bears can cross into British Columbia, a Canadian province with a legal grizzly hunt.

After the Craigheads' studies in Yellowstone and the streak of grizzly attacks in American national parks, grizzly bear research received more funding. The first formal international bear management conference was held in Calgary, Alberta, in 1970, and subsequent meetings attracted many Canadian and American researchers. In Glacier National Park tremendous interest and numerous research opportunities followed the passage of the U.S. Endangered Species Act and its designation of grizzlies as a threatened species in 1975. Glacier National Park joined the Interagency Grizzly Bear Committee, which worked on a recovery plan with state, local, and national agencies to ensure the survival of the grizzlies. However, perhaps because the Endangered Species Act was national legislation without a Canadian federal counterpart at the time, most funding and formal agreements were tied to the United States even though many protected bears roamed between both nations.[57]

Waterton Lakes National Park continued to cooperate with Glacier, mostly in informal but nevertheless valuable ways. The park, along with Canadian provincial fish and wildlife divisions, were considered "members-at-large" in the Northern Continental Divide Grizzly Bear Managers Subcommittee.[58] Waterton also benefited indirectly from Glacier's funding to study their joint grizzly population. Canadian staff contributed to American research projects by monitoring huckleberry plots in late summer and early fall to assess the impact of the berry crop on bear activity, and by participating in cross-border meetings and inquests into bear incidents.[59] Parks Canada libraries retain many reports and theses about bears in Glacier National Park, and bibliographies of bear studies often include literature from both sides of the border.

Although Americans initiated most grizzly research projects, in the 1990s a small Canadian bear DNA study in British Columbia sparked the interest of U.S. Geological Survey biologist Kate Kendall. The Canadian project set up bait stations with barbed wire, from which researchers collected grizzly bear hair samples and analyzed their DNA. Kendall had long been interested in finding more reliable ways to measure bear populations in the Glacier National Park region. In 1985 bear biologist Stephen Herrero had quipped, "The grizzly could vanish from the contiguous United States while the bear biologists are still debating how to count them."[60] In 1998 Kendall received American funding for a study modeled in part on the Canadian one, but much larger in scope. Her team conducted DNA surveys on 2 million acres of land in Glacier National Park and surrounding areas. Kendall's team used scent-baited hair traps modeled on the Canadian study, but they also placed barbed-wire collection points on natural rub trees, making the collection methodology even less invasive. Kendall's results confirmed some of what wardens and rangers already suspected from anecdotal evidence, but at least one Waterton warden was surprised at how often bears were using some high-traffic areas, and at how skilled they were at avoiding encounters with people. The project ran from 1997 to 2002 and produced the first reliable grizzly population estimate for the region. Kendall led a second similar but even larger project from 2002 to 2008, which covered almost 8 million acres of the Northern Continental Divide ecosystem in Montana.[61]

Conclusion: The Importance of the Local

National parks remain at once federal projects and deeply local places. Cross-border ties remain strong in Waterton and Glacier National Parks. Staff have recently worked together not only on grizzly bear issues but also on projects concerning plants, fish, bats, aquatic and terrestrial invasive species, and grassland and fescue restoration.[62]

National parks around the world have long grappled with regional issues, often involving complicated, contentious relationships between roving humans and animals.[63] The American and Canadian national park systems now extend far beyond the world-renowned mountainous areas such as Yellowstone or Glacier or Banff, and wildlife regulations vary from place to place. In Canada, for example, Native communities legally hunt and harvest resources in national parks that have been created in their traditional territories through recent land-claim negotiations. Two eastern Canadian parks have introduced a licensed moose hunt as an alternative to staff culling.[64] Such initiatives, often developed through local activism or collaboration with park employees, can lead to official policy changes that spread to other regions.

Histories of borderlands often focus on the moments where tensions explode, boundaries harden, or contradictions are laid bare. These remain important stories to tell. Chris Conte's and Steve Rodriguez's chapters on East African and

Indonesian parks in this volume, or Mark David Spence's work on the relationship of the Blackfeet Nation to Glacier National Park, investigate incidents where state and local interests have painfully collided with serious consequences.[65] Historians Pekka Hämäläinen and Samuel Truett, however, have warned against the temptation to place state and nonstate actors in constant opposition to each other, pointing out that national and local narratives are often interdependent.[66] Waterton Lakes and Glacier park employees provide a good example of this overlap. Many staff members proudly represented their own national park service while working on local cross-border initiatives. Although national and regional priorities sometimes came into conflict, park officials were often able to navigate between them. Grizzly bear management in Waterton-Glacier International Peace Park points to a more flexible type of borderlands experience: the often unrecorded stretches of everyday life in which many individuals are able to productively juggle multiple identities and allegiances.

Notes

1. Waterton Lakes National Park, "Superintendent's Annual Report 1972," Parks Canada Library, Western and Northern Service Centre, Calgary, Alberta (hereafter WNSC).

2. Earl D. Smith to Secretary Udall, August 14, 1967, U.S. National Archives and Records Administration, College Park, Maryland (hereafter NARA), Record Group 79, Entry 11, Administrative Files, 1949–71, Box 370, Folder: "Complaints About Service and Personnel, Glacier National Park [GLAC], 1949–69." For more citizen opinions see the section below on the 1967 attacks.

3. For examples of how groups of people have functioned well and often cooperatively within a locally developed set of rules, see Robert Ellickson, *Order without Law: How Neighbors Settle Disputes* (Cambridge, Mass.: Harvard University Press, 1991). For a case where transborder cooperation failed due to divergent ideals of conservation, see Emily Wakild, "Border Chasm: International Boundary Parks and Mexican Conservation, 1935–1945," *Environmental History* 14, no. 3 (July 2009): 453–75. On another organism—weeds—shaping community relationships, land use, and laws, see Mark Fiege, "The Weedy West: Mobile Nature, Boundaries, and Common Space in the Montana Landscape," *Western Historical Quarterly* 36 (Spring 2005): 22–47.

4. John George Brown to J. B. Harkin [?], June 1, 1913, Library and Archives Canada, Ottawa, Ontario (hereafter LAC), RG 84, A-2-a, Volume 2167, W2–1a, "Waterton Lakes National Park—Park Property—Encroaching on a Game Preserve," Part 1. See also Sergeant Allen Chamberlain to [?], undated but ca. 1913, LAC, RG 84, A-2-a, Vol. 2167, W2–1a, "Waterton Lakes National Park—Park Property—Encroaching on a Game Preserve," Part 1.

5. James B. Harkin to William Wallace Cory, Deputy Minister of the Interior, 24 November 1913, LAC, RG 84, A-2-a, Volume 2167, W2–1a, "Waterton Lakes National Park—Park Property—Encroaching on a Game Preserve," Part 1.

6. Waterton is 505 square kilometers (195 square miles), while Glacier is 4,144 square kilometers (1,600 square miles). National Park Service, "General Information about Waterton and Glacier International Parks," http://www.nps.gov/glac/forteachers/general_info.htm; Parks Canada, "Waterton: A Quick Guide," http://www.pc.gc.ca/pn-np/ab/waterton/visit/visit6.aspx (both accessed May 1, 2014).

7. James Harkin, as cited in Catriona Mortimer-Sandilands, "'The Geology Recognizes No Boundaries': Shifting Borders in Waterton Lakes National Park," in *The Borderlands of the American and Canadian Wests: Essays on the Regional History of the Forty-ninth Parallel,* ed. Sterling Evans (Lincoln: University of Nebraska Press, 2006), 309–33, quotation on 319.

8. Graham A. MacDonald, *Where the Mountains Meet the Prairies: A History of Waterton Country* (Calgary: University of Calgary Press, 2000), 12–13, 61–63 (lumbering), 65 (grazing), 74 (dam), 82 (hotels), 93 (roads), 97 (roads), 109 (irrigation), 136 (elk, fire), 142 (roads and expanding Waterton), 156 (biosphere reserve). For training programs, see Waterton Lakes National Park, "35th Annual Report," 1948; Waterton Lakes National Park, "28th Annual Report," 1941; Waterton Lakes National Park, "Superintendent's Monthly Report," September 1952. For fires, see Waterton Lakes National Park, "22nd Annual Report," 1936; Waterton Lakes National Park, "23rd Annual Report," 1937. All of these primary sources are found in Waterton Lakes National Park Warden Compound files, Waterton Park, Alberta (hereafter WLNP), Box 207.

9. Waterton Lakes National Park, "28th Annual Report," 1941; Waterton Lakes National Park, "35th Annual Report," 1948; both in WLNP, Box 207.

10. Waterton Lakes National Park, "42nd Annual Report," 1955, WLNP, Box 207.

11. Glacier National Park [Superintendent?] to the Director of the National Park Service, September 21, 1937, NARA, Record Group 79, Entry 10: Classified Central Files, 1907–49, Box 2960, "Proposed National Parks, Waterton-Glacier International Peace Park, 1932–49, 1." From the 1960s onward there would also be collaboration on naturalist and interpretive programs. See, for example, William J. Briggle (Glacier Superintendent) to Thomas J. Ross (Waterton superintendent), April 20, 1970, in "Bears 1951–1970," WLNP, File 212 v. 1, Box 226.

12. Commissioner of National Parks [J. B. Harkin] to Honourable G. Hoadley [Minister of Agriculture for Alberta], April 11, 1922, LAC, Record Group 84, A-2-a, Vol. 173, "Waterton Lake National Park—Bears 1918–1946," file w212, part 1 (hereafter "Bears 1918–1946").

13. A. T. Cleaver to Superintendent of Waterton, May 22, 1930. See also Cleaver to Superintendent, September 19, 1929, both in "Bears 1918–1946."

14. Assistant Controller of National Parks to Waterton Superintendent, July 10, 1944, in "Bears 1918–1946." See also Smith Bros Co. Ltd. et al. [petition] to the Dominion Government of Canada, July 11, 1925; the Commissioner of National Parks [J. B. Harkin] to the Acting Superintendent of Waterton Lakes National Park, July 7, 1925; Acting Superintendent of Waterton Lakes National Park to "Sir" [copied to J. B. Harkin], August 8, 1925, all in "Bears 1918–1946." For mentions of grizzlies shot outside the park boundaries, see, for example, Waterton Lakes National Park, "29th Annual Report" [1942], WLNP, Box 206; Waterton Lakes National Park, "Superintendent's Monthly Report," May 1951, WLNP, Box 208.

15. See, for example, Waterton Lakes National Park, "Superintendent's Monthly Report for July 1952" (bears as wildlife); "Superintendent's Monthly Report for June 1959" (wildlife); "Superintendent's Monthly Report for December 1937" (game); "33rd Annual Report," 1946 (predators); "32nd Annual Report," 1945 (predators); "31st Annual Report," 1944 (predators); "41st Annual Report," 1954 (not listed under predators); all from WLNP, Boxes 206 and 207. On changes to bear management language in Glacier National Park in the 1960s, see Theodore Catton (Principal Investigator), "Protecting the Crown: a Century of Resource Management in Glacier National Park," June 2011, 128 http://www.cfc.umt.edu/CESU/Reports/NPS/UMT/2008/08Catton_GLAC_History_Resource%20Mgmt_final.pdf (accessed May 1, 2014).

16. Bob Thompson, Interview with Maryalice Stewart, Tape S23/3–15, Side C, Parks Canada Fonds, Whyte Museum of the Canadian Rockies, Banff, Alberta. Bob Thompson was a warden in Waterton and several other Canadian national parks. In 1943 a Canadian parks official claimed that "not one tourist in ten thousand ever gets a glimpse of a grizzly, whereas almost all tourists encounter black bears." C. H. D. Clarke to Mr. Lloyd, November 10, 1943, "Bears Original File," WLNP, Box 226, File 212.

17. "Bear Turns Bandit and Robs Home of Senator," *Denver Post,* July 7, 1931.

18. Charles J. Kraebel (Glacier Superintendent) to the Director of the National Park Service, December 19, 1924, NARA, Record Group 79, Entry 10: Classified Central Files, 1907–49, Box 253, Folder "[Glacier National Park], Predatory Animals, 1918–1932." Kraebel added that he personally felt there was a "scientific or biological value in maintaining the species [cougars] in existence."

19. Bear policy was similar in other U.S. national parks from 1916 to 1929. See Richard Sellars, *Preserving Nature in the National Parks* (New Haven, Conn.: Yale University Press, 1997, 2009), 79. See also Alice Wondrak Biel, *Do (Not) Feed the Bears: The Fitful History of Wildlife and Tourists in Yellowstone* (Lawrence: University of Kansas Press, 2006). For more details about bear policy in Glacier National Park, see Catton, "Protecting the Crown," esp. 123–136, 189–194.

20. Arno B. Cammerer to the Superintendent of Glacier National Park, March 24, 1938, NARA, Record Group 79, Entry 10: Central Classified Files, 1907–49, Box 968, Folder: "[Glacier National Park] Bears, 1930–49"; F. H. H. Williamson to Superintendent of Waterton, August 24, 1938, in "Bears 1918–1946."

21. National Parks Bureau of Canada, "Don't Blame the Bears," ca. 1944. Copies in both "Bears Original File," Box 226, WLNP, and in NARA, Record Group 79, Entry 10: Central Classified Files, 1907–49, Box 968, Folder: "[Glacier National Park] Bears, 1930–49."

22. Jerome C. Miller, Acting Director Region Two, to the Superintendent of Glacier National Park, August 2, 1946, NARA, RG 79, Entry 10: Central Classified Files, 1907–49, Box 968, Folder: "[Glacier National Park] Bears, 1930–49." Miller wanted the superintendent to obtain more copies from Waterton for him.

23. D. S. Libbey (Glacier superintendent) to the Regional Director of Region Two, May 13, 1943, NARA, Central Plains Region, Record Group 79, Region #CCF, 1936–52, Box 68, File 715_02; Superintendent of Glacier National Park to Regional Director Region Two, September 26, 1950, ibid. See also

Glacier National Park, "Can Bears Read?" [Press Release?], June 3, 1949; E. N. Fladmark to Regional Director Region Two, October 4, 1948, both in NARA, Record Group 79, Entry 10: Central Classified Files, 1907–49, Box 968, Folder: "[Glacier National Park] Bears, 1930–49."

24. "The Bear Problem," NARA, Record Group 79, Entry 10: Central Classified Files, 1907–49, Box 968, Folder: "[Glacier National Park] Bears, 1930–49."

25. Waterton Lakes National Park, "40th Annual Report," 1953, WLNP, Box 207.

26. Catton, "Protecting the Crown," 125–27 (summary of Glacier bear policy in this period); Jeanne N. Clarke, "Grizzlies and Tourists," *Society,* January/February 1990, 26 (cull numbers); Keith Brady, "A Report on the Distribution and Population of Black and Grizzly Bear in Waterton Lakes National Park [1972]," 7, WNSC (Waterton grizzly control); Charles R. Wasem to Chief Park Ranger of Glacier National Park, August 19, 1963, Glacier National Park Archives [hereafter GNPA], Administrative Files, 1910–1984, Box 258 Folder 4 (fewer beggar bears).

27. Keith Neilson to B. I. M. Strong, December 14, 1964, NARA, Rocky Mountains Region, Record Group 79, Numerical Subject Files 1949–1965, Box 24, N16.

28. For more on bears and tourists in Canadian and American national parks in the postwar period, see George Colpitts, "Films, Tourists, and Bears in the National Parks: Managing Park Use and the Problematic 'Highway Bum' Bear in the 1970s," in *A Century of Parks Canada 1911–2011,* ed. Claire Elizabeth Campbell (Calgary: University of Calgary Press, 2011), 164; Susan Sessions Rugh, *Are We There Yet? The Golden Age of Family Vacations* (Lawrence: University Press of Kansas, 2008), chap. 5; and Wondrak Biel, *Do (Not) Feed the Bears.*

29. Clarke, "Grizzlies and Tourists," 27 (Glacier); Rugh, *Are We There Yet?* 131–32 (U.S. national parks); Alan MacEachern, *Natural Selections: National Parks in Atlantic Canada, 1935–1970* (Montréal: McGill–Queen's University Press, 2001), 162 (Canadian national parks).

30. Waterton Lakes National Park, "42nd Annual Report," 1955, WLNP, Box 207.

31. Stephen Herrero, *Bear Attacks: Their Causes and Avoidance* (New York: Nick Lyons, 1985), 56–61.

32. Adrian Maas to Secretary Udall, August 15, 1967, NARA, Record Group 79, Entry 11, Administrative Files, 1949–71, Box 370, Folder: "Complaints About Service and Personnel, Glacier National Park [GLAC], 1949–69" (hereafter "Complaints About Service and Personnel, 1949–69.")

33. Janis Wilson to Secretary Udall, August 15, 1967, in "Complaints About Service and Personnel, 1949–69." Bear biologist Stephen Herrero and his family avoided Glacier in 1967 for the same reason, although he certainly did not advocate bear extermination. Herrero, *Bear Attacks,* xii.

34. John Franklin Donahoo to Secretary Udall, August 15, 1967, in "Complaints About Service and Personnel, 1949–69."

35. Nick Carter, Former Chief Ranger of Glacier (1925–31), to the editor of *Hungry Horse News,* September 15, 1967, copied in WLNP, Box 226, File 212 v. 1.

36. A. Carl Ellis to the editor of *Hungry Horse News* [Columbia Falls, MT], copied in WLNP, Box 226, File 212 v. 1.

37. Tony Hoyt to Secretary Udall, undated, received August 30, 1967, in "Complaints About Service and Personnel, 1949–69."

38. See n. 36.

39. Waterton Lakes National Park, "Superintendent's Report for the Three Months ending September 1967," 5–6, WNSC.

40. W. J. Lunney (Waterton Superintendent) to Regional Director of the Canadian National Parks Service, September 18, 1967; Lunney to Superintendent of Glacier National Park, May 22, 1968, both in "Bears 1951–1970," File 212 v. 1, Box 226, WLNP. Note that I have only seen the letters written to the Department of the Interior in Washington, D.C., which are more heavily weighted toward grizzly extermination than the letters Lunney describes.

41. Copies of the report were sent out to dozens of people over several days. Superintendent Lunney was the first name on the list of copies sent out. That first day, the park also sent copies to ten others, including the director of the National Parks Service, the superintendent of Yellowstone, and the manager of concessionaire and hotel operator Glacier Park, Inc. [Untitled list of recipients of copies of Grizzly Bear Attacks Report], GNPA, Administrative files, 1910–1984, Box 247, Folder 1.

42. "Enjoying Bear Country," ca. 1967–1968, WLNP, Box 226, File 212 v. 1.

43. W. J. Lunney to Keith Neilson, June 21, 1968, in ibid.

44. "Bear Facts" [Canadian Pamphlet] and "Enjoying Bear Country" [U.S. pamphlet], WLNP, Box 226, File 212 v. 1.

45. Canada National Parks Act (S.C. 2000, c. 32). http://laws-lois.justice.gc.ca/eng/acts/N-14.01/page-2.html (accessed May 1, 2014). The (U.S.) Act to Establish a National Park Service reads, "which purpose is to conserve the scenery and the natural and historic objects and the wild life therein and to provide for the enjoyment of the same in such manner and by such means as will leave them unimpaired for the enjoyment of future generations." http://www.nps.gov/parkhistory/online_books/anps/anps_1i.htm (accessed July 24, 2015).

46. In 1959 Joseph Williams, a young Glacier National Park hotel waiter on his first day of work, went for a hike and was mauled for forty-five minutes until a ranger arrived and shot the bear. He claimed workman's compensation for the incident, which the district court granted him, and a higher court overturned. *Williams v. Glacier Park Co.,* 140 Mont. 440, 445, 373 P.2d 517, 520 (Mont.1962). For Canadian concern about lawsuits, see B. I. M. Strong, circular to western Canadian national parks, August 1, 1962, WLNP, Box 226, File 212 v. 1.

47. Glacier National Park, "Bear Management Plan," NARA, Record Group 79, Entry 11: Administrative Files, 1949–1971, Box 2331, Folder: "Wildlife Management, including Predator Control, Glacier National Park, 1952–67, 4." The standard Department of the Interior reply to letters about the 1967 maulings listed many of these policies. See, for example, Karl T. Gilbert to Tony Hoyt, September 19, 1967, in "Complaints About Service and Personnel, 1949–69."

48. Virginia Bacher to Director of the National Park Service, September 18, 1967, in "Complaints About Service and Personnel, 1949–69." See also Herrero, *Bear Attacks,* 57.

49. Lunney to Acting Regional Director of Canadian National Parks Service, May 13, 1968, in "Bears 1951–1970," WLNP, Box 226, File 212 v. 1.

50. Herrero, *Bear Attacks,* 15–16. See also Steve J. Gniadek, "Bear Management in Glacier National Park, 1960–1994," Glacier National Park, 1995 Central Files; Catton, "Protecting the Crown," 123–30.

51. Waterton Lakes National Park, "Superintendent's Annual Report 1977–78," WNSC.

52. Warden Service, Waterton Lakes National Park, *Bear Management Plan (Waterton Lakes National Park, 1984), 7, 20, WNSC.*

53. Charles B. Sigler, Chief Park Ranger, to Mr. and Mrs. William M. Stevens, January 26, 1977, GNPA, Administrative Files, 1910–1984, Box 260, Folder 3.

54. Cliff Martinka, as cited in Tristan Scott, "Glacier Park Has History of Scientific Discovery," *Missoulian,* October 31, 2010.

55. Adolph Murie, *The Grizzlies of Mount McKinley* (Seattle: University of Washington Press, 1985, 2000). On the Craigsheads, see R. Gerald Wright, *Wildlife Research and Management in National Parks* (Chicago: University of Illinois Press, 1992), 112; Wondrak Biel, *Do (Not) Feed the Bears,* 86–112; Etienne Benson, *Wired Wilderness: Technologies of Tracking and the Making of Modern Wildlife* (Baltimore: Johns Hopkins University Press, 2010). People had been debating the effectiveness of existing national parks for protecting grizzlies for decades. See, for example, Aldo Leopold, *A Sand County Almanac, and Sketches Here and There* (Oxford: Oxford University Press, 1949, 1989), 198; Aldo Leopold to Charles Kraebel (Glacier Superintendent), January 18, 1927, NARA, Record Group 79, Entry 10: Central Classified Files, 1907–49, Box 253, Folder: "[Glacier National Park], Predatory Animals, 1918–32"; Sellars, *Preserving Nature,* 92–93; A. E. Demaray, NPS Acting Director, to the Superintendent of Glacier National Park, September 24, 1936, NARA, Record Group 79, Entry 10: Central Classified Files, 1907–49, Box 968, Folder: "[Glacier National Park] Bears, 1930–49."

56. Ted Catton, personal communication, November 2011. For a study demonstrating the varied nature of another international border, see Geraldo Cadava, "Borderlands of Modernity and Abandonment: The Lines within Ambos Nogales and the Tohono O'odham Nation," *Journal of American History* 98, no. 2 (2011): 363.

57. C. J. Martinka and K. L. McArthur, in U.S. Bureau of Land Management, *Bears: Their Biology and Management: Papers of the Fourth International Conference on Bear Research and Management Held at Kalispell, MT, USA, February 1977* (Bear Biology Association, 1980) n.p., in preface (about meetings); Catton, "Protecting the Crown," 130–36, 189–94 (the Endangered Species Act and its effects); Ted Catton, personal communication, fall 2011.

58. Northern Continental Divide Grizzly Bear Managers Subcommittee, "Preliminary Working Copy of the Conservation Strategy for the Grizzly Bear in the Northern Continental Divide Ecosystem, December 1995," GNPA, Central Files Fiscal Year 1996.

59. Rob Watt, personal communication, July 18, 2011. See also Waterton Lakes National Park, "Superintendent's Annual Report 1975–76," 43, WNSC; Warden Service, Waterton Lakes National Park, *Bear Management Plan,* 11, WNSC; William J. Briggle to Thomas J. Ross, April 20, 1970, in WLNP, Box 226, Bear management files 1943–1969, Bears 1951–1970, File 212 v. 1.

60. Herrero, *Bear Attacks,* 257.

61. Catton, "Protecting the Crown," 191–92; Rob Watt (former warden), personal communication, July 18, 2011; Scott, "Glacier Park Has History of Scientific

Discovery"; John C. Woods et al., "Genetic Tagging of Free-Ranging Black and Brown Bears," *Wildlife Society Bulletin* 27, no. 3 (2009): 616–27 (Canadian study); Katherine C. Kendall et al., "Grizzly Bear Density in Glacier National Park, Montana," *Journal of Wildlife Management* 72, no. 8 (November 2008): 1693–1705. For summaries of Kendall's projects, see http://www.nrmsc.usgs.gov/rescarch/glac_beardnagb.htm and http://nrmsc.usgs.gov/research/NCDE-beardna.htm (accessed August 4, 2015).

62. Dennis Madsen, personal communication, summer 2011. Retired warden Rob Watt (personal communication, July 18, 2011) confirmed that there have long been significant alliances with Glacier, at least since he arrived in 1978. Another recent Waterton employee explained that Waterton and Glacier staff have frequent contact with their counterparts by email and telephone, and that many individuals also cross the border to collaborate and participate in joint events.

63. See, for example, William M. Adams, *Against Extinction: The Story of Conservation* (London: Earthscan, 2004), especially chap. 1 about the transnational management of gorillas; Jane Carruthers, this volume and "Conservation and Wildlife Management in South African National Parks, 1930s–1960s," *Journal of the History of Biology* 41, no. 2 (2008): 203–26; Karen R. Jones, *Wolf Mountains: A History of Wolves along the Great Divide* (Calgary: University of Calgary Press, 2002). For a case study of an Alberta provincial park that adds dead animals (dinosaur fossils) to the mix, see Sterling Evans, "Badlands and Bones: Towards a Conservation and Social History of Dinosaur Provincial Park, Alberta," in *Place and Replace: Essays on Western Canada,* ed. Adele Perry, Esyllt W. Jones, and Leah Morton (Winnipeg: University of Manitoba Press, 2013), 250–70.

64. The U.S. NPS currently administers more than twenty different types of sites, ranging from national battlefields to national seashores to the National Mall in Washington, D.C. For a list see "Find a Park," http://www.nps.gov/findapark/index.htm (accessed July 24, 2015). As of 2015, Parks Canada's main holdings consist of 167 national historic sites, 44 national parks, and 4 national marine conservation areas. For more information, see http://www.pc.gc.ca/eng/index.aspx (accessed July 24, 2015). On moose, see "Moose Hunting Starts in National Parks," http://www.cbc.ca/news/canada/newfoundland-labrador/story/2011/10/12/nl-park-hunting-112.html (accessed May 1, 2014). For the story of the establishment of Canada's first national park created out of a modern land-claim agreement, see Brad Martin, "Negotiating a Partnership of Interests: Inuvialuit Land Claims and the Establishment of Northern Yukon (Ivvavik) National Park," in Campbell, *Century of Parks Canada.*

65. Mark David Spence, "Crown of the Continent, Backbone of the World: The American Wilderness Ideal and Blackfeet Exclusion from Glacier National Park," *Environmental History* 1 no. 3 (1996): 29–49.

66. Pekka Hämäläinen and Samuel Truett, "On Borderlands," *Journal of American History* 98, no. 2 (2011): 349.

| # From Randomness to Planning
The 1979 Plan for Brazilian National Parks

JOSÉ DRUMMOND

THE GOAL OF THIS CHAPTER is to examine how Brazilian national park policy evolved from an enduring pattern of randomness to the status of a solidly planned effort. Randomness prevailed in protected area policies even before the creation of Brazil's first national park in 1937, something that could be expected of such a new policy area, but it remained the dominant feature all the way up to 1979. An encompassing plan for national parks and other types of protected areas was launched during this year. It was based on an original blend of ecological, biogeographical, and policy criteria that helped move Brazilian protected areas into a new and enduring era of systematic growth and diversification.

The analysis of the rationale, the origins, and the initial workings of the 1979 plan is the major aim of this chapter, through examination of the plan itself and the documents that were instrumental in its drafting. A secondary goal is to identify the degree to which the Brazilian park system was shaped by national concepts and policies or by foreign ones (particularly the U.S. concept of national parks), or by a combination of both. These two goals require the examination of protected area and park proposals and initiatives during a rather long period between the mid-nineteenth century and the mid-1980s. However, the major focus of the chapter falls on the 1979 plan.

Findings show that Brazilian parks were shaped by a mix of local and foreign influences, concepts, and policies. The U.S. concept of national parks was indeed present at the creation of the Brazilian park system in the 1930s. However, it was gradually diluted by the aforementioned randomness, in the forms of loose managerial standards, haphazard creation criteria, and the quest for different types of protected areas. The combination of these three traits shaped the Brazilian park system between its inception (1937) and 1979.

Starting in 1979, a concerted effort by the park service and Brazilian scientists steered park policy and the more encompassing protected area policy in the direction of a complex, diversified, and planned model quite different from that of the United States and probably unique in global terms. Innovative criteria—locally created and/or adapted from several sources—for plotting new parks generated a more systematic and eclectic policy approach that significantly overhauled the preexisting park system. Basic park management goals (protection, visitation, leisure, education) remained in line with those of the U.S. model. Overall, however, after 1979 Brazilian park policy gained an increasingly original or national content, but aesthetic and scientific values that are not necessarily Brazilian maintained a degree of influence over park policy and the park system.

The matter of national or foreign influences (or a combination of both) is relevant for several reasons. Brazil has one of the world's most extensive park and protected area systems. Being the largest tropical country on the planet and having a mostly humid climate, Brazil hosts what is probably the world's richest terrestrial biodiversity under the sovereignty of a single country. Additionally, Brazil has vast amounts of some of the most widely used natural resources, such as wood, fresh water, arable soils, ores, oil, natural gas, and so on. Brazilians used these resources intensively to help the national economy grow faster than that of any other country during the twentieth century. As results of the oldest environmental policy applied in Brazil, parks and other protected areas are obviously a matter of national and international relevance. Brazil's vast natural endowment and the competence of a new generation of park planners made its park policy stray considerably from the U.S. model or that of other foreign countries. Therefore, the analysis herein does not approach the matter with a rigid dichotomy between national and nonnational, but rather considers a continuum of conceptual origins and a specific effort on systematic planning, with special attention given to the 1979 plan.

The starting point of the period under analysis (1862) was chosen because it marks the inception of a major conservation effort undertaken in the city of Rio de Janeiro, which a century later led to the creation of a national park. The closing point (1979) corresponds to the drafting and initial execution of an encompassing planning document—a master plan—that deeply changed park policy and the profile of park system.

The chapter begins by summarizing the current status of the Brazilian park system. Second, in order to highlight the matter of randomness, it examines briefly a series of episodes, concepts, and processes concerning Brazilian protected areas in general and national parks in particular, from 1862 to the late 1960s. Third, the text discusses new concepts introduced in Brazilian park planning documents dating from the late 1960s and mid-1970s. This leads to a fourth section that examines the 1979 planning document, which launched a new era for park policy in Brazil. A fifth section presents final remarks. Major sources include policy and legal documents, inventory data on national parks, and technical and analytical texts.

National Parks in Brazil Today

Brazil has a world-class status in the matter of national parks and other types of protected areas. It created its first national park in 1937. As of mid-2011 it had sixty-six parks, a figure that excludes two cancelled ones and another park converted into a national monument. Parks expanded remarkably since 1937— but so did several other types of federal conservation units (such as national forests, biological reserves, and extractive reserves).[1] The Brazilian Ministry of the Environment reports that currently Brazil has the fourth largest expanse of protected areas in the world, behind the United States, Russia, and China.[2]

As of mid-2011, 9.02 percent (76,848,771.30 ha) of the Brazilian national territory was covered by 305 federal protected areas, belonging to twelve distinct types. The sixty-six national parks covered about 2.9 percent of Brazil's territory (24,658,349.29 ha), or 32.08 percent of the area affected by all types of federal conservation units. National parks led in terms of the share of the total area of these units, followed by national forests (24.99 percent) and extractive reserves (15.97 percent). Parks are also the most numerous type of federal area, compared to sixty-five national forests and lesser numbers of other types of units. Additionally, national parks are the most widely distributed type of protected area—they are plotted in twenty-three of the twenty-six Brazilian states and in the Federal District, in all five major geographical regions, and in all six terrestrial biomes (there are also parks in the marine biome). Besides federal protected areas, 615 state protected areas add 75,540,950.48 ha (8.87 percent of the Brazilian territory) to the country's protected area system. Of the 615, 195 (31.70 percent) are state parks, covering of 9,063,804.27 ha (11.99 percent of the combined area of state protected areas).[3] These figures show that national parks are the "jewels in the crown" of Brazil's extensive and highly diversified federal protected area system.

From 1862 to the late 1960s

This section presents brief sketches concerning selected concepts, policies, and decisions related to Brazilian protected areas, between 1862 and the late 1960s.[4] The purpose is to illustrate two points: (i) the aforementioned randomness of events related to Brazilian protected areas—national parks in particular—and (ii) the long "prehistory" (prior to 1937) and equally long "early history" (1937–79) of protected areas during which that randomness prevailed. This background will provide a sense of how innovative the 1979 plan was.

The earliest fact to be considered is the replanting of the Tijuca Forest, in Rio de Janeiro, between 1862 and 1889. The country's first recorded large conservation effort reclaimed a watershed in the nation's capital, prompted by repeated water shortages in Rio. The area, after decades of intensive deforestation and coffee cultivation, was reforested mostly with native species (maybe a purely technical decision, maybe an intentional "nationalistic" statement). It recovered its water-producing capabilities, which served the city well for almost an additional century. Over the years, the reclaimed area was used also for leisure (hiking, horseback riding, mountain climbing, picnics), landscape painting, and so on. In 1961 it became a national park (Tijuca). Manuel Gomes Archer (1821–1905), in charge of the effort, knew about German reforestation and forest management efforts. Emperor Pedro II, Archer's immediate boss, proudly invited foreigners to visit the replanted forest. Nonetheless, this was a reclamation project imposed by resource scarcity, not a nature preservation project. Also, the reclaimed area was but a tiny fraction of deforested areas around the city, not to mention the country.[5]

The first recorded reference to national parks made by a Brazilian dates from 1876. The military engineer André Pinto Rebouças (1838–98) proposed parks for two remote sites: Bananal Island (the world's second-largest fluvial island, on the Araguaia River) and the Sete Quedas rapids (on the Paraná River). He cited Yellowstone National Park as the inspiration for his proposal and included the expression "national park" in the title of his text, an obscure, self-financed essay, with no recorded impact on future park policies and concepts. Rebouças hoped that parks would generate income by attracting tourists, an early argument made also in favor of U.S. parks. As a well-traveled black engineer and journalist, however, Rebouças is remembered for his feats in road, waterworks, and port construction, but even more for his inflamed antislavery activism. His "correct" premonitions—many years later parks were indeed created in those two locations—did not make him an engaged conservationist. He has gained limited fame as a conservation pioneer, but after 1876 he probably knew and cared little about the particulars of Yellowstone and other protected areas of the world.[6]

A failed conservation episode in Brazil probably indicates the role of nationalistic motivation in park prehistory. In 1891 Federal Decree 8,843 created a huge "forest reserve" (2.8 million hectares) in lands located in the future Amazonian state of Acre, but it was never implemented. Its precise location remains uncertain until this day, as most of Acre at that time was actually part of Bolivian territory. In the 1890s Brazil was initiating a diplomatic effort to consolidate its international borders in the Amazon basin, and most likely this elusive reserve was an overt nationalistic statement made to serve this effort. Not coincidentally, what is now the state of Acre was taken by Brazil from Bolivia in 1904, practically at gunpoint.

Two examples of foreign initiative in the creation of protected areas—but not national parks—illustrate another force acting in favor of protected areas—non-Brazilian natural scientists. Two foreign scientists, residing temporarily in Brazil, independently created two "biological stations" (the similar name was probably coincidental). In 1914 the Swedish botanist Johan Albert Constantin Löefgren (1854–1918), active in scientific expeditions and institutions in Brazil, helped create the Biological Station of Itatiaia, precisely where Brazil's first national park (Itatiaia) was established in 1937. The area had been regularly crisscrossed by expeditions made by Löefgren and Brazilian scientist colleagues from Rio's botanical garden. The German scientist Hermann Friedrich Albrecht von Ihering (1850–1930) used his own money to buy the area of the Alto da Serra Biological Station, in the state of São Paulo. In 1909 he donated it to his employer, the Museu Paulista, a state natural history museum. This area was also regularly explored by scientific expeditions.[7] Both units supported continuing field research. That these initiatives were taken by foreigners is somewhat diluted by the fact that both scientists worked in Brazil for long periods, employed by Brazilian institutions.[8]

The 1934 Forest Code was Brazil's first encompassing regulation on forests and parks. Still another force was emerging to shape protected areas policy: a nationalistic, interventionist central government seeking to control the use of the nation's natural resources. The code was one among several items of a "surge" of conservation laws issued in the mid-1930s—for water, mining, hunting, fishing, food safety, irrigation, and so on.[9] Article 1 stated flatly that all Brazilian "forests" were considered to be of the "common interest" of all Brazilians. Since 1934, therefore, no forest in Brazil has been fully considered to be legally sanctioned private property. Given the historically weak control over public and private lands, however, this concept was hard to enforce, although not without impact. This code was at heart a forestry regulation, but it also contained the first legal reference to national parks in Brazil. This legal initiative for the conservation of the native flora was a new variable. Article 9 defined national parks plainly as "natural monuments" worthy of "permanent protection," in the form of "remnant" forests, distinct from "productive," "model," and "protective" forests.

Articles 1 and 9 had lasting influence over forestry and park policies. Although this classification seems to be uniquely Brazilian, the commission that drafted the decree was well informed about European and North American forestry laws and practices. Productive forests, for example, were those located on private lands. They were open to exploitation, but only by means of specific permits. Additionally, productive forests could be created on public lands. These "national forests" became the numerous and expansive conservation units of the present, although a bidding system for exploiting them has yet to be established. Model forests were those replanted for commercial purposes, quite rare at the time in Brazil but fairly extensive today.

Little happened in terms of conservation laws or measures after the 1934–37 surge. When the 1934 code was reformed, in 1965,[10] several important changes occurred. Now, all forms of native vegetation—and not only "forests"—were considered to be of the common interest of Brazilians. "Protective forests" evolved into "permanently protected areas" (APPs—steep slopes, hilltops, riverbanks and associated wetlands, watersheds, mesa edges, etc.) and "legal reserves" (RLs, mandatory reserves of native vegetation). Both are defined as existing in every private and public rural property in Brazil and must be spared from clear-cutting. This is relevant because all landowners became legally required to spare considerable parts of the native vegetation on their properties. Although these provisions of the new code are highly unpopular among farmers, their motivation is less the protection of nature and more the improvement of agricultural practices by means of conserving resources at the property level. The relevance of mentioning the opposition to APPs and RLs is that it overspills to national parks and other protected areas, even though the number of landowners directly affected by parks is minimal in comparison with those affected by APPs and RLs.[11]

Article 5 of the 1965 code renewed the concept of national parks, a reemergence of the 1934 code's concern with nature conservation. They were redefined (jointly with biological reserves, a new type of protected area) as areas designed to "protect exceptional natural attributes, reconciling the full protection of flora, fauna and natural beauties with use for educational, recreational and scientific purposes." This definition combines "exceptionalism," protection of natural features, public use and scientific relevance, coming closer to the U.S. model of national parks than the more succinct definition of the 1934 code.

A lasting consequence of the 1934 and 1965 forest codes is that, distinct from the United States and several other countries, Brazilian national parks and all other conservation units are created by executive decrees. Legislative initiative or approval is not required. Thus, conservation units have had weak support in Congress, but executive privilege also "buffers" them from changes initiated by Congress. Indeed, there is a continuous stream of bills presented by congressmen proposing to reduce or even cancel national parks. These bills usually die in committees, but they express tangible attritions between parks and local interests.

Another durable consequence of the 1934 and 1965 codes is that parks have been under the responsibility of the Ministry of Agriculture, together with national forests. This differs sharply from the U.S. solution of dividing parks and forests between two departments (Interior and Agriculture). This changed only in 2007, when Brazil finally created a park agency and placed it under the Ministry of the Environment.[12] However, in contrast with the United States, there has not been a historically sharp divide in Brazil between conservation and preservation in the minds of most environmentally concerned Brazilians. The terms were and are used almost always interchangeably and indicate both the rational use of resources and the preservation of resources for scientific and esthetic purposes.[13]

The wording used to define national parks in the 1934 and 1965 forest codes suggests the topic of "monumentalism" (or "exceptionalism") as a criterion for selecting areas for Brazilian national parks. "Monumentalism," not necessarily a foreign concept, was indeed a driver of Brazilian national park policy until 1979, and it lingers today, although weakened. Examples start with the first four Brazilian parks: Itatiaia (1937) was created around a peak believed at the time to be the highest in Brazil; Serra dos Órgãos (1939) contains a set of prominent peaks visible from downtown Rio de Janeiro; Iguaçu (1939) showcases the impressive waterfalls of the Iguaçu River; and the now defunct Paulo Afonso Park (1948) was also created around a major waterfall on the São Francisco River.

There are many other examples of the influence of "monumentalism" in the Brazilian park system. Ubajara (1959) is a rare, moist, densely forested mountaintop in the mostly semi-arid state of Ceará. Aparados da Serra (1959) protects a section of Brazil's deepest canyon (called Itaimbezinho). Araguaia (1959) was plotted on the world's largest fluvial island. Caparaó (1961) protects another peak once deemed to be Brazil's highest. Sete Cidades (1961) protects unique erosional rock formations. São Joaquim (1961) was selected in part for being the only place in Brazil with regular annual snowfalls. Sete Quedas (1961) protected beautiful rapids of the Paraná River, before the lake formed behind the gigantic Itaipu hydroelectric dam engulfed the rapids, leading to the closing of the park. Tijuca (1961) hosts the famous statue of Christ the Redeemer, from which visitors have spectacular views of the city of Rio de Janeiro. Serra da Canastra (1972) protects the headwaters and a major waterfall of the mighty São Francisco River. Even Pico da Neblina (1979), created when the "monumentalist" criterion became secondary, hosts what is now considered Brazil's highest peak.[14]

In 1947 "exceptionalism" was strong in the mind of Wanderbilt Duarte de Barros, the first director of Brazil's first national park and for many years a ranking staff member of the national park agency. In the first book ever written about Brazilian national parks, he suggested fourteen sites for new national parks, almost all of them on account of their "scenic value." He considered "exceptionalism" to be the most important criterion for locating parks.[15]

Paulo Afonso (1948) and Sete Quedas (1961) are prime examples of "monumentalism." They illustrate also a different, unfortunate phenomenon: both parks were terminated. National parks are universally defined as permanent units, in Brazil and abroad, but Brazil violated this precept. Hydroelectric dam lakes victimized both parks, created for the purpose of providing vistas of exceptional waterfalls and rapids.[16]

One trait of early park creation is particularly revealing of the randomness under examination. Many parks were created in areas previously subjected to several types of public management or control. As mentioned, scientists from Rio's botanical garden and national museum regularly collected specimens in sections of the future Itatiaia National Park. Before 1939 the area of Iguaçú National Park was set aside by the Paraná state government as an ill-defined reserve; it was later donated to the federal government. Ubajara National Park was originally one of the Department of Agriculture's many "experimental forestry stations."

Aparados da Serra National Park was partially acquired by the state government of Rio Grande do Sul in 1957 and donated to the federal government. In 1952 a federal "protective forest" was created in its vicinity and later added to the park. The area of the Araguaia National Park was donated to the federal government by the state government of Goiás in 1958. Monte Pascoal National Park was originally a state park, created in 1943 and later donated to the federal government. Tijuca National Park was created by joining nine federal "protective" forests (created in the 1930s) with the replanted Tijuca Forest. Caparaó National Park was originally a Minas Gerais state forest preserve, created in 1948. Serra do Cipó National Park was originally a Minas Gerais state park, created in 1978. São Joaquim was nominated for park status in 1958 by the Department of Agriculture's Forest Service (the agency in charge of forest management and logging), as representative of forest formations dominated by *Araucaria angustifolia*, one of Brazil's two only native conifers, intensively logged at that time. Brasília National Park, also nominated by the Forest Service, was created to protect the watershed that supplied the new national capital.[17] It is thus obvious that many parks owe their location and status to a number of disconnected federal or state research and/or conservation initiatives. This strong and durable trend made the park system evolve in different and simultaneous directions and shows that the system was built on the basis of haphazard, dispersed initiatives.

This leads to the distinct but related topic of the geographical/ecological distribution of Brazil's national parks. By the late 1970s, after forty years of national park policy, the park system had a markedly coastal, urban/industrial profile. Most parks were plotted in intensively farmed areas, not far from major cities. This was in sharp contrast with the park systems of Canada, the United States, Chile, and Argentina, among other countries, which planted parks first in remote regions (in Mexico and Australia, however, many parks were plotted close to cities). This distribution did not result from a deliberate Brazilian project or concept but from three loosely connected trends, two of them already mentioned: (i) monumentalism, (ii) the random inclusion of lands subject to previous research and conservation efforts, and (iii) a bias in favor of coastal forested areas of the Brazilian Atlantic Forest biome, in which most Brazilians still live. This profile had a "demographic" logic, but it also demonstrates that a shortsighted park policy ignored for many decades the country's vast interior and its biological, hydrographic, and geological features.

Indeed, nine of the eighteen parks in existence in 1978, just before the 1979 plan was drafted, were located in the Atlantic Forest biome. They were plotted mostly in settled areas, coastal or quasi-coastal, near the country's largest cities; they were easily accessible from these cities; most had "monumental" features as their strongest appeal; and some supported research projects conducted by scientists employed in those large cities. Not incidentally, several of the eight biological reserves extant in 1978 had similar features and origins.

A few parks plotted in the interior were not necessarily exceptions to this coastal bias. Emas, Chapada dos Veadeiros, Brasília, and Araguaia, although located deep in the country's interior, were created as potential leisure areas for the numerous government workers scheduled to be transferred from coastal Rio de Janeiro to the new national capital, Brasília, inaugurated in 1960. Actually, Brasília National Park lies within the urban area of Brasília. Tapajós National Park, the first Brazilian park in the Amazon region, was created in 1974 as a side effect of aggressive frontier occupation policies; not incidentally, it was designed to be cut by the famous Trans-Amazonian Highway. None of these "backland" parks resulted from an autonomous decision to change the system's dominant profile: coastal, forested, and plotted in settled areas. This profile translated into an equally biased distribution of parks among Brazilian biomes or ecoregions. The 1979 reform plan radically changed this profile and the poor representation of several biomes and ecoregions in the park system.

One last matter concerning the randomness of the growth of Brazil's park system remains to be discussed: the proliferation of other types of protected areas. Although this is not necessarily unique to Brazil, it is difficult to write about Brazilian national parks without dealing simultaneously with other types of protected areas, some actually created, others only proposed. Other types of protected areas affected the identity and public visibility of national parks and sometimes helped scatter protection efforts. Also, some of these other units embody foreign influences of interest to this text. Some of these different types of protected areas were proposed before 1979, while others appeared after the 1979 plan to be examined later.

By the mid-1980s Brazil had no fewer than twelve types of protected areas, defined by different regulations.[18] National parks and national forests were the oldest, defined by the 1934 and 1965 forest codes. Parks were supposed to promote protection, education, recreation, and scientific research in areas with exceptional natural features. National forests were slated for multiple use— wood production, watershed management, scientific research, and recreation. Biological reserves, the third type, based on Law 5,197 of January 3, 1967, were created only after 1974. They aimed to protect faunal species and their habitats. Hunting parks, the fourth type, based on the same law, were for sports hunting. Sources indicate that they were created in a single state, Rio Grande do Sul. In contrast with first three types, the few hunting parks that were created died out and were eventually cancelled, for reasons that remain undetermined. The Instituto Brasileiro de Desenvolvimento Florestal (IBDF), a federal forestry and conservation agency created in 1967, under the Department of Agriculture, managed these four types of protected areas.

Ecological stations, a fifth type, based on Law 6,902 of April 27, 1981, were created only in 1981 but had been proposed and widely discussed since the mid-1970s. They combine preservation of intact and ecologically representative areas with the scientific monitoring of intentionally induced changes in small portions of them.[19] Another agency managed them—the Secretaria Especial do

Meio Ambiente, a federal regulatory office that competed with IBDF in conservation efforts. Law 6,938 of August 31, 1981, allowed SEMA to create types six, seven and eight: (i) ecological preserves (similar to biological preserves), (ii) environmental protection areas (integrated planning units based on the zoning of productive activities), and (iii) areas of special ecological interest (which could be speedily established to allow for the subsequent creation of permanent conservation units). There were also two types (nine and ten) of tourism-oriented protected areas (managed by a federal tourism company), besides ill-defined protective forests (eleven) and forest reserves (twelve), under the authority of the Department of Agriculture.

Two more types appeared in the late 1980s and 1990s. Extractive reserves, the thirteenth type, appeared only in 1987. Initially they were a type of land reform settlement, based on Ruling 627 of Brazil's land reform agency (INCRA), an outgrowth of the activism of rubber tappers led by Francisco "Chico" Mendes. In 1990, Decree 98,897 officially recognized them as protected areas.[20] After several inconclusive initiatives in the 1970s and 1980s by the IBDF and its successor agency, the Instituto Brasileiro do Meio Ambiente e dos Recursos Naturais (IBAMA), Decree 1,992 (1996) instituted a fourteenth type of protected area—private natural patrimony reserves.[21] They are created on preserved private lands by the initiative of their owners, who are entitled to mild tax breaks and certified by IBAMA.

This remarkably varied list of protected areas includes only the types for which units were actually created—about a dozen other types of areas found in several databases were excluded here. This reflected interagency rivalry and different interpretations of conservation and preservation concepts, but also different foreign influences. National parks were loosely based on the U.S. concept, but national forests had both German and U.S. inspiration. Biological preserves (and ecological reserves) were inspired mostly by U.S. wildlife preserves or refuges. Hunting parks were modeled on those of several Western European countries. Ecological stations were inspired by field research stations of unspecified U.S. universities visited in the late 1970s by Paulo Nogueira Neto, longtime SEMA director, and his staff. Environmental protection areas were inspired by French natural parks and Italian national parks, both of which include all sorts of productive activities, rural and urban, besides towns, roads, dams, and waterworks. Extractive reserves were in part adapted from Mexican *ejidos* and other types of communal land holdings in South and Central America.[22]

Of course, the sheer number of types of protected areas hindered coherent planning. The listing reveals also that foreign conservationist influences went beyond national parks; they spilled over into other types of protected areas. One cause for proliferation of protected areas was the difficulty of creating national parks. Most of these other types have management goals that are more flexible than those of national parks, and several of them do not demand public ownership of the lands. Parks lacked both funding (for acquiring private lands and/or indemnifying investments made in them) and popular support, besides having a limited constituency of visitors and friends; many parks lacked basic implementation measures.

The 1979 Plan

In the mid-1970s, after almost forty years of national park policy, Brazil had created only eighteen parks—small, underfunded, understaffed, unevenly distributed among regions and biomes, plotted mostly in settled areas, and scantily visited. This was mainly because they were created in a haphazard manner.[23] It was a meager heritage.

As early as 1969, the small staff of four assigned to the young Department of National Parks and Equivalent Reserves, a powerless third-echelon IBDF subunit, was well aware of this meager legacy and of all the disparate forces that had shaped the park system over several decades. Its leader was the young IBDF staff member Maria Teresa Jorge Pádua, whose major collaborator was the seasoned Alceo Magnanini, a staff member of the Department of Agriculture.[24] They led the successful effort to turn national parks into the central components of an expanded, planned, and renovated "conservation unit" system.

This effort had the outside help of scientific teams linked to a major environmental NGO, the Fundação Brasileira de Conservação da Natureza (FBCN), founded in 1958, of which Magnanini had been an active member since the mid-1960s. Pádua later joined the organization.[25] This "park agency + local environmental NGO" format, crucial for park policy reform, was new in Brazil, although it probably appeared in other countries at the same time. The Brazilian NGO joined the International Union for the Conservation of Nature (IUCN) in 1962. Its leaders were active in IUCN's technical committees and international board and, after 1972, in UNESCO's "Man and the Biosphere" program. FBCN members attended the first three international meetings on national parks—Seattle (1962), Yellowstone (1972), and Bali (1982)—and park management courses offered by the U.S. National Park Service; they also engaged in many exchanges with environmental NGOs in several countries.[26] Thus, the FBCN ushered several international influences into IBDF's initially weak and unworldly park policy/experience.

Pádua recognizes that the national park idea has a non-Brazilian origin. However, she maintains that in the 1930s "international rivalry" spurred Brazilian conservation policies more than the pursuit of local or foreign models. Neighboring Argentina had large areas designated as national parks. Both Pádua and the North American environmental historian Warren Dean mention that the Argentinian Parque Nacional Iguazu (created in 1934) motivated the creation of Brazil's neighboring Iguaçu National Park in 1939. In 1960 Jair Tovar, a park service employee, recorded that Brazilian parks lagged behind Argentina's—although three times smaller than Brazil, Argentina had six times more park area. He chauvinistically stated that Brazil was "making a comeback" and could "take the lead" soon, thanks to its extensive territory and geographical diversity.[27] By the late 1960s there was a perception inside and outside IBDF that the Brazilian park system was weak, but that it could grow in quantity and quality if adequately planned.

Magnanini and Pádua built on this feeling. In 1969 they initiated a planning "wave" that yielded several documents and activities that opened the path to the 1979 plan.[28] Magnanini's unassuming 1970 document (drafted and published for internal discussion only) spelled out for the first time the single most important principle of the future 1979 plan: for him, a sound Brazilian park system must include sections that were representative of the entire variety of Brazil's rich vegetational regions and formations. This was not unheard of as a directive for park systems in other countries. Actually, in September 1968 a group of 300 scientists from sixty countries met in Paris for a conference (the Biosphere Conference), called by UNESCO (with the support of the United Nations, IUCN, World Health Organization (WHO), and the Food and Agriculture Organization of the United Nations (FAO) to design and launch the Man and the Biosphere Program (MAB). The basic concept of this program was the establishment of protected areas in each "ecotype" of each country. Ecotype was the name given to every distinctive floral-ecological formation within a territory and even within the same biome in the same territory. Targeting varied ecotypes guaranteed that protected area systems would encompass the richest sampling possible of natural features and diversity. This concept gained a relatively large following over the succeeding years, inside and outside of the MAB program.[29]

Magnanini, a soils engineer and self-taught biogeographer, was most probably aware of this meeting and its recommendations. He soon became a maverick defender of the concept of ecotype protection inside the Ministry of Agriculture and its forestry agency, the IBDF. For him, the concept was more than a theoretical directive recommended by an international meeting or entity, because he based his view on extensive travels and field research all over Brazil. At that moment he was probably the Brazilian with the most extensive firsthand knowledge of the country's ecotypes. In his 1970 document he gave the names and locations of all ecotypes that should be protected in the entire Brazilian territory.

In 1975 Pádua and Magnanini set up a long-term working agreement between IBDF and FBCN. A joint task force of IBDF personnel and twenty-five scientists rounded up by the FBCN worked continuously to produce the 1979 plan, besides preparatory and follow-up documents. The first task force publication came out in 1975.[30] Although preliminary, it was ambitious, realistic, highly technical, and well informed about park problems all over the world. It pointed out Brazil's very low ranking in the international park scene—only .28 percent of its territory was protected by parks and reserves (compared to 15.95 and 7.30 percent in the United States and Japan, respectively). It reviewed all major managerial problems and listed/prioritized measures to address them. One proposal was the drafting of management plans for each unit, reportedly mirrored on U.S. parks.

A second task force document, issued in 1978, sharpened the list of problems and priorities, such as personnel shortage, training, private land acquisition, and the financial shortcomings of Brazil's three most visited parks. It should be added that the preparation of the 1979 plan was not a strictly national effort, as the all-Brazilian task force received support from foreign organizations, such as the IUCN, World Wildlife Fund, and New York Zoological Society.[31]

In early 1979 the plan was published and put into execution. It turned out to be an unprecedented blueprint for park and reserve classification, creation, management, research, and consolidation. The basic concept of national parks was subsumed in a vast set of conceptual and policy variables and principles, including proposals for other types of protected areas, all this with a distinctive Brazilian blend.

The 1979 plan deserves an entire article by itself, as the major turning point in Brazil's park policy.[32] The two points that matter most for the purposes of this chapter are its break with randomness and the originality of its content. The plan resulted from a long-term Brazilian planning effort, although it was well informed about foreign and international issues of park policy and science and was partly supported by non-Brazilian funds. The end result, in terms of concepts, design, and sources, was both eclectic and original in global terms.[33]

First, it is important to reemphasize that the 1979 plan resulted from five years of formal collaboration between a government agency (IBDF) and an environmental NGO (FBCN). This original arrangement is explained by the small number of IBDF personnel deployed in conservation and by the FBCN's superior expertise in the matter. Related to this, FBCN's participation brought in the expertise of university and research institute scientists, who until then had been practically aloof to park policy. Knowledge acquired by university-based FBCN members in their research efforts and in the aforementioned international forums was thus channeled directly into park policy.

The second point is conceptual. The plan's first paragraph shows how even IBDF personnel deeply involved with parks were amenable to conservation—as distinct from preservation—principles. Written by Pádua, this passage could have come directly from the pen of Gifford Pinchot: "the basic concept of conservation is translated by the rational use of natural resources, aiming at the continuous production of the renewable ones—air, water, soil, flora, and fauna—and at the maximum yield of nonrenewable ones." The next phrase states that "nature conservation . . . thus involves both the use . . . and the preservation of nature, in cases when it is best not to use resources directly but to obtain indirect benefits from them."[34] The document thus claimed full compatibility between use and preservation, something not to be found, for example, in the U.S. experience of placing resource use in one agency/policy area and preservation in another.

The plan's methodology, the heart of its originality and effectiveness, is a third point to be highlighted. It followed five basic steps: (1) select areas with potential conservation or preservation value; (2) point out "gaps" to be filled in the park system; (3) use technical and scientific criteria for selecting areas for new units and new types of units; (4) revise management goals; and (5) propose priority locations for plotting new protected areas.[35] The "secret," however, was the marshaling of abundant information, organized in thematic maps. Data were plotted on twenty-two photo-laminated thematic maps drawn in the same scale. Successive map overlaps led to the identification of locations suited for protected areas (on account of the desired traits) and to the exclusion of unsuitable locations (with unfavorable traits). Proposed parks and preserves were

TABLE 1   National-scale maps used in the drafting of the 1979 IBDF-FBCN park policy plan

CONTENTS	AUTHOR(S)	NATIONALITY
Phytogeographical regions	C. T. Rizzini	Brazilian
Biogeographical provinces	Miklos Udvardy	Hungarian
Morphoclimatic and phytogeographical domains	Aziz Ab'Saber	Brazilian
Climatic zones	Edmon Nimer	Brazilian
Roads	Brazilian Department of Transportation	Brazilian
Geopolitical divisions	Instituto Brasileiro de Geografia e Estatística (IBGE)—Brazilian census agency	Brazilian
River basins	IBGE	Brazilian
Existing and planned protected areas	Instituto Brasileiro de Desenvolvimento Florestal (IBDF)—Brazilian forestry agency	Brazilian
Geological formations	Brazilian Mining and Energy Department	Brazilian
Areas proposed for protection by the Projeto Radambrasil	Projeto Radambrasil	Brazilian

Source: Instituto Brasileiro de Desenvolvimento Florestal and Fundação Brasileira de Conservação da Natureza, *Plano do Sistema de Unidades de Conservação do Brasil* (Brasília: 1979) [mimeographed], 20–31.

TABLE 2 Regional-scale maps used in the drafting of the 1979 IBDF-FBCN park policy plan

CONTENTS	AUTHOR(S)	NATIONALITY
Indigenous lands in the Amazon region	Fundação Nacional do Índio (FUNAI), Brazil's indigenous peoples agency	Brazilian
"Development poles" of the Second National Development Plan	Brazilian Department of Planning	Brazilian
Limits of the "Legal Amazonia" region	Instituto Brasileiro de Geografia e Estatística (IBGE)—Brazilian census agency	Brazilian
Amazonian phytogeography	Ghillean Prance	English

Source: Instituto Brasileiro de Desenvolvimento Florestal and Fundação Brasileira de Conservação da Natureza, *Plano do Sistema de Unidades de Conservação do Brasil* (Brasília: 1979) [mimeographed], 20–31.

plotted between and away from areas that might spark resistance, but the goal was to include as many areas as possible with the desirable vegetational-ecological features. All this was done in a painstakingly manual manner, as the age of satellite imagery combined with image interpretation software, electronic map overlay and geographical information systems had not arrived yet. Pádua explains that the plan's final proposals met "no resistance" from state governments and federal agencies connected to road building, indigenous peoples' lands, mining, land-reform settlements, and so on.[36] Tables 1 and 2 summarize the national- and regional-scale thematic maps produced by the team that drafted the 1979 plan and the authors of the respective databases.

Eight additional regional-scale maps were used. Four of them recorded the location of Amazonian "Pleistocene refuges," identified in part by foreign scientists. The other four recorded (i) all official Brazilian urban areas (by the Brazilian census agency), (ii) all areas of historical interest (by the IPHAN, the Brazilian historical heritage agency), (iii) coastal ecosystems (by the Brazilian navy), and (iv) faunal concentrations (by the IBDF, the Brazilian forestry agency). The plan also used firsthand data collected at thirty-four locations during five field expeditions made in 1977–78 by four teams of task force members. These trips were supported by several Brazilian agencies and universities, and by WWF, IUCN, and the New York Zoological Society.[37]

With the exception of Udvardy and Prance (see tables 1 and 2) and some of the Pleistocene refuge researchers, all databases were produced by Brazilian scientists and technicians, working in Brazil. Radambrasil's maps, texts, and inventories alone provided a wealth of pertinent information. This was a pioneering resource inventory project conducted by the Brazilian government. Begun in the early 1970s using remote sensing, it employed the cutting-edge technology of side-looking aperture radar (SLAR) to record geology, geomorphology, topography, water bodies, soils, ores, vegetation, and so on. It covered the entire Brazilian territory and was the most comprehensive natural resource survey of its time. Its results continue to be widely used for many planning and study purposes besides parks. So successful was this inventory that all texts and images of the lengthy thirty-four printed volumes (published in the 1970s) generated by the project were digitized and posted on the Internet.[38]

Therefore, the task force operated and reached its proposals on the basis of an overwhelming amount of nationally generated information. Such a quantity and variety of socioeconomic and natural data had never been assembled in Brazil. Its availability was in itself a novelty; its application to park planning was unheard of. Nonetheless, the entire effort went back to Magnanini's original idea of having all vegetational formations represented in the Brazilian park system. Conceived in the 1960s by a self-taught biogeographer, informed about nonmandatory IUCN and MAB directives, Magnanini's ideas were well attuned with future definitions of and debates about biodiversity and bioregional management.

As a result, the 1979 plan made viable proposals for new protected areas. The plan allowed even preexisting protected areas to be better managed. The task force's suggestions were based also on an independent but related 1976 document coauthored by Pádua, partially funded by the United Nations Development Program, spelling out specific priorities for protected areas in the Brazilian Amazon region.[39] This UNDP funding can be considered an additional foreign input received by the 1979 plan.

The plan selected thirteen areas for immediate creation of new protected areas—five parks and eight biological preserves. Ten of them were in the Amazon region. About a dozen other areas were preselected but required further studies to confirm their viability as conservation unit sites.[40] Not all of these thirteen units were created immediately, but nonetheless the short-term results were remarkable. Between May 1979 and September 1984, nine new national parks pushed the total number of units from eighteen to twenty-seven and expanded park area 3.8 times, from 22,292 km² to 85,990 km². Eight new biological reserves (pushing the total to sixteen) expanded reserve area 53.3 times, from 372 km² to 19,844 km².[41]

The resulting system, besides growing rapidly and diversifying sharply in terms of protected vegetational forms and types of protected areas, gained legitimacy in the eyes of other sectors of government. It became much more representative of the ecological diversity of the Brazilian territory, particularly its Amazonian sections. The profile of the protected area system changed radically. Now there were larger numbers of units, their individual sizes were much larger, and they belonged to more diversified categories. Most new areas were plotted in unsettled or isolated areas, deep in the country's interior. Many native floral formations were included in the system for the first time, together with sections of all Brazilian biomes and major ecosystems.

A fourth relevant aspect of the 1979 plan is the priority given to the Brazilian Amazon region. In the late 1970s Brazil was under fire from a recently emerged international environmental public opinion, composed of scientists, environmentalists, and multilateral organizations. Criticism focused on the high rates of deforestation in the Amazon region. Could this qualify as foreign influence on the 1979 plan? In part, yes. The IBDF-FBCN task force was well aware of this criticism before the general Brazilian public and politicians, on account of its deepening engagement in conservation issues. Foreign criticism is not discussed explicitly in the 1979 plan, but it probably spurred the task force to give special attention to the Amazon region, for which it proposed numerous parks, biological reserves, and ecological stations. However, priority given to Amazonian lands was determined first by the task force's technical and "endogenous" criteria of including all types of Amazonian floral formations (recall that before 1979 there was only one Amazonian park), its focus on public lands located in unsettled areas, and its goal of protecting large continuous areas.[42]

A remarkable aspect of the 1979 plan is the decision to include "Pleistocene refuges" as a criterion for park location. This gave the plan a special "ecological" blend and reinforced the shift away from monumentalism as the basic criterion for creating parks. Refuges were plotted on maps based on recent research conducted by biologists not linked to the IBDF or FBCN: three of them are foreign and one is Brazilian, Paulo Vanzolini). These areas were identified as major forested holdouts during climatic fluctuations in which Amazonian rain forests cyclically shrank and expanded. These "forested islands" of high biological diversity survived adversities and recolonized the Amazon region to its present-day levels. Refuges were defined in relation to four different sets of species that saved themselves from extinction (birds, lizards, butterflies, and plant communities). They were deemed to host a large proportion of endemic species. Parks and reserves were therefore plotted in or around them in order to maximize biodiversity preservation.[43] This was the first time in the world that park areas were selected on the basis of this type of criteria and evidence. However, again this was compatible with the 1979 plan's major directive—Magnanini's concern with the representation of all vegetational forms. Other countries followed suit and included Pleistocene refuges in their protected areas, among them Peru.[44]

The 1982 stage of the plan consolidated proposals for twelve additional parks, sixteen biological preserves, and twenty-one ecological stations for the Amazon region, many of them created only in the 1990s and 2000s. In biological terms, the "ideal" for the Amazon region was stated to be the preservation of "three grand samples" of each of its phytogeographic regions, each with 5,000 km^2, plus twenty-four units for "unique microhabitats," each with 1,000 km^2.[45]

Brazilian national parks took thirty-seven years (from 1937 to 1974) to reach the Amazon region, but after 1979 they arrived on a massive scale and on the basis of sophisticated ecological criteria. Before 1979 only one national park existed there, but by 1989 six new parks (and several biological reserves and ecological stations) had been added. Today, twenty-one of the sixty-six (31.81 percent) national parks are located in the region and make up an overwhelming 87 percent of Brazil's total area protected by national parks. The emphasis on the region was in part sparked by international pressure, but the Brazilian response was part of an endogenous effort of overhauling its park policy and park system.

Three additional aspects of the 1979 plan deserve a few words. First, older parks were much smaller than younger ones. The average size of the eighteen national parks created before 1979 was 1,485 km^2. The next eleven parks, created under the influence of the 1979 plan, had an average size of 6,070 km^2. Today the average size for all Brazilian parks is 3,823 km^2; for all Amazonian parks, it is 10,767 km^2. These expanded figures reflect the focus on public lands located in unsettled sections of the country and the adoption of the precept of the nascent science of conservation biology in favor of large reserves capable of better preserving biological diversity. Conservation biology was not widely known in Brazil at the time of the 1979 plan, but Brazilian scientists were aware of some of its emerging principles and concepts.

Second, the 1979 plan suggested eight new types of protected areas, in addition to the fourteen types mentioned earlier. Although some of these eight types were never implemented, this part of the plan did add to the relative vagueness of the identity of protected areas. The eight suggested new types were natural monuments, wildlife refuges, faunal reserves, scenic rivers, road parks, resource reserves, natural parks, and cultural monuments. At least the first four were taken from U.S. models or proposals. Therefore, the quest for types of conservation units more flexible than national parks continued after the 1979 plan, but only the first three types listed above have been legally sanctioned since then.[46]

The third and last matter concerning the 1979 plan relates to management plans. Reportedly inspired by U.S. national parks, IBDF and FBCN, by 1982, had drafted and published twenty individual management plans for old and new parks and reserves. Researched and written by almost a hundred scientists and managers, they were a novelty for the management of Brazil's conservation units.

After 1986, the 1979 plan "ran out of steam," due mainly to Brazil's runaway inflation and budgetary chaos. Several parks, biological preserves, and ecological stations were created in the 1990s and 2000s in areas selected by the plan. However, the overall emphasis of park policy since the 1990s has shifted away from biodiversity protection to "socially oriented" protected areas, such as extractive reserves, not contemplated in the 1979 plan.

Conclusion

The overall finding is that the Brazilian national park system, which eventually ranked as one of the most outstanding internationally, was, until the late 1970s, built on a random mix of Brazilian and non-Brazilian initiatives and concepts. Findings indicate, however, that locally based randomness prevailed over any clear-cut foreign influences between 1937 and 1979.

Brazilians involved in park policy, especially since the early 1970s, were well informed about park matters on a global scale and open to nonlocal concepts, experiences, and solutions. The original U.S. concept of national parks and "orthodox" interpretations of it were considerably "diluted" in Brazil, as time went on, by the addition of new criteria and other types of protected areas. More importantly, the 1979 plan expressed a successful effort to reduce the effects of randomness in the configuration of the national park system.

In the Brazilian context, national parks, "America's best idea" (as Ken Burns might say), became something like the stone in a stone soup—it is present, but it is not the main course. Early on, Brazilian policy makers absorbed and selectively applied foreign influences in regard to parks and other types of protected areas, but in 1979 they came up with an original, endogenous, systematic set of concepts and policies. The end result is that the Brazilian national park system does not display clear-cut traces of U.S. or any other foreign concepts.

Notes

1. The expression "conservation unit" (*unidade de conservação*) seems to be used only in Brazil to designate publicly protected areas for the purpose of nature conservation or preservation. The common English language expression, adopted whenever proper in this text, is "protected area," but "national parks" is used when this type of protected area is addressed explicitly.
2. Ministério de Meio Ambiente, *O Sistema Nacional de Unidades de Conservação da Natureza* (Brasília: MMA, 2011), 4, 5.
3. All data come from the official directory of Brazilian protected areas, http://www.mma.gov.br/sitio/index.php?ido=conteudo.monta&idEstrutura=119 (accessed October 15, 2011). Other types of federal conservation units, state parks, and other state areas are not studied in this text, although data about them is available in this registry.
4. Some points in this section are discussed in earlier publications: José Augusto Drummond, José Luiz de Andrade Franco, and Alessandra Bortoni Ninis, *O Estado das Áreas Protegidas no Brasil—2005* (Brasília: Centro de Desenvolvimento Sustentável, 2006), available at https://www.academia.edu/3307044/O_Estado_das_Areas_Protegidas_do_Brasil_-_2005 (accessed July 3, 2015); José Augusto Drummond, José Luiz de Andrade Franco, and Alessandra Bortoni Ninis, "Brazilian Federal Conservation Units: A Historical Overview of their Creation and of their Current Status," *Environment and History* 15 (2009):

463–91; José Augusto Drummond, *Devastação e Preservação Ambiental no Rio de Janeiro—Os Parques Nacionais do Estado do Rio de Janeiro* (Niterói: EDUFF, 1997); José Augusto Drummond, *O Sistema Brasileiro de Parques Nacionais—Análise dos Resultados de uma Política Ambiental* (Niterói: EDUFF, 1997).

5. José Augusto Drummond, "The Garden in the Machine: An Environmental History of the Tijuca Forest (Rio de Janeiro, Brazil), 1862–1889," *Journal of Environmental History* 1, no. 1 (January 1996): 83–104; Claudia Heyneman, *A Floresta da Tijuca: Natureza e Civilização—Século XIX,* Série "Biblioteca Carioca" (Rio de Janeiro: Prefeitura Municipal do Rio de Janeiro, 1995); C. M. Bandeira et al., *Pesquisas e Escavações Arqueológicas em Sítios Históricos do Parque Nacional da Tijuca e Arredores* (Rio de Janeiro: Fundação Brasileira de Conservação da Natureza, 1984) [typewritten]; P. B. Cezar and R. R. Oliveira, *A Floresta da Tijuca e a Cidade do Rio de Janeiro* (Rio de Janeiro: Nova Fronteira, 1992). There is no published biography of Archer. Cezar and Oliveira present biographical data that place Archer in close proximity to Emperor Pedro II's family and court.

6. See André Rebouças, *Excursão ao Salto do Guaíra: O Parque Nacional* (Rio de Janeiro: 1876). Several biographies of Rebouças are available: Roberto Ruiz, *André Rebouças, sua vida, sua obra, seus ideais* (Brasília: Ministério dos Transportes, 1973); Sydney M. G. dos Santos, *André Rebouças e o seu tempo* (1985); Maria Alice Rezende Carvalho, *André Rebouças e a construção do Brasil* (Rio de Janeiro: Revan, 1998).

7. Wanderbilt Duarte de Barros, *Parques Nacionais do Brasil* (Rio de Janeiro: Ministério da Agricultura, 1952); José Luiz de Andrade Franco and José Augusto Drummond, "Frederico Carlos Hoehne: A Atualidade de um Pioneiro no Campo da Proteção à Natureza no Brasil," *Ambiente & Sociedade* 8, no. 1 (Jan.–June 2005): 1–26; Warren Dean, *With Broadax and Firebrand—The Destruction of the Brazilian Atlantic Forest* (Berkeley: University of California Press, 1995).

8. During the nineteenth century, many foreign scientists made long field trips and extensive natural history collections all over Brazil. Back in their homelands, they helped spread among scientists a diffuse concern with the sources of their specimens. Löefgren and von Ihering inherited this concern. The importance of foreign naturalists in the earliest research efforts on Brazilian fauna, flora, minerals, rivers, and so on, particularly in the nineteenth century, can be gauged in two excellent publications: João Meirelles Filho, *Grandes Expedições à Amazônia Brasileira—1500–1930* (São Paulo: Metalivros, 2009); and Anna Maria de Moraes Belluzzo, *O Brasil dos Viajantes* (São Paulo: Objetiva, 2000).

9. The code was issued by Decree 23,793, on January 23, 1934. See José Drummond and Ana Flávia Barros-Platiau, "Brazilian Environmental Laws and Policies, 1934–2002: A Critical Overview," *Law & Policy* 28, no. 1 (January 2006): 83–108.

10. The new code was issued by Law 4,771 on September 15, 1965.

11. Curiously, current pro-agriculture and pro–protected areas lobbyists, although in disagreement about everything else, agree that APPs and RLs are uniquely Brazilian provisions, not to be found in forestry or conservation laws anywhere

in the world. Pro-agriculture activists argue that their uniqueness makes them more objectionable, while preservationists praise them exactly for their uniqueness. I have not found research that confirms this uniqueness, but the matter suggests the possible influence of agricultural conservation concepts on park and protected area policies in Brazil.

12. This agency is the Instituto Chico Mendes para a Conservação da Biodiversidade, created by Law 11,516, on August 28, 2007. Unfortunately, the contrast with the United States persists, as the agency also manages national forests.

13. Barros, *Parques Nacionais do Brasil;* Wanderbilt Duarte de Barros, "Legislação de Conservação da Natureza," *Boletim FBCN* 9 (1974): 28–35; J. L. Bélart, "Por uma Política Nacional de Conservação," *FBCN—Boletim Informativo* 11(1976): 47–48; Alceo Magnanini, "Conceitos de Conservação," *FBCN—Boletim Informativo* 1 (1966): 13–22. See also J. L. A. Franco and J. A. L. Drummond, "Wilderness and the Brazilian Mind (II): The First Brazilian Conference on Nature Protection (Rio de Janeiro, 1934)," *Environmental History* 14 (2009): 82–102; José Drummond, *O Sistema Brasileiro de Parques Nacionais: Análise dos Resultados de uma Política Ambiental* (Niterói: Editora da Universidade Federal Fluminense, 1997).

14. All information was retrieved from the decrees creating each park and from personal visits to most of these parks. All decrees are transcribed in Carmen Moretzsohn Rocha, ed., *Legislação de Conservação da Natureza*. 4th rev. ed. (São Paulo: Centrais Elétricas de São Paulo; Rio de Janeiro: Fundação Brasileira de Conservação da Natureza, 1986). The momentum of monumentalism is strongly illustrated by the curious fact that three parks were created in succession (1937, 1961, and 1979) to protect Brazil's highest measured peaks, with a new park being created each time a "new" mountain was proclaimed the highest.

15. Barros, *Parques Nacionais do Brasil*, 26, 30–35.

16. Paulo Afonso was cancelled by Decree-Law 605, on June 2, 1965; Sete Quedas by Decree 86,071, on June 4, 1981. The trend continues. On August 15, 2011, president Dilma Roussef signed an executive order altering the areas of three national parks in the Amazon region (Amazônia, Mapinguari, and Campos Amazônicos), also for the purpose of accommodating lakes of hydroelectric dams. See Oswaldo Braga de Souza, "Redução de Unidades de Conservação abre precedente perigoso," http://www.socioambiental.org/nsa/detalhe?id=3393, (accessed October 15, 2011). Pioneer environmental activist Teresa Urban, in *Missão Quase Impossível—aventuras e desventuras do movimento ambientalista no Brasil* (Uberaba: Fundação Peirópolis, 2011), examined citizens' resistance to the flooding of Sete Quedas National Park; actually, she considers this resistance to be one of the founding events of Brazil's modern environmental movement.

17. All information was taken from park creation decrees; see also Instituto Brasileiro de Desenvolvimento Florestal, untitled typewritten report, 1979, available at IBDF archives, Brasília, Brazil; IBDF, *Projeto de Implantação e Consolidação de Parques Nacionais, Reservas Equivalentes e Proteção à Natureza* (Brasília, 1975); Instituto Brasileiro de Desenvolvimento Florestal and Fundação Brasileira de Conservação da Natureza, *Plano do Sistema de Unidades de Conservação no Brasil* (Brasília, 1979); Instituto Brasileiro de Desenvolvimento Florestal and

Fundação Brasileira de Conservação da Natureza, *Plano do Sistema de Unidades de Conservação no Brasil. II Etapa* (Brasília, 1982); Secretaria Especial de Meio Ambiente, *Projeto Nacional do Meio Ambiente* (Brasília: 1988), [typewritten]; Rocha, *Legislação de Conservação.*

18. José Augusto Drummond, *National Parks in Brazil: A Study of 50 Years of Environmental Policy (with Case Studies of the National Parks of the State of Rio de Janeiro)* (master's thesis, Evergreen State College, Olympia, Washington, 1988), chap. 4, table 2. See also Rocha, *Legislação de Conservação;* Maria Tereza Jorge Pádua, interview with author (Brasília, May 18, 1988); Maria Tereza Jorge Pádua, *Unidades de Conservação* (Brasília, 1988), [typewritten]; Departamento de Parques Nacionais e Reservas Equivalentes, IBDF, unpublished report [ca. 1989]; Harold Edgard Strang et al., "Parques Estaduais do Brasil: Sua Caracterização e Essências Nativas Mais Importantes," Tese Apresentada ao Congresso Nacional sobre Essências Nativas (Campos do Jordão [SP], September 12–18, 1982), 8–10.

19. The aforementioned biological stations, created earlier in the century, were designed strictly for collection expeditions and fieldwork.

20. Extractive reserves, framed as protected areas, opened an entirely new chapter in Brazilian conservation policies, aimed at the "sustainable use" of natural resources. This matter will not be examined here.

21. One such reserve has gained international visibility: RPPN Fazenda Bulcão, owned by the famous Brazilian photographer Sebastião Salgado. The unit, rather than protecting natural vegetation, is dedicated to the reclamation of deforested Atlantic Forest lands in the states of Minas Gerais and Espírito Santo. See http://www.institutoterra.org/eng/conteudosLinks.php?id=22&tl=QWJvdXQgdXM=&sb=NQ==#.VeHksbTtKfQ (accessed July 24, 2015).

22. Pádua, interview with author; Pádua, *Unidades de Conservação*; Paulo Nogueira Neto, personal communication with author. 2001, 2004; Paulo Nogueira Neto, *Uma Trajetória Ambientalista—Diários de Paulo Nogueira Neto* (São Paulo: Empresa das Artes, 2011); Drummond, *National Parks in Brazil.*

23. There was also a string of eight biological preserves and fifteen national forests.

24. Since the 1970s Pádua has written extensively about conservation issues and lobbied prominently for national parks. She briefly presided over the IBDF in the 1980s and was active for many years in the NGO Funatura. Pádua summarized her rich and informed views on Brazilian conservation issues in Maria Tereza Jorge Pádua and Marc Dourojeani, *Biodiversidade: a Hora Decisiva,* 2nd ed. (Curitiba: Editora da UFPR, 2007). More recently, she published her memoirs: Maria Tereza Jorge Pádua, *Conservando a Natureza do Basil* (Curitiba, Paraná, Brazil: editor of UFPR, 2015).

25. FBCN's role in park policy and other conservation efforts is studied in José Luiz de Andrade Franco and José Augusto Drummond, "Nature Protection: The FBCN and Conservation Initiatives in Brazil, 1958–1992," *História Ambiental Latinoamericana y Caribeña* 2 (2013): 338–67.

26. Franco and Drummond, *Nature Protection;* Drummond, *Brazilian National Parks,* chap. 3; José Cândido de Melo Carvalho, "Dia Mundial do Meio Ambiente: Considerações sobre a Participação do Brasil em Organizações, Programas

e Convenções Relativas à Conservação do Meio Ambiente," *Boletim FBCN* 14 (1979): 5–16; José Cândido de Melo Carvalho, "O Papel das Entidades Não-Governamentais na Conservação da Natureza," *FBCN—Boletim Informativo* 12 (1977): 49–53; FBCN, "Currículo Sintético da FBCN" [Rio de Janeiro, ca. 1982], [typewritten]. Available at FBCN's archives, currently closed to the public.

27. Pádua, interview with author; Warren Dean, "Forest Conservation in Southeastern Brazil, 1900 to 1955," *Environmental Review* 9 (1985): 63; Jair Tovar, "Parques Nacionais," *Revista do Serviço Público* 87, no. 3 (1960): 126; Ângela B. Quintão Tresinari, "Evolução do Conceito de Parques Nacionais e sua Relação com o Processo de Desenvolvimento," *Brasil Florestal* 13, no. 54 (1983): 14; Barros, *Parques Nacionais,* 21, also mentions Argentina's leadership in parks. Barros examines park policies in the United States, Canada, Mexico, Argentina, Chile, Cuba, Santo Domingo, Uruguay, and Venezuela. He states that Brazilian national parks were created in part as a result of "cultural exchange" with these and other nations.

28. See Alceo Magnanini, *Política e Diretrizes dos Parques Nacionais do Brasil* (Rio de Janeiro: Instituto Brasileiro de Desenvolvimento Florestal, 1970) [mimeographed], 3, 6–10. Other early planning documents were Instituto Brasileiro de Reforma Agrária and Instituto Brasileiro de Desenvolvimento Florestal, *Parques Nacionais e Reservas Equivalentes no Brasil: Relatório com Vistas a uma Revisão da Política Nacional nesse Campo* (Rio de Janeiro: IBRA-IBDF, 1969) [mimeographed]; Alceo Magnanini and Maria Teresa Jorge Pádua, "Situação dos Parques Nacionais no Brasil," *FBCN—Boletim Informativo* 4 (1969): 28–58.

29. See UNESCO, *The Biosphere Conference 25 Years Later* (2003), available at http://unesdoc.unesco.org/images/0014/001471/147152eo.pdf (accessed May 3, 2014).

30. Instituto Brasileiro de Desenvolvimento Florestal, *Projeto de Implantação e Consolidação de Parques Nacionais, Reservas Equivalentes e Proteção à Natureza* (Brasília, 1975), 1–8, 20–24, 28, Anexo 1-A.

31. Instituto Brasileiro de Desenvolvimento Florestal, *Diagnóstico do Subsistema de Conservação e Preservação de Recursos Naturais Renováveis* (Brasília, 1978), introduction, 53–56, 61–65, 75–77, 111–33, tables 23–25. The meager numbers of visitors in almost all Brazilian national parks is a relevant comparative variable, but it not will be explored here. Even today, many parks attract only a few hundred or a few thousand visitors per year. Many other parks have never been opened to visitation, or were opened to the public many years after their creation.

32. The plan was published in a mimeographed format as Instituto Brasileiro de Desenvolvimento Florestal and Fundação Brasileira de Conservação da Natureza, *Plano do Sistema de Unidades de Conservação do Brasil* (Brasília, 1979). It had a "second stage," also mimeographed, published as Instituto Brasileiro de Desenvolvimento Florestal and Fundação Brasileira de Conservação da Natureza, *Plano do Sistema de Unidades de Conservação do Brasil—II Etapa* (Brasília, 1982). I will use mostly the first document (cited henceforth as IBDF/

FBCN, *Plano do Sistema*) because the 1982 version was an abridged updating of the 1979 document. Drummond, *National Parks in Brazil,* chap. 3, has a broader evaluation of the 1979 plan. It is important to note that the plan was never turned into law or decree; despite its deep and lasting influence, it remained an internal IBDF policy document.

33. Steve Rodriguez's essay in this volume shows a strong correlation between the Indonesian military dictatorship of Suharto and the creation of a large number of national parks in that country. The temptation to correlate Brazil's park policy with its military dictatorship (1964–85) yields results contrary to those from Rodriguez. Between 1964 and 1978, the peak years of Brazil's dictatorship, only three national parks were created. Between 1979 and 1985, however, ten parks and eight biological preserves were created—a very strong spree, particularly in terms of the huge areas of these new units. However, 1979 to 1985 were the exact years of the waning of the dictatorship, which practically fell apart in 1982. Another difference between the two situations is that the younger Brazilian parks have survived well past the end of the dictatorship, while in Indonesia the parks created during the Suharto regime were to a great extent destroyed by logging. Additionally, numerous additional parks and other protected areas anticipated by the 1979 plan were created during the several civilian democratic governments that followed the end of the Brazilian military dictatorship.

34. IBDF/FBCN, *Plano do Sistema*, 6.

35. Ibid., 6–18, 20–31.

36. Maria Teresa Jorge Pádua, "Situação Atual do Sistema de Parques Nacionais e Reservas Biológicas," *Boletim FBCN* 16 (1981): 35. It is surprising that the task force ignored hydroelectric dams and lakes as potential "enemies" of national parks. In 1979 Paulo Afonso had already been flooded, and Sete Quedas was about to be flooded. Besides, task force members were most likely aware of the Hetch Hetchy dam controversy (in Yosemite National Park) and of a similar dispute about a dam that would affect the Grand Canyon National Park.

37. IBDF/FBCN, *Plano do Sistema*, 20–31.

38. http://biblioteca.ibge.gov.br/pt/biblioteca-catalogo?acervo=todos&campo=todos &opeqry=&texto=Projeto%20Radam&digital=true&fraseexata= (accessed August 5, 2015).

39. Gary B. Wetterberg et al., *Uma Análise das Prioridades em Conservação da Natureza na Amazônia,* PNUD/FAO/IBDF/BRA 45. Série Técnica, Número 8 (Brasília, 1976).

40. IBDF/FBCN, *Plano do Sistema*, 27–33. As the task force continued to work beyond 1979, dozens of additional parks, reserves, and ecological stations were proposed; they appeared in the plan's 1982 update.

41. Data taken from the "second stage" of IBDF/FBCN, *Plano do Sistema*, "Prefácio," 12.

42. In *Amazon Conservation in the Age of Development: The Limits of Providence* (Gainesville: University Press of Florida, 1991), Ronald A. Foresta captured the intersection of national conservation projects, foreign pressures, and development policies for the Amazon region exactly at the time when the 1979 plan was being drafted and put into execution.

43. These findings in the field of paleontology and their application to preserve policy were controversial, of course, but this matter will not be explored here. The four "refuge" authors cited by the plan are Jürgen Haffer, Paulo Vanzolini, Ghillean T. Prance, and K. S. Brown, Jr.

44. IBDF/FBCN, *Plano do Sistema,* 26–27; Pádua, interview with author.

45. IBDF/FBCN, *Plano do Sistema, II Etapa,* 39–40, 42.

46. In 2000, partially under the influence of the proposals contained in the 1979 plan, the Brazilian Congress voted into law a plan that instituted a "conservation unit system" composed of twelve types of protected areas. Some were reminiscent of the 1979 proposals, while others were not. See Drummond et al., "Brazilian Federal Conservation Units: A Historical Overview of Their Creation and of Their Current Status," *Environment and History* 15 (2009): 463–91.

ELEVEN

Conquering Sacred Ground?
Climbing Uluru and Devils Tower

ANN MCGRATH

TWO LANDMARK SITES OF NATIONAL SIGNIFICANCE, Uluru (formerly Ayers Rock) in Central Australia and Devils Tower in Wyoming, have sparked controversies over climbing, and primarily over whether it should be permitted at all.[1] These huge monoliths are located in landscapes considered worthy of inclusion in the national estates of two settler-colonizer nations: Uluru as a national park in Australia, and Devils Tower as a national monument administered by the National Park Service of the United States. Despite the protests of indigenous custodians over the last decades, both rocks remain popular climbing destinations. Uluru is accessible to anyone with a reasonable degree of fitness, while the sheer rock faces of Devils Tower prevent any but skilled and experienced climbers from scaling its heights. In Australia, Uluru has taken on the status of a crucial national icon. Although Devils Tower is well known, it does not hold the same central place in the American imagination. In this chapter, I examine how climbers and other visitors have engaged in contested national and transnational history stories. I will also explore how the histories of national parks have helped render indigenous landscapes a shade of colonizer white, while simultaneously creating theaters for defining, redefining, and contesting nation.

In settler-colonizer states, "national parks" and "national monuments" take on historical meanings that reveal the basis of their respective national foundations. As national parks are ostensibly for the public and for the nation, a person or a group's distinctive relationship to nation becomes highly significant. Not everyone feels an insider in the national polity. Historical thinking reinforces such perceptions. For example, national parks have been dubbed a "gift of the new world to the old."[2] While settler-colonizer nations have long been designated as the New World, they were obviously someone else's Old World. Hence, that "gift" of national parks inevitably signifies a loss. What colonizers might see as a "benefit" may not appear so altruistic to others, notably indigenous traditional landowners. Rather, the legal status of a national park may serve to lock them out of their own land. With land appropriation a core enterprise of colonialism, the topic of indigenous dispossession is thus vital to the history of national parks.

The nations of the United States and Australia were founded more than a century apart, with quite different indigenous cultures, colonizing histories, and emergent collective national identities. In both cases, however, it was the later colonized, relatively undeveloped zones that attracted national parks designations. Under the auspices of two settler-colonizer states, their recognition as significant geographical places enabled protection through special legislation. As sites left underdeveloped by colonizer industries, and valued for antiquity and "wilderness" attributes, Devils Tower and Uluru enabled some people to imagine that these places stood outside colonialism—if not outside modernity and history.[3] The concept of wilderness has often designated a place without human imprints, and without human occupation—past or present.[4]

As spectacular sites in nonurban locations, they have also become focal points for competing assertions of indigenous and colonizer sovereignty. While they might offer the welcome relief of pristine natural vistas far from the ravages of industrialization, settler-colonizers combine this with an impulse to "preserve" and "conquer" for modern nationalistic purposes. At the same time, they are used as decolonizing sites for transnational negotiations between colonizer and colonized. Here, I refer to indigenous "nations" and the settler-colonizer nation as entering "transnational" negotiations. I do not use the term "transnational" to indicate connections, but rather as a way of looking *across* nations or transnationally, and also *between* and across nations within settler-colonizer states.[5]

At a cosmetic level, Uluru and Devils Tower appear to have commonalities that make them worthy of comparison. Both stand in sharp contrast to the surrounding flat country, their lofty peaks and massive size visible from many miles distant. Both are tourist destinations, and at both, climbing is a central attraction. Yet there is much that divides them. Uluru, along with the adjacent Kata Tjuta range (previously known as the Olgas), are the principal features of the Uluru–Kata Tjuta National Park, which has been a UNESCO World Heritage site since 1987 and is one of the few Australian sites listed for both cultural and aesthetic values. It is somewhat unusual in being managed by the federal government rather than by state governments, which manage most other Australian national parks. Uluru–Kata Tjuta is located in the Northern Territory, which, although self-governing, does not hold the status or independence of a state and has a low population. Federal or Commonwealth refusal to give up control to the Northern Territory government is another indicator of Uluru's vital national status. "The Rock," as it is colloquially known, is a destination for large numbers of Australian and international visitors. They are drawn to the Red Centre to experience the heart and essence of that vast continent. It is the iconic Australian place known all over the world for its remarkable geology and coloration, a place that has, since it became more widely known to nonindigenous peoples in the early twentieth century, attracted adventurers, travelers and sightseers. Visitors, including the European Ernest Giles, who arrived in 1873, found it both "ancient and sublime": "Its appearance and outline is most imposing, for it is simply a mammoth monolith that rises out of the sandy desert soil around, and stands with a perpendicular and totally inaccessible face at all points, except one slope near the north-west end, and that at least is but a precarious climbing ground to a height of more than 1,100 feet."[6]

Although Devils Tower is not nearly as well known to Americans or internationally, its fame grew after the release of Stephen Spielberg's blockbuster movie *Close Encounters of the Third Kind,* in November 1977. This film entrenched its status as eerie and alien, extraplanetary, and exuding strange energies attractive to New Age spiritual seekers. Nonetheless, the early impetus for the recognition of this "lofty and isolated rock in the state of Wyoming" as a national monument arose from being "such an extraordinary example of the effect of erosion in the high mountains as to be a natural wonder and an object of great historical and scientific interest."[7] By 1936 the National Park Service was borrowing imperial imagery to encourage Americans to participate in the "great Wonder" of this majestic natural spectacle, to "pause for a moment in reflection of this royal member of the world's prodigious marvels and realize that it is more than a National Monument. As we view it resting in silent solemnity on its imposing throne, we sense that it bespeaks a commemorative message from the directing force of all nature."[8]

Including an analogy to a "royal member" and a "throne" is somewhat unusual in an avowedly republican nationalist discourse. Unlike Uluru, Devils Tower is neither a World Heritage Site nor a national park, and although managed by the U.S. National Park Service, it is technically classified as a national monument of the United States. Significantly, however, it was the first national monument, proclaimed under the federal Antiquities Act of 1906, which gave the president of the United States executive powers to protect "historic landmarks, historic and prehistoric structures, and other objects of historic and scientific interest" without the need for congressional approval. The level of protection for a national monument is not as great as for a U.S. national park, which may demand the protection, for example, of wildlife and biodiversity. Consequently, Devils Tower is not protected as a cultural landscape but rather, in keeping with the aims of the legislation, as a monument "confined to the smallest area compatible with the proper care and management of the objects to be protected."[9]

The word "national" in parks and national monuments may be taken for granted, but it encapsulates a deeper semiotic meaning, in that it indicates a particular collective and nationalistic form of political and cultural connection to wilderness landscapes. Belonging to "nation" suggests something larger than individual self-interest, an implied membership in a supposed entity glued together by sentiment. "National landscapes" is a new designation that Australia's national parks establishment is applying to regionally themed tourist destinations, whereas the much earlier American term "national monuments" metaphorically heightens the sense of drama in the notion of a "landscape." Like the man-made Washington Monument, Devils Tower rises above flatness to memorialize a nation. National parks, national monuments and national landscapes are not just about management; they make emotive assertions of nation, ownership and sovereignty.

Inspired by the American romanticist movement and transcendentalism, national park landscapes are often valorized not only as offering a historic "first" gaze but as something pure, an unmediated connection between man and nature. When humans connect with the beauty of nature, it is as if they are elevated. However, national parks are experienced cross-culturally and historically in different ways. Marilyn Lake and Henry Reynolds demonstrated how, during the late nineteenth and early decades of the twentieth century, Australia and the United States were both explicitly conceptualized as "white men's countries."[10] As Melissa Harper and Richard White suggest, in contrast to transcendentalism, Australia's national parks activists stressed leisure, recreation, and a favorable climate's capacity to forge a healthy race.[11]

Despite national differences, both Uluru and Devils Tower reveal important things about the way protected areas, including national parks and monuments, may be used by nations to approach questions of the rights of indigenous peoples, and in the process, to illuminate issues of national identity and competing uses of natural spaces. Both places are of high significance to Native custodians as sacred places. In the case of Uluru, this was crucial to the successful struggle of Anangu traditional owners to regain control of their land under the Australian Federal Land Rights Act (1976). Similarly, for Devils Tower traditional custodians, the Arapaho, Cheyenne, Kiowa, Lakota, Shoshone, and Crow, it is the sacred nature of the rock that has underwritten attempts to restrict climbing. In 1996 this desire was reflected in the National Park Service Final Climbing Management Plan, which proposed a voluntary suspension of climbing during the month of June "in deference to Indian sacred ceremonies."[12]

In both places, the climbing practices of visitors have ignited assertions of indigenous religious freedom and rights. Their pleas to control an activity that attracts strong support in the wider community serves to crystallize larger issues of indigenous rights in the management of public lands, and it highlights the uneasy relationship of indigenous minorities to the nation-state. Both the Anangu traditional owners of Uluru and the Plains Indian groups with a spiritual connection to Devils Tower have prominently expressed their wishes that climbing should be curtailed. Under two very different legal and management regimes, both parks authorities have endeavored to introduce voluntary climbing protocols. Although they have advised visitors that they object to climbing, until recently the Anangu have lacked the backing to enforce their right to stop climbing. Similarly, while Plains Indian tribes have gained the support of the U.S. National Park Service for a voluntary cessation of climbing for one month of the year, the power of the service to control such activity on Devils Tower has not been supported by U.S. courts.[13] Nevertheless, there is evidence that voluntary bans have had an impact, with both places recording substantial compliance amongst visitors, suggesting a desire by many to respect the wishes of the traditional custodians.

Layering the Transnational

For the purposes of this paper, artificially placing Devils Tower next to Uluru is a practice of transnational juxtaposition that enables us to consider issues in fresh ways, and to benefit from a wider scholarship and contrasting historical trajectories.[14] Rather than evaluating only difference and likeness, I will examine thematic links within and between networks of meaning. While several transnational studies have compared the Anglophone, British imperial–settled nations, there has been little attention to the extra layering of the transnational as a contest over sovereignty within settler-colonizer societies, that is, between Native nations and the colonizer nation-state. Yet focusing on this dynamic offers potential for some decolonizing critique. The two nation-states are generally defined according to end population—with most settler immigrants arriving from the British Isles and Europe. However, in other analyses, the settler-colonizer state refers to the hegemonic power structures of colonialism, and a subsequent polity that came about through dispossessing indigenous peoples of their land.[15]

In comparing these sites in national contexts, my approach takes the opportunity to compare climbing in discrete "spaces of nation," and in two settler-colonizer nation-states. Australia and the United States have different geographical, demographic, and sociopolitical characters, and their nation-state chronologies and key policies also differed. Significantly, in the United States, American Indian groups have long been officially recognized as a type of nation. Given the outright warfare conducted during the eighteenth and nineteenth centuries, the Department of War saw them as an enemy or at least as a competing polity; under a treaty system, they were diplomatically dealt with as, or akin to, foreign nations. The 1831 Marshall judgment defined them as "domestic dependent nations."[16]

In Australia no treaties were negotiated, and with settlement legally based on *terra nullius* or wasteland rather than conquest by war, indigenous sovereignty and ownership of lands were not recognized. After the 1970s, however, the introduction of limited land rights shifted this dynamic. Since the *Mabo* judgment of 1992, more Aboriginal Australians are entitled to claim native title, and larger language-sharing groups increasingly describe themselves as nations.[17]

The evolution of national parks in each country has certainly differed. For one thing, while Australian parks are called "national," almost all are funded and managed by individual states within the federal system. Various state parks are also classified as World Heritage—for example, Willandra Lakes, Greater Blue Mountains—twelve of them on the basis of "natural" significance. The only three classified exclusively for "cultural" significance are British-Australian built sites, including convict sites and the Sydney Opera House. Those with both natural and cultural values include indigenous-owned Kakadu National Park (1981) and Uluru–Kata Tjuta National Park (1987).[18] In these areas, Aboriginal people have continued to live and carry out culture on their traditional lands, with nonintensive European economic activities leaving a relatively minimal colonial footprint.

Parks Australia, a federal authority, has oversight of World Heritage listed sites, with the parks within its primary authority located in Australian Territories, including the Northern Territory. When the Australian federal government moved belatedly, in 1979, to establish national parks in the Northern Territory, Kakadu was developed with explicit reference to and help from the International Union for the Conservation of Nature. In negotiating indigenous majority management in Uluru–Kata Tjuta and Kakadu National Parks and the handover of Uluru to indigenous owners in 1985, Australia exercised international leadership in developing protocols for indigenous or joint control of national parks. As Catton's chapter in this volume reveals, neighboring New Zealand's early national parks were informed by Māori concepts of custodianship. This reflected a contrasting history marked by Māori wars that ended in the Treaty of Waitangi, which is celebrated, albeit often controversially, as New Zealand's day of national unification. In a land that had been designated as terra nullius, or vacant wasteland, Australian joint management of the Northern Territory parks constituted an important, even radical, recognition of indigenous land rights. It followed in the wake of the Northern Territory Land Rights Act of 1976. The handback of Uluru/Ayers Rock preempted recognition of native title by the High Court of Australia, which did not take place until 1992.[19]

The prior history of national parks lands is a vital factor in this story. While each different era will reinvent its own conceptions of the parks mission, in the United States the 1963 Leopold and National Academy reports envisaged that each national park should be an "illusion of primitive America." Their mission was thus to preserve and create the "mood of wild America." Parks historian Richard West Sellars implies that this was therefore a patriotic, ethnocentric goal—to maintain some "landscape remnants of a pioneer past" as they were "when first visited by the white man" or "viewed by the first European visitors."[20] Such a preservation mindset reifies the "first moment" of the colonizer gaze, meaning that it was primarily settler-colonizer traditions that became codified in land, heritage, parks, and antiquities law.[21]

As historian Mark Spence explains, parks history became an arm of settler-colonizer land alienation and, as his title sums up, was a means of "dispossessing the wilderness."[22] After all, their resumption showed that even "waste land" incapable of "improvement" had some meaning to colonizers and could be alienated by the nation-state.[23] Taking this further, Australian and Pacific historian Tracey Banivanua Mar concludes that the declaration of national parks was an endpoint of dispossession and the ultimate triumph of private property.[24]

Because parks designations encoded race segregation, heritage expert Denis Byrne describes the archaeologies of designated Aboriginal reserves as "nervous landscapes."[25] While "national monuments" and "national landscapes" may be viewed similarly as landscapes of colonizer anxiety, Byrne and Maria Nugent argue more optimistically for the possibility of "shared landscapes," which include parks.[26] Indigenous people in both nations are using national parks venues to assert their ongoing sovereignty claims. They also use them as sites over which to test the sincerity of national reconciliation goals, although levels of state interest in accommodating them may differ. In Australia the 1992 Mabo decision and the 2008 prime ministerial apology raised national consciousness of indigenous history and ways to compensate for past injustice. While Americans have long been cognizant of frontier warfare, many historians still depict America's "colonial" era as something that ended with the anti-British revolution.[27] The powerful legacies of African American slavery and a more multifaceted European colonizing history detract attention from the singularity of any foundational act of indigenous dispossession.

National Monoliths

Despite the treaties enjoyed by many Native Americans, the several American Indian tribes with links to Devils Tower lack formal ownership or officially recognized rights in regard to the site. Before it became a national monument, its surroundings were a ranching area held by private landowners. It would appear that the "protection of lands" declared in conservation enactments have been just as effective as other styles of landholding in diminishing indigenous title rights.

Furthermore, national parks lands are understood as publicly accessible. While special conservation values or "wilderness" classifications create exceptions, in order to justify government land takeover and maintenance costs, generally a national park must be used and visited. By promoting a voluntary June climbing closure to coincide with Sundance ceremonies, the National Park Service has attempted a compromise in relation to Devils Tower. This ensures that it is not totally ignoring one of the nation's constituencies. Yet despite such gestures, the invitation to climb remains ubiquitous. With some safety warnings, this also applies at Uluru, where climbing remains the much-anticipated visitor activity.

In the 1980s Ayers Rock, as it was then known, became the centerpiece in a heated controversy in which political conservatives declared that conferring indigenous land rights to the rock would "divide the nation." Politicians of the newly self-governing Northern Territory staged an aggressive campaign to assert masculinist "outback" tropes of white entitlement and colonizer sovereignty. However, the federal Labor government, backed by Prime Minister Bob Hawke and Minister for Aboriginal Affairs Clyde Holding, appreciated the powerful unifying and healing symbolism of "handing back the Rock."[28] The Anangu thus gained rights of veto, joint management, and income rights from the national park. The handback ceremony involved dance and song rituals in which the Anangu leadership, and various tiers of the white national leadership, including the queen's representative, Governor General Sir Ninian Stephen, officiated. The Anangu valued this deeply, as they appreciated that it was the British government, with its head the queen, that had originally divested them of sovereignty, and without treaties.[29] Ninian Stephen was admired for having made landmark public statements in favor of Aboriginal rights. Dealing with the regal and vice regal was in keeping with Aboriginal sovereignty thinking, as through history they had nurtured a relationship with and had petitioned the kings and queens of England for redress for their lands. While the Northern Territory government absented itself, the Anangu saw the event as a breakthrough marking a hopeful future for their children. The Anangu now held statutory and traditional title to Uluru. This recognition of title rights, however, came with a big catch, for it was conditional upon Uluru being leased back as a national park for ninety-nine years.

Both Uluru and Devils Tower have long been singled out as suitable locations for nationalist performances, reinforcing their status as icons of nation and of settler ownership. President Theodore Roosevelt's declaration of Devils Tower as the first U.S. national monument inscribed a European notion of antiquity as a national value. Thus the ancient "old" was incorporated into the modern nation. Before that, the Lakota knew the mountain as Mato Tipila, while other Plains Indian groups used their own language and storied names with distinctive meanings. It was and remains a ceremonial and gathering site for the Arapaho, Cheyenne, Kiowa, Lakota, Shoshone, and Crow, who each narrate and reenact legends about the creation of Devils Tower and the profound power of the bear, which made the place sacred and thus incompatible with recreational climbing.[30] Many consider the official name offensive, explaining that "Bear Lodge" is a more accurate translation of indigenous names for the place than Devils Tower or "Bad God's Tower."

Similarly, while Uluru is legally recognized as belonging to the Anangu people, a wider range of Pitjanjanjara and Aranda people hold it to be a significant feature of their stories, histories, and cultural landscapes. Its contested status is reflected in its changing names. In 1873 the surveyor William Gosse named it Ayers Rock after a senior government official of South Australia. A dual naming system of Ayers Rock and/or Uluru applied from the 1980s, and then in 2002 its official name reverted to the preexisting name of Uluru. Today, many Australians familiarly refer to it simply as "The Rock," and many tourism promotions still call it Ayers Rock.

The Anangu have opposed climbing because the climb route is a sacred journey route of Mala the hare-wallaby man and is exclusively reserved for ascent by specified elders conducting seasonal ceremonies.[31] The Anangu see the visitors' climbing practices as senseless. Because of the way their long lines appear from a distance, they call them *minga* or "crazy ants." Furthermore, and this may be equally important, the Anangu are concerned that they cannot protect those who walk on their country as they feel obligated to do. Deaths on the rock are therefore especially troubling, and the Anangu are particularly concerned about their responsibility for these events, and for the protocols to be followed regarding departed strangers.[32]

Uluru and Devils Tower continue to hold indigenous sacred associations and meanings as ceremony sites within a broader cultural landscape, one which carries inscribed histories of creation, long journeys, large gatherings, kin and interpersonal connection, marriage, trade, and feasting. However, until recently, National Park Service literature about Devils Tower's sacred significance to Indians was minimal. While a recent leaflet documents the significance of the rock to six of the tribes, the main information section is vague, stating that Devils Tower is a sacred site and that Native Americans would like to see the climb banned due to their "beliefs."[33] Climbing blogs use the tower's "sacredness to all" to justify their conviction, arguing that this made climbing even more "special": a place where climbers could exercise their rightful, even God-given, privilege.[34]

Discourses of sacredness have been used to support both indigenous and nonindigenous claims to decision-making rights over the activities allowed on these sites.[35] In my title, the words "conquering" and "sacred" may seem an awkward pair, but they go to the heart of contestations over these sites. Nationalism imbues an emotional connection that draws on secular and sacred iconography. Its narratives are built up, becoming increasingly layered and naturalized over time. In this case, the big rocks in focus stand for many meanings beyond themselves, including complex sets of values associated with their respective settler-colonizer nations' self-image and sense of prerogative.[36]

When Uluru was declared a national park in 1975, this only heightened competing assertions of sovereignty. Indeed, when the Northern Territory Land Rights Act was introduced in 1976, the Anangu peoples of Uluru lost any immediate chance of having their land rights recognized. Formerly classed as an Aboriginal reserve (reserved by the Crown or the government, not owned by Aborigines), its declaration as a national park officially alienated the land from the Crown and thus meant it could not be claimed under the Land Rights act. Anangu elders could not comprehend why white Australians assumed possession; they asserted that whites visiting Uluru had no *tjukurrpa*—a word for dreaming stories or creation ontologies. Indeed, whites "saw nothing" other than a big lump. For Anangu, each crag was proof of ancient epic events, as well as intimate present-day connections that laid down their system of law. What is more, Aboriginal people continued to live there, while very few whites did so on a temporary basis, let alone permanently.

In 1977, when indigenous owner and artist Wenten Rabuntja was informed that Uluru's national park status precluded it from being claimed under the new Northern Territory Land Rights Act (1976), he explained:

> That's my place out there . . . said Nipper Winmati, the white fella knows no stories out there. That's my place, I own, my father was born there and I was born there. White fellas haven't got their stories or dreamings out there—I don't try and take away Melbourne or Adelaide. . . . We want the white men to come and visit our place, not take it away. They must know it belongs to us. Now I am left with just my tent and broken promise.

Discovery and Expeditions

Settler-colonizer narratives involve an "invention of tradition" that includes inscribing moments of "firstness"—of origins, primacy, originality, preeminence, and naming rights.[37] As Gosse's 1873 journey to Central Australia was for a colonial purpose and backed by government authorities, he was not dubbed a traveler, wanderer, or rambler. As the first white male discoverer and climber, he was glorified with the valorized label of explorer. As John Weaver has demonstrated, the activities of surveying, mapping, and journeying were all part of the Great Land Rush of the New World, especially the search for pastoral land, and in Australia's case, for inland seas.[38] As with many other famous explorers, Gosse was not alone when he climbed the Rock. With him was camel-driver Kamran, in bare feet. An Aboriginal guide, whom Gosse knew as Moses, led the party—his nickname probably an abbreviation of his Aboriginal name, and perhaps also a humorous allusion to the biblical tale of Exodus.

In 1894 a scientific "expedition"—another loaded category of colonial purposefulness[39]—climbed the Rock. Only twenty-five Europeans climbed it in the next sixteen years. However, after a chain was erected on stanchions driven into the rock, the slippery, steep parts of the climb became more accessible to the wider public. From the 1940s and 1950s, with the advent of sturdy trucks and light aircraft, tourism expanded. Illustrated magazines such as *Walkabout* and *Smith's Weekly*, as well as visits by talented photographers using the new arts of aerial photography, adventurous journalists, writers and novelists, contributed to Ayers Rock being firmly etched into the Australian national consciousness.

With the outback already laden with cultural meaning, Ayers Rock's central desert location was more "out" the "back" than most—an immense distance from the east coast metropolitan centers. Devils Tower is similarly distant from large urban concentrations and in an iconic region associated with nostalgia for the Wild West. Since the 1960s, with the development of roads and the growing appeal of mass and remote tourism, the two monoliths have become increasingly accessible to tourists. Both can be accessed by good roads and highways, and they are readily visible without requiring visitors to leave their cars. This accessibility also makes them more reachable climbing destinations. In more recent decades, the rise of adventure tourism and "extreme" experiences only added to their appeal.

The National Ascent

Devils Tower was officially surveyed by the U.S. Geological Survey party in 1875, a couple of years after Gosse's survey of Uluru; however, the party found its steep slopes too forbidding to climb. In 1906 President Theodore Roosevelt's imprimatur on Devils Tower as the first national monument firmly established its special national status, but it had been earning its place as a symbol of nation beforehand. Local rancher William Rogers, the attributed "first climber" of Devils Tower, staged a nationalistic spectacle of the event. Accompanied by his wife and by fellow rancher Willard Ripley and climbing via a wooden ladder, they chose the Fourth of July, Independence Day, 1893, to make the ascent. This was also the year of the World's Columbian Exposition (or World's Fair) in Chicago, designed to celebrate the 400th anniversary of Columbus's arrival in the New World and to declare America's modernity to the wider world. And fittingly, at this exposition, historian Frederick Jackson Turner delivered his influential paper on "the significance of the frontier in American history," with its encounters-with-Indians nostalgia, outdoorsman values, and articulation of a national narrative of a modern, white colonizing manhood.[40] In this heyday of nationalist sentiment, America's progress was being watched by the wider world.

One way of asserting authority over landscapes is by naming or renaming. Another is by conquering geography—trekking to a locale, or in this case, climbing to the top. As we have seen, the official name of Devils Tower does not take indigenous sensibility into account. Renaming may present special problems as the custodial tribes have different names for the site, with divergent meanings. Historical timelines, with their positivist authority, tend to emphasize 1893, when the local rancher William Rogers made the first climb.[41] Demonstrating something to be shared with and circulated among his fellow Americans, Rogers's climb exhibited an iconography of masculine performance emphasizing courage, action, and strength. Dressed in the evocative representational garb of an Uncle Sam outfit, when Rogers reached the summit on the Fourth of July, he draped a huge stars-and-stripes flag over the monument. Local white women had crafted the flag and the suit.[42] Presumably Theodore Roosevelt would have approved; he was fond of projecting himself in the iconography of nationalist dress, being photographed in the fringed deerskin attire that drew on the hybrid garb of Native Americans and the pioneering frontiersman or backwoodsman.

In 1936 the Devils Tower climb was closed for safety reasons, although by then only about twelve people had succeeded in climbing it. The first "technical climb" took place in 1937. By 1954 only thirty-six people had climbed it, although within five years the figure climbed to 143. Due to safety concerns, the National Park Service conceded that climbing was becoming a "public relations problem." Subsequently, the service recruited skilled mountaineers and even started to encourage the activity, promoting a Mountaineers Week event in 1956. Soon the climb gathered further momentum.[43]

Following in the patriotic footsteps of Roosevelt and Rogers, white residents continued to deploy Devils Tower as a site of national celebration. In recent decades, climbs have taken place regularly on Independence Day. People dressed in historic clothing gather at the site to reenact George Washington's proclamations and the Declaration of Independence—moments that explicitly take pride in and memorialize American-born, white settler-colonizer sovereignty. The anniversary performances glorify the virtues of independence and liberty, and especially the gaining of New World manhood by ousting the British Old World imperialists. With the Fourth of July marking the independence of the American Republic and its sacred "manifest destiny," such bedecked climbs valorize moments of ascendant national sovereignty.

Similarly, Uluru has been frequently used as a setting for special ceremonies of nation. In the lead-up to the Australian Bicentenary in 1988, which celebrated the anniversary of the arrival of the First Fleet of British convicts (at faraway Port Jackson, Sydney), Uluru was chosen as the backdrop for a choir representing the diversity and inclusivity of the multicultural Australian nation. It was the starting point of the Olympic torch relays for the 2000 Sydney Olympics, and it has also been a conspicuous setting for advertisements for Qantas Airways' "the Spirit of Australia" theme and for a private loan company. Uluru's distinctive profile is constantly rehashed in tourism brochures, souvenirs, and artifacts.[44] As a complex signifier, it represents many things—the outback, a vast continent, emptiness, loneliness, the mystical, and the sacred. Since writers started to engage with the place, they have inscribed this monolith with a special role in the anthropomorphic body of the nation. As "the red heart" of Australia, and in the geographic heart of the country, it was as if Uluru had come alive, representing the life force, survival, a beating pulse, and as a symbol of courage, love, and sentiment for a solid and permanent nation.[45]

Sites of Pilgrimage or Sacrilege?

As monumental sites of memorialization, the national landscapes of Uluru and Devils Tower rely heavily on the creation of new chronologies of settler-colonizer histories of connection. In *Pilgrims of the Vertical,* American author Jay Taylor has explored pilgrimage as a theme taken up by Yosemite rock climbers.[46] At the same time, the mystique of the longer indigenous presence enhances Uluru and Devils Tower as tourist attractions. The indigeneity of the landscape purportedly adds magic, power, aura, and mystery, offering in situ ruminations on an almost lost, or a frozen, static moment of "primitive" history. Ninian Stephen's reference to Uluru as the "spiritual heart of Australia" in 1985 concurred perfectly with what historian and anthropologist Nicholas Thomas referred to as the "primitivist renovation of white identity via indigenous culture."[47] In American culture the theme of pilgrimage has powerful historical resonances; by the early nineteenth century, historical and religious pilgrim themes were drawn upon to create a new colonizer history for New England, as well as for the wider nation.[48]

Between the 1940s and 1960s, Australian writers and journalists such as Frank Clune, Beryl Miles, Arthur Groom, and Bill Harney wrote popular books that whetted the public appetite for tourism. Hedley Herbert Finlayson popularized the catchy metaphor of the "Red Centre."[49] Not only did such writers self-consciously follow each other's journeys and climbs, they repeatedly quoted each other, their fellow members of the "Brotherhood of the Rock," as poet Rex Ingamells had put it.[50] Frank Clune thus wrote: "Like Explorer Gosse . . . I 'had a view which repaid me for my trouble'; and, like him also, my feet 'were all in blisters, and it was as much as I could do to stand.'" When Clune descended, he blandly stated that it was "with our mountain-climbing ambitions satisfied."[51] Lacking deep historical meanings, connections with place could be shallow and comments disappointing. Travel writer Arthur Groom mentions Gosse's ascent and his own interest in timing his party's "race" to the top. But, like Clune, he desperately searches for a deeper significance: "I have been on many a mountain summit, and seen many a cairn of stones and bottle filled with names; but none excited me more than those accounts of the past few who have travelled hundreds and in some instances, thousands of miles, to ascend Ayers Rock. In this lonely land it seemed to give the names written in ink and pencil definite reality and personal presence."[52]

Another intrepid traveler, Beryl Miles, imagined her encounter with the Rock as the discovery of "a lost scroll," an image connecting the site with Northern Hemisphere antiquity, its history of writing, and the Bible. Describing her anticipation of the cairn of stones on the summit, she declared: "It was a proud moment when I added my name to those in the bottle."[53] Through the on-site archive of this commemorative cairn, it was as if individual experience had been elevated and at the same time sublimated to a collective ethos. In their names and signatures, visitors had thus left part of "themselves" behind. In schools, such entertaining tales delivered colonizer history lessons featuring heroes who conquered "a hostile landscape." Due to such literature and the growing visual record, climbing the Rock became a national, and increasingly well-documented, aspiration.

From time to time, Anglo-Australian notions of sacredness, and convergent ones merging with understandings of the Aboriginal sacred, have also been enacted around the Rock. Bill Harney, appointed the first ranger of the area in 1958, remembered "the holy expressions in white people's eyes as they came down after going to the top of the mountain."[54] In 1959 the legendary "Flynn of the Inland"—the Flying Doctor whose services assisted white women to settle and give birth in the "wilderness"—recorded a ritual conducted on slabs from the rock:

> with the magnificent view all around —the Olgas standing up like a medieval castle . . . all Australia stretching away on every side—the Sacrifice of Praise, of Gratitude, of Apology, of Petition, began. Father Brown was the first to raise the Host over Australia and over the men and boys kneeling on the Rock. . . . They were offered, these Masses, for the "complete Australia," both the living and the dead of our nation, all of white skin and all of black.[55]

Thus was celebrated an inclusive moment in which black and white were welcomed into the arms of a renewed, benign colonizing nation. Rex Ingamells and Miles had their own sacred missions that embraced reconciliation, albeit imagined exclusively from the colonizer perspective. Ingamells was instrumental in setting up the Jindyworobaks, a white nativist movement that aspired to create an authentic Australian culture by integrating Aranda/Central Australian Aboriginal language, song, and meaning into art, poetry, and literature. However, the sacred pilgrimage could also be secular. More recently, poet Barry Hill wrote of walking not in Aboriginal tracks but in Bill Harney's tracks, and more generally in "history's tracks."[56] To Hill, such makers of written national history themselves became the "congregation of pilgrims" who had gone before him.

Few non-Aboriginal people ever see the rock without the aid of cars, buses, and planes, yet they tend to consider Aboriginal people who travel in Toyotas and who live in towns and communities around the Rock as "inauthentic." Some tour operators have been reported to refer to contemporary Uluru residents as an "eyesore," contributing to a "degraded" rather than "pure" Aboriginal vista. For others, Aboriginal people at Uluru emphasize disparities between the social status of black and white, constituting an unwelcome reminder that all is not well with the nation. Devils Tower's associations with the wonder of the Wild West are celebrated in its souvenir shops and fort-style architecture, yet less evident is the role of several American Indian tribes in the ranching industry and other modern economic pursuits, because they do not fit as neatly with the designated "authentic" image.[57] Dream catchers and spiritual iconography sell better, too, as the Indian link is more comfortably associated with the spiritual, and with "lost cultures," rather than land development and the ranching economy.

So What Is It about Climbing?

Climbing is an activity heavily laden with cultural values. It offers a powerful poetic: the individual arises above the masses, pitting body and mind to meet a great challenge. After the ascent, he or she achieves a new vantage point on the world. Especially associated with manhood, it enables the individual to prove himself—his mind and body can master a mountain. The virtues of perseverance, determination, planning, intelligence, self-reliance, and skill are essential. The ascent takes grit, but the reward—getting to the top—becomes a metaphor for reaching the toughest of human goals.

Mountains have featured powerfully in humankind's assertion of dominance over the natural world. In the twentieth century, this reached its ultimate expression in the response to the news that a New Zealand "man of the Empire," Edmund Hillary, and his Nepalese companion, Sherpa Tenzing Norgay, had reached the summit of Mount Everest, the world's tallest mountain, in 1953. This was reified as a "conquering" moment of empire, with the queen of England conferring the highest imperial honors upon Hillary and the second highest upon Tenzing. Although not staking any claim to land, reaching the summit, the "roof of the world," was nonetheless considered a symbolic victory for the British Empire. For much of the twentieth century, Boys' Own adventure books promoted ideas of empire and conquering the world. The language of domination was slotted into an imperial tradition of landscape mastery. Explorers symbolically conquered strangeness, distance, drought, wilderness, and "hostile natives." Although they followed their tracks and were usually assisted by indigenous trackers in Australia and Indian "scouts" in the United States, the main heroes were white men. By the 1970s, however, with so many decolonizing movements around the world, the lure of the imperial adventure was losing its hold.[58]

Climbing was increasingly associated with the spiritual and the "rising above" implicit in the transcendent experience.[59] While climbing pits man against nature, to use a common phrasing of mastery and domination, at the same time the climber ends up with nature, in an almost religious state of being.[60] The climber's sense of achievement is inevitably in connection with landscape, nature, place—and, as we have seen, this can also include another kind of transcendence—beyond the individual, toward a sense of belonging to the larger collectivity of nation.

At certain sites, reaching the summit can function as an endlessly repeated individual assertion of an earlier chronology of national dominance and conquest, a moment of arrival in a strange place. The moment of conquering comes to reside in a timeless zone of endless colonizer repetition. That is, every climber can win the prize and conquer. For the individual, it can always be a first. Each later climber walks or climbs in the pathways of earlier climbers, and as we have seen, people often self-consciously try to reconnect with the moment of the primeval white gaze. As a landed communication with the wilderness and the modern, climbing promises a white national moment belonging to all. Yet the seemingly innocuous "open to all" inevitably generates contests of power and control.[61]

As national parks authorities have the role of managing the national estate, they place controls for ecological or environmental reasons. They may ban or restrict certain kinds of accommodation or activities, for example, the establishment of tents and camping grounds near environmentally sensitive sites. They may also ban indigenous people from hunting or fishing, or at least using modern technologies to do so. In other words, while a prescribed set of activities can be followed, for conservation or other reasons, people may be locked out of activities that would enable them to maintain the very land connections that add deep cultural value to the sites.

At the same time, many visitors to sites value indigeneity as the star attraction. As historian James Clifford has pointed out, the "indigenised landscape" is often imagined as an outdoor museum, "a fixed heritage," like a photograph or a museum collection designed to entertain the visitor.[62] Some climbers are on a specific quest for a more ancient, primeval imagining of what it is to be indigenous. The rhetoric of heritage reinforces the notion that indigenous people have heritage but not necessarily history, which is the preserve of white settler-colonizers. Urbanized indigenous people also travel to such places simply as tourists, or searching for a connection with "pure" or "real" indigenous culture prior to colonizer destructiveness.[63] For others, it is simply to get into the spirit of an adventurous school field trip.

Many tourists cherish a sense of entitlement to do whatever they please. At Uluru, a common activity young men use to mark arrival at the top is urination, which, in its after-stench, can make the journey's climax somewhat unpleasant. People scratch their name on the rock. Hitting a golf ball off the summit to see how far it might go across the Plains is also a favorite visitor activity. Advertising agencies have asked permission to use Uluru to shoot scenes of paint splashing over it or planes crashing into it, suggesting that many people have little insight into indigenous sensibilities or the basics of parks management. At least such proposals have been tightly vetted by parks and by indigenous owners and disallowed. Recently a female French performer did a striptease atop a windy Uluru that proved a hit on YouTube. When the Anangu male elders stated that they considered her actions offensive (it was a men's site), she retorted that Aboriginal people had traditionally been naked. Primitivist ideas have wide circulation far beyond the nations in which these iconic geographies are situated.

Devils Tower has had its own set of bizarre events, including the story of the adventurer George Hopkins, who in 1941 parachuted out of a plane onto the mountain and could not get down.[64] In the 1980s and 1990s, bolts were inserted into the granite to facilitate and improve the safety of crack climbs. With abseiling gaining popularity, climber numbers increased. Every year, hundreds of brave climbers attempt the very steep ascent, some of whom strap themselves with cameras as well as ropes to leave a full record on YouTube. At the summit, successful climbers have erected small and large stone cairns. One contains an encased visitor's book in which climbers record their comments.[65]

Souvenirs reveal changing meanings. One American pro-climbing organization uses the name "Devils Tower Sacred to Many People," and assures its website customers that its sales of souvenirs and T-shirts will assist Native Americans.[66] In the 1980s, T-shirts declaring in large letters "I Climbed Ayers Rock" were conspicuously worn around the site. That T-shirt now competes for sales with another one proclaiming, "I Didn't Climb Ayers Rock." Visitors to Devils Tower make strained gestures toward respectfulness. As well as purchasing recognizable "Indian" motifs such as feathered headdresses and Wild West imagery, Devils Tower T-shirts glorify its associations with *Close Encounters of the Third Kind*—and with believers in UFOs and aliens generally. While popular culture can absorb both aliens and Indians into a mystique of otherness, this majestic rock's association with outer space "aliens" lends it a mysterious, ominous aura that can entirely wash away indigenous meanings.[67]

Sacred Nations

White American nationalism and Native nationalism have had some transnational moments arguing over rocks. As James Clifford explains, local and global "contact zones" are sites of "identity making and transculturation, containment and excess," that suggest an ambiguous future for "cultural difference."[68] With legislative endorsement, national parks function to serve particular communities that are "inside nation." Not necessarily a stable space, these can variously include indigenous and other peoples, as Catton's chapter has also demonstrated for New Zealand.[69] As the exchange, the transaction, however, involves more than "gifting," it will be constantly contested—creating sites for sovereignty struggles that go to the foundations of settler-colonizer nations.

At Uluru and Devils Tower, we have observed how a conquering, settler-colonizer sacred vies with a modern, decolonizing, indigenous sacred. The Anangu close the Rock when an important elder dies. While they have no rights to close the climb at Devils Tower, Plains Indians tie prayer ribbons and meet to conduct gatherings and ceremonies annually. Some nonindigenous peoples also view the two monoliths as magic places—sites of potential good and evil power.

The symbolics of these two national monuments seem to be enhanced by the continuing "Indian" and Aboriginal presence and their mythologies. For Ayers Rock, the Aboriginal presence became ubiquitous on postcards, ashtrays, plates, and other souvenirs from the 1960s, adding a picturesque touch of the "primitive" and the unique. Apparently, such popular iconography of Aboriginal people was compatible with notions of emptiness, ancientness, discovery, conquest, or wilderness. The Aboriginal and the Indian were part of nature and the original white panorama. European accounts of visits to the Rock by scientists such as Walter Baldwin Spencer and Charles Mountford brought Aboriginal culture to the forefront.[70] National parks literature has increasingly emphasized the indigenous significance of Uluru and Kata Tjuta in an attempt to educate visiting tourists and thus broaden their appeal as tourist destinations.

Similarly, the significance of Devils Tower to Plains Indian groups has long been recognized by the National Park Service, albeit in a deracinated, deperson-alized format that presents the traditional stories of the place as mythologies, even as "colorful fantasies."[71] Largely absent in the official record are the voices and presence of contemporary Plains Indian peoples. Their interpretations of Devils Tower are sometimes represented as if they belong in the past, conve-niently dislocated from present-day issues and concerns that stress continued connection and belonging, and from a desire to be involved in the management of the park.

Both Devils Tower and Uluru host Western-style New Age and Christian spiritual gatherings, including spiritual quests to "be Indian" or "Aboriginal." The rocks around Devils Tower are used in Plains Indian ceremonies, as is the site itself. Ideas about power centers of special magnetic force draw people into cosmic earth journeys. New Agers and spiritual seekers have been found on various occasions chanting in off-limits sites such as sacred women-only caves. At both sites, climbers have died, imbuing the place with additional memorial significance. At both, tourists constantly pilfer rocks and, after taking them home, fall into the belief that runs of bad luck were attributable to the rocks being in their possession. They send them back in such quantities that the parks authorities do not know what to do with them.[72]

We have now discussed how settler-colonizer states deploy Uluru and Devils Tower as modern ways of asserting new meanings imbued with nationalistic sacredness, and how Native people use these sites to assert ongoing prior sover-eignties often based on spiritual significance. Yet beyond the binaries—which are essentially blurred through long histories of coexistence, intermarriage, and a mutual engagement with modernity and the larger nation—the situation is complex and convergent. Climbers, too, may be indigenous and may not neces-sarily follow tribal protocols.

Significantly, however, numerous indigenous people are currently using national parks as a forum in which to assert their knowledge, to tell stories, and to teach deeper meanings and longer histories. In Australia and North America, many indigenous rangers work in parks interpretation. Many sit on key committees and spend hours each week pursuing their cultural obligations. In Uluru and other Australian World Heritage areas, Aboriginal people have formed land councils, elders councils, and local, regional, and state committees to which parks turn for advice. Although the formation of parks often led to the dispossession of indigenous people, national parks have now become a key venue for sharing histories. Park interpretation may serve as the "pedagogical arm of reconciliation." For nationals associated with the respective sites, some use key national parks sites to imaginatively jump the gap between the perceived binaries of colonizer and colonized.

This paper has suggested that, when looking at "national parks" in settler-col-onizer nations, the term "transnational" takes on new layers.[73] The appellation "national" for landscapes denotes a strategy of ongoing spatial colonization that potentially creates categories of indigenous exclusion and inclusion. Yet national parks and national monuments have created spaces for the assertion of indigenous and nonindigenous sovereignty over lands, as well as for negotiation and enactment of the ideals of reconciliation. Following a long period of consultation with traditional owners, tourist access to the climb was scheduled to cease at Uluru in 2012. The date appears to have been extended. Who steals the higher moral ground, if any, is yet to be seen. Parks are elastic signifiers of the sacred that can become venues for imagining a more benign colonization, and possibly a more just settlement, between Native peoples and colonizers. Nonetheless, in this metaphoric national climb, the summit always seems to loom slightly out of reach.

Notes

1. This paper refers to Uluru and Devils Tower according to contemporary official terminology. Where relevant, I refer to Uluru as Ayers Rock. The appellation "Devils Tower" has been a lingering source of controversy, and there have been periodic but unsuccessful attempts to adopt a second official name that better reflects its religious significance to Plains Indian groups. See Jared Farmer, "Devils Tower May Get a Second Name," *High Country News,* September 2, 1996, available at https://www.hcn.org/issues/89/2740 (accessed August 26, 2015). The place is sometimes referred to as Bear Lodge (Mateo Tepee), or by its Lakota name Mato Tipila, both of which refer to the place's sacred association with the bear. Recognizing that different names apply across different tribal groups and the contested nature of place names, in this paper I somewhat reluctantly adhere to its official name.

2. Melissa Harper and Richard White, "How National Were the First National Parks? Comparative Perspectives from the British Settler Societies," in *Civilizing Nature: National Parks in Global Historical Perspective,* ed. Bernard Gissibl, Sabine Höhler, and Patrick Kupper, (New York: Berghahn, 2012), 50.

3. Haydn G. Washington, "The "Wilderness Knot," USDA Forest Service Proceedings (2007), Rocky Mountain Research Station, www.fs.fed.us/rm/pubs/rmrs_p049/rmrs_p049_441_446.pdf (accessed August 26, 2015).

4. William Cronon, "The Trouble with Wilderness: or, Getting Back to the Wrong Nature," in *Uncommon Ground: Rethinking the Human Place in Nature,* ed. William Cronon (New York: W. W. Norton, 1995), 69–90.

5. I. Tyrrell, "Comparative and Transnational History," *Australian Feminist Studies* 22, no. 52 (March 2007): 49–54; Joseph E. Taylor III, "Boundary Terminology," *Environmental History* 13, no. 3 (2008): 454–81.

6. Ernest Giles, *Australia Twice Traversed: The Romance of Exploration: Being a Narrative Compiled from the Journals of Five Exploring Expeditions into and through Central South Australia and Western Australia, from 1872 to 1876* (1889; reprint, Perth, Australia: Hesperian Press, 1995), May 21–July 20, 1874; see also Jay Arthur, "Natural Beauty, Man-Made" and Michael Cathcart, "Uluru," in *Words*

for Country: Landscape and Language in Australia, ed. Tim Bonyhady and Tom Griffiths (Sydney: UNSW Press, 2002), 191–221, 207–21.

7. Devils Tower National Monument, "A Proclamation by the President of the United States of America," Record Group 79 (National Park Service), Entry 10: Central Classified Files, 1907–49, U.S. National Archives, College Park, Maryland (hereafter cited as RG 79), Box 584, Folder "[Devil's Tower National Monument] History, General, 1892–1942" (location: RG 79/150/32/24/3).

8. Devils Tower National Monument, "Glimpses of Our National Parks, 1932–35," RG 79, Box 2155 (location: RG 79/150/34/8/3).

9. http://www.law.cornell.edu/uscode/html/uscode16/usc_sec_16_00000431----000-notes.html (accessed January 25, 2012).

10. M. Lake and H. Reynolds, *Drawing the Global Colour Line: White Men's Countries and the International Challenge of Racial Equality* (Cambridge: Cambridge University Press, 2008)

11. Harper and White, "How National Were the First Parks?" 59.

12. George Linge, 2000, "Ensuring the Full Freedom of Religion on Public Lands: Devils Tower and the Protection of Indian Sacred Sites," 27 B.C. Envtl. Aff. L. Rev. 307, p. 311, http://lawdigitalcommons.bc.edu/ealr/vol27/iss2/5/ (accessed January 20, 2012).

13. Ibid., 312–14.

14. See Hilary du Cros and Chris Johnston, 2001, "Tourism Tracks and Sacred Places: Pashupatinath and Uluru. Case Studies from Nepal and Australia," in Conference Proceedings, Australia ICOMOS 2001, *Making Tracks,* http://www.aicomos.com/wp-content/uploads/Tourism-tracks-and-sacred-places-Pashupatinath-and-Uluru.-Case-studies-from-Nepal-and-Australia.pdf (accessed August 27, 2015). For another insightful transnational comparison, see Jane Carruthers, "South Africa and Australia: Comparing the Significance of Kalahari Gemsbok and Uluru–Kata Tjuta National Parks" in *Disputed Territories: Land, Culture Identity in Settler Societies,* ed. D. Trigger and G. Griffiths (Hong Kong: Hong Kong University Press, 2003), 233–68. See also Taylor, "Boundary Terminology."

15. Patrick Wolfe, "Land, Labor, and Difference: Elementary Structures of Race," *American Historical Review* 106 (2001): 867.

16. Andrew Boxer, 2009, "Native Americans and the Federal Government," *History Review* 64 (September 2009), available at http://www.historytoday.com/andrew-boxer/native-americans-and-federal-government (accessed January 30, 2012).

17. Sarah Maddison and Morgan Brigg, *Unsettling the Settler State: Creativity and Resistance in Indigenous Settler-state Governance* (Sydney: Federation Press, 2011).

18. See World Heritage Convention website on Willandra Lakes Region, http://whc.unesco.org/en/list/167 (accessed July 10, 2015).

19. For a summary of these developments, see Ann McGrath, *Contested Ground: Aborigines under the British Crown* (St. Leonards, NSW: Allen & Unwin, 1995), introduction and chap. 1.

20. Richard Sellars, *Preserving Nature in the National Parks* (New Haven, Conn.: Yale University Press, 2009), 214, 256.

21. Ibid., 13.

22. Mark David Spence, *Dispossessing the Wilderness: Indian Removal and the Making of the National Parks* (New York: Oxford University Press, 1999), 214.

23. Harper and White, "How National Were the First Parks?" 52.

24. T. Banivanua Mar and P. Edmonds, eds., *Making Settler Colonial Space: Perspectives on Race, Place, and Identity* (New York: Palgrave Macmillan, 2010), 76.

25. Denis R. Byrne, "Nervous Landscapes: Race and Space in Australia," *Journal of Social Archaeology* 3, no. 2 (June 2003): 170.

26. D. Byrne and M. Nugent, *Mapping Attachment: A Spatial Approach to Aboriginal Post-contact Heritage* (Hurstville, Australia: Department of Environment and Conservation, 2004).

27. McGrath, *Contested Ground*.

28. "Courageous MP Fought for Social Justice," *Canberra Times,* August 6, 2011.

29. The term *terra nullius* was often applied as the rationale behind colonizing takeover. This was overturned by the *Mabo* judgment of 1992, and later by native title legislation. For a summary of Australian government policy and legislation, see McGrath, *Contested Ground*, chap. 1.

30. George L. San Miguel, 1994, "How Is Devils Tower a Sacred Site to American Indians?" National Park Service, U.S. Department of the Interior, http://www.nps.gov/deto/index.htm (accessed April 17, 2011).

31. Robert Layton, *Uluru: An Aboriginal History of Ayers Rock* (Canberra: Aboriginal Studies Press, 1986), 3–16; Stanley Breeden, *Uluru: Looking after Uluru-Kata Tjuta the Anangu Way* (Marleston, South Australia: J. B. Books Australia, 1997), 95.

32. Barry Hill, *The Rock: Travelling to Uluru* (Sydney: Allen & Unwin, 1994), 96.

33. A range of valuable new resources are now posted on the NPS website. See National Park Service, "Devils Tower National Monument: First Stories," http://www.nps.gov/deto/historyculture/first-stories.htm (accessed April 13, 2011); http://www.nps.gov/deto/learn/historyculture/first-stories.htm (accessed August 17, 2015).

34. See, for example, http://www.billandcori.com/devilstower/devils_tower.htm (accessed July 9, 2015); http://travelingringo.com/2011/02/the-power-of-devils-tower (accessed April 14, 2011).

35. A. McGrath, "Travels to a Distant Past: Mythologies of the Outback," *Australian Cultural History* 10 (1991): 113–24; J. Marcus, "The Journey Out to the Centre: The Cultural Appropriation of Ayers Rock," in *Aboriginal Culture Today/Kunapipi, ed. Anna Rutherford* (Sydney: Dangaroo Press, 1998), 254–74; Elvi Whittaker, "Public Discourse to Sacredness: The Transfer of Ayers Rock to Aboriginal Ownership," *American Ethnologist* 21, no. 2 (1994): 310–34.

36. McGrath, "Travels to a Distant Past."

37. Nicholas Thomas, *Colonialism's Culture: Anthropology, Travel and Government* (Melbourne: Melbourne University Press, 1994), 2. See also Harper and White, "How National Were the First Parks?" 60; Jean O'Brien, *Firsting and Lasting: Writing Indians Out of Existence in New England* (Minneapolis: University of Minnesota Press, 2010).

38. John Weaver, *The Great Land Rush and the Making of the Modern World, 1650–1900* (Montreal: McGill–Queen's University Press, 2003).

39. For example, see Martin Thomas and Margo Neale, eds., *Exploring the Legacy of the 1948 Arnhem Land Expedition* (Canberra: ANU E-Press, 2011).
40. See Ray Allen Billington, ed., *Frederick Jackson Turner: Historian, Scholar, Teacher* (New York: Oxford University Press, 1973).
41. Devils Tower National Monument, RG 79/150/34/8/3.
42. John B. Dorst, *Looking West* (Philadelphia: University of Pennsylvania Press, 1999), 201–2. See also Frederick Augustus Fidfaddy, *The Adventures of Uncle Sam in Search after His Lost Honor* (Middletown, Conn.: Seth Richards, 1816), which is attributed as launching this allegory.
43. "Devils Tower National Monument; History, General, 1892–1942," RG 79/150/32/24/3; "Devils Tower National Monument; Newspaper Articles, Press Notices," RG 79, Box 2155, RG 79/150/34/8/3; National Park Service, U.S. Department of the Interior.
44. Hill, *The Rock*, 14.
45. McGrath, "Travels to a Distant Past."
46. Joseph E. Taylor III, *Pilgrims of the Vertical: Yosemite Rock Climbers and Nature at Risk* (Cambridge, Mass.: Harvard University Press, 2010).
47. Thomas, *Colonialism's Culture, 171.*
48. For an in-depth discussion of pilgrimage and history making, see O'Brien, *Firsting and Lasting*; Joseph Conforti, *Imagining New England: Explorations of Regional Identity from the Pilgrims to the Mid-twentieth Century* (Chapel Hill: University of North Carolina Press, 2001); Joseph Conforti, *Jonathan Edwards, Religious Tradition, and American Culture* (Chapel Hill: University of North Carolina Press, 2005).
49. H. H. Finlayson, *The Red Centre: Man and Beast in the Heart of Australia* (Sydney: Angus & Robertson, 1935).
50. Rex Ingamells, 1953, "Journey to the Rock," in *Walkabout*, 119, 120. See also Hannah Hueneke, "To Climb or Not to Climb? The Sacred Deed Done at Australia's Mighty Heart," unpublished honors thesis, Australian National University, 2006, 17–18.
51. Frank Clune, 1941, "Ayers Rock," in *Walkabout, 7*, 11–15.
52. A. Groom, *I Saw a Strange Land: Journeys in Central Australia* (Sydney: Angus & Robertson, 1950), 166.
53. Beryl Miles, *The Stars My Blanket* (London: John Murray, 1954), 18, cited in Hueneke, "To Climb or Not to Climb?" 72.
54. Bill Harney, 1963, *To Ayers Rock and Beyond* (London: Robert Hale, 1963), 92.
55. Frank Flynn and Keith Willey, *Northern Frontiers* (Sydney: F. P. Leopard, 1964), 101, 19.
56. Hill, *The Rock, 17.*
57. See Ann McGrath, *Born in the Cattle: Aborigines in Cattle Country* (Sydney: Allen & Unwin, 1987).
58. Marie Louise Pratt, *Imperial Eyes: Travel Writing and Transculturation* (London: Routledge, 1992), 1–11 and part 3.
59. David Craig, *Native Stones: A Book About Climbing* (London: Pimlico, 1995), ix.
60. Taylor, *Pilgrims of the Vertical, 13–14.*
61. Craig, *Native Stones*; Taylor, *Pilgrims of the Vertical.*

62. James Clifford, *Routes: Travels and Translation in the Late Twentieth Century* (Cambridge, Mass.: Harvard University Press, 1997), 217.

63. Ruby Langford, *Don't Take Your Love to Town* (Ringwood, Australia: Penguin, 1988), 222–26.

64. Devils Tower National Monument, RG 79, Box 2154, RG 79/150/34/8/3.

65. The book appears to be held in a cylinder. See also "Devils Tower National Monument; History, General, 1892–1942," RG 79/150/32/24/3; U.S. National Archives, "Devils Tower National Monument; Newspaper Articles, Press Notices," "Devils Tower National Monument; Publications, General, 1899–1948," and "Devils Tower National Monument; Glimpses of Our National Parks, 1932–35," all in RG 79/150/34/8/3.

66. http://www.devilstower-sacredtomanypeople.org/shop.html (accessed January 28, 2012; online shop no longer exists).

67. See, for example, M. Sturma, "Aliens and Indians: A Comparison of Abduction and Captivity Narratives," *Journal of Popular Culture* 36 (2002): 318–34.

68. Clifford, *Routes,* 219.

69. Sellars, *Preserving Nature in the National Parks*; Spence, *Dispossessing the Wilderness,* 214.

70. Breedon, *Uluru,* 140–41.

71. U.S. National Park Service, "Devils Tower National Monument: A Brief Description," March 1938, RG 79, Box 2154, RG 79/150/34/8/3.

72. Craig, *Native Stones,* 6. See also David Craig, *Landmarks: An Exploration of Great Rocks* (London: Pimlico, 1996); Sarah James, "Negotiating the Climb: Uluru, a Site of Struggle or Shared Space?" (unpublished research paper, University of Melbourne, 2005); Hueneke, "To Climb or Not to Climb?" 59.

73. Tyrrell, "Comparative and Transnational History"; Taylor, "Boundary Terminology."

| # The Trouble with Climate Change and National Parks

MARK CAREY

GLOBAL WARMING IS THREATENING what many see as the world's most pristine environments: national parks from Glacier in the United States to Kilimanjaro in Tanzania, Sagarmatha (Mount Everest) in Nepal, and Huascarán in Peru, among many others. Worse, many see global warming as driven by the same nefarious forces that people hope to escape in parks—industrialization, natural resource exploitation, development and sprawl, consumerism, and rampant capitalism—thereby making climate change particularly offensive when it tarnishes the parkland sanctuary. Melting glaciers, forest fires, hungry polar bears, species migration and extinctions, diminishing water supplies, depleted scenery, declining tourism economies—these are what the media and environmental groups report as the most severe consequences of climate change in the world's national parks. As the Rocky Mountain Climate Organization and Natural Resources Defense Council recently asserted, "Human disruption of the climate is the greatest threat ever to our national parks."[1]

Media accounts and environmental groups' reports about climate change in national parks reveal a great deal not only about climate impacts but also about embedded ideas of wilderness, human-nature interactions, and the place of parks in national narratives of nature.[2] Journalists and nongovernmental organizations (NGOs) typically have to—or at least choose to—simplify their coverage of climate change in national parks. What they decide to convey to readers and, more importantly, what they leave out actually reveals a great deal about their values, ideas, and perceptions of both climate change and national parks. In many cases, the climate change discourse on national parks resembles the declensionist narrative of environmental degradation that has long been at the center of the environmental and wilderness preservation movements—a narrative that scholars have also critiqued.[3] When news stories and climate reports simplify human-environment dynamics, exaggerate global warming impacts, or avoid discussion of complex human-nature dynamics, they can also perpetuate a certain "traditional" view of national parks.[4] I thus argue that the last fifteen years of media accounts and environmental groups' reports on climate change in national parks actually divulge as much about popular perceptions of parks and the relationship between people and nature as they do about climate change impacts. What's more, these portrayals that depict climate change as such a tragic threat may even impede more effective efforts in climate change adaptation, because they emphasize tourist wishes and the aesthetics of park landscapes, or they portray people as passive victims, instead of delving deeper into ecosystem processes, social-ecological systems, human livelihoods, natural hazards, resource management, social justice, and many other critical issues affected by global climate change.[5]

In addition to uncovering these narratives of national parks embedded in climate discourse, this essay also strives to break out of the pervasive and highly restrictive believer/skeptic dichotomy that characterizes much climate research. After studying climate change and glacier retreat impacts in the Andes for more than a decade, I can attest to the fact that climate change has already unleashed catastrophic consequences that killed thousands of residents around Huascarán National Park. Further, my research underscores how more marginalized populations often suffer disproportionately from climate-related hazards, resource conflicts, and biophysical changes stemming from climate change.[6] But just because I understand the deadly effects of climate change does not mean I should step into the mold that pits global warming "believers" against "skeptics." Clinging to one of these two climate "camps," I believe, has derailed scholarship and made it nearly impossible for scholars to critique any aspect of climate discourse without coming across as a skeptic.[7] What's more, the language that refers to believers and skeptics gives the discussion strong religious overtones, thereby shutting down open, critical analysis of climate change. Finally, an uncritical adherence to the "believers" camp out of fear of providing ammunition for skeptics runs the risk of potentially doing social science and humanities research in the service of Western science or environmentalist groups' priorities, both of which scholars have long critiqued for being socially constructed and having embedded agendas of their own.[8]

The goal in this essay is to critically examine media accounts and NGO reports of climate change impacts in Huascarán National Park in Peru and Glacier National Park in Montana. Both Huascarán and Glacier are located in high mountain glaciated environments, are globally high-profile parks, and have received significant international attention related to climate change. This analysis involved an exhaustive search for news articles examining these parks written during the last fifteen years. Media accounts and news articles influence public views and also offer helpful insights into popular perceptions. Interestingly, most journalists tended to cite and quote environmental organizations, rather than scientists or even park officials, as the climate experts in their news stories. Therefore, this research also involved analysis of reports by influential environmental NGOs—not park officials or scientists—because they are the ones overwhelmingly represented in the media. Media reports in Spanish from Peru and in English from the United States and the United Kingdom were consistent in their portrayal of parks and climate change: they all linked people and parks in Peru. The Spanish-language sources, however, rarely mentioned Glacier National Park; therefore, a divergent view between Glacier and Huascarán was not possible to detect from this research. Of approximately 100 news articles and a dozen NGO reports found on national parks and climate change, about one-third of them were devoted entirely or significantly to Huascarán and Glacier. This research also included historical analysis to help juxtapose past depictions of Huascarán and Glacier against more recent accounts. The historical perspectives show not only how perceptions changed over time, but also how the issues identified as global warming impacts also reveal long-standing cultural values and ideas about national parks.

Parks and People in Peru

The iconic Huascarán National Park in the Peruvian Andes has attracted significant international attention related to climate change. In fact, international NGOs have petitioned the World Heritage Committee more than once to designate Huascarán—along with a handful of other parks such as Glacier and Sagarmatha (Mount Everest)—as an "endangered" site precisely because of the effects of global warming. The climate discourse about Huascarán reveals several things about perceptions of national parks in Peru. First, nature is generally portrayed, or idealized, as static scenery, and Huascarán is often cast as a place primarily for recreation and tourism. Second, environmental processes and socioenvironmental dynamics are usually simplified, while certain societal or environmental changes are misattributed to climate change. Third, deterministic predictions about future climatic/environmental changes tend to minimize human agency and ignore past adaptation accomplishments, thereby turning Peruvians into passive victims. Despite many similarities in the depiction of climate change in Huascarán and Glacier National Parks, the conceptualizations of climate impacts in the two parks also illuminate a fundamental difference in the perceptions and meanings of U.S. and Peruvian national parks. In Peru, the parks are not discursively separated from people as they are in the United States. Rather, Huascarán National Park is inextricably connected to human populations through natural hazards, natural resources, the tourism economy, and water supplies.

Huascarán National Park was created in 1975, and it remains one of the country's preeminent parks. It protects a significant part of the upper elevations within the Cordillera Blanca, a 200-kilometer-long mountain range that includes twenty-seven peaks above 6,000 meters, including Peru's highest, Mount Huascarán (6,768 meters). Huascarán became a UNESCO biosphere reserve in 1977 and was first inscribed on the World Heritage list in 1985. The Cordillera Blanca is the highest and most glaciated tropical mountain range in the world, and parkland varies from 2,500 meters to 6,768 meters above sea level, offering a host of ecological life zones, ecosystems, plant and animal diversity, and climatic variation within the park. The park also protects the *puya raimondi,* the world's largest bromeliad, and is home to animals such as the spectacled bear and Andean condors.[9] Yet the park also has strong foundations in recreation, tourism, and mountaineering. The roots of the park's creation, in fact, lie in tourism promotion as much as plant and animal conservation.[10] Interestingly, the early proposals and recommendations for the park barely mentioned the park's glaciers; instead, they focused on its lakes, flora and fauna, forests, high mountains, and geology.[11] Today park descriptions focus on the glaciers themselves, as in Glacier National Park.

With shrinking glaciers highlighted as the principal climate change impact in Huascarán National Park, media accounts imply that glaciers and other aspects of the natural environment in national parks should remain as static scenery primarily for tourists to enjoy. Many articles focus on the effects that glacier retreat will have or has had on tourism and mountaineering, and they cite a 20 to 35 percent loss of Cordillera Blanca ice since the late 1970s as evidence of this impact.[12] Some also lament the loss of the ice caves, which were a beautiful feature of Huascarán's Pastoruri Glacier before they melted.[13] Another complained, "The glacier looks like a patient dying of a virus." But the real problem expressed in this news story is that the glacier terrain is "not normal" because it is unstable and problematic for mountain climbers.[14] Accounts also tend to exaggerate the rate of future glacier retreat, thereby making a statement about how changing park scenery is lamentable. Some accounts suggest twenty years for the disappearance of glaciers, and one journalist lamenting ice loss in Huascarán National Park and other World Heritage sites noted that Peru will lose "almost all [glaciers] within the next 7 years."[15] The claim that Cordillera Blanca glaciers will disappear even in fifty years cannot be found in scientific literature, and informally I have heard glaciologists say total ice loss in the Cordillera Blanca would likely be on the scale of centuries, not decades.[16]

Iconic species also appear prominently in climate discourse, suggesting how the nature in national parks is often identified as charismatic flora and fauna— even if news stories provide little evidence of climate impacts on these species. As one representative news article mentions for Huascarán, "the Andes are home to many rare species. The mountains are populated by llamas which can be found living at high altitudes, predominantly in Peru and Bolivia. The South American condor, the largest bird of its kind in the Western hemisphere, is also found here as are pumas, camelids, partridges, parinas, huallatas (geese) and coots."[17] The article's focus on climate change impacts leads the reader to believe these species are actually affected by climate, but there is no evidence provided. Moreover, the statement that llamas are a rare species would be like suggesting that white-tailed deer are rare in Massachusetts or cows unusual in Iowa. Of course, climate change does threaten species inhabiting national parks, and in some cases species might migrate outside park boundaries, thus creating new dilemmas for park managers.[18] But when climate change news stories claim that climate change affects bears, llamas, condors, and other symbolic species without offering any evidence of those impacts, then the articles reveal a tendency to associate national parks with iconic species and charismatic fauna rather than other lesser-known species or ecosystem processes.

Most media accounts of climate change impacts in Huascarán National Park reveal a very different perspective than exists for Glacier National Park in the United States: local people in Peru are shown to be intimately tied to the national park, unlike the common portrayal of U.S. parks that (mis)depict them as standing in isolation from all surrounding populations. The two contrasting popular narratives of people connected to parks in Peru and of vacant wilderness parks devoid of nearby populations in the United States are thus affecting the ways in which people learn about climate change impacts—and vice versa. One way the link between people and nature in national parks comes up for Huascarán is through an emphasis on natural hazards associated with climate-induced glacier retreat. In the United States, in contrast, the disaster narrative for parks often centers on the natural environment: disappearing glaciers or threatened plant and animal species that must migrate or go extinct. In Peru, climate change has already caused catastrophic consequences in Huascarán from glacial lake outburst floods (GLOFs) that have killed thousands of residents.[19] Glacial lakes formed at the foot of retreating glaciers after the end of the Little Ice Age in the mid-nineteenth century. As the ice retreated, lakes formed that were precariously dammed behind unstable moraines. In 1941 Lake Palcacocha burst through its moraine dam and killed 5,000 people in the city of Huaraz. Two additional GLOFs occurred in 1945 and 1950 that killed nearly 1,000 people and destroyed one of the country's largest hydroelectric stations at Cañón del Pato. Other types of glacier disasters originating in the park have caused even more deaths, including the 1962 glacier avalanche that killed 4,000 people in Ranrahirca and the 1970 earthquake-triggered avalanche in Yungay that killed at least 6,000 people (based on studies completed since official documents put the death toll at 15,000).

News stories about climate change in the park generally mention this history of disasters and the potential for more in the future, though they sometimes exaggerate the potential impacts. A 2007 report written by a variety of international NGOs and published by UNESCO exemplifies these views. The section of the case study report on Huascarán asserted that "the livelihood of two million people living within the immediate vicinity of the Huascarán National Park is threatened by high-altitude glacial lakes with the combination of climate change, local seismic activity, and increased glacier and hill slope instability."[20] Another news story offers a similarly high number, suggesting that "millions of people in the Peruvian Andes live under threat from catastrophic floods caused by global warming."[21] Recent census data and GIS spatial analysis, however, reveals that fewer than a half million people live in all the surrounding areas. And the portion of those people exposed to potential glacier and glacial lake hazards is much smaller.[22] Media reports make other exaggerations by claiming, for example, that 70,000 people died in the Yungay glacier avalanche in 1970.[23]

Another way news stories connect the national park to surrounding populations—and thus blur nature and culture together in ways that discourse about Glacier does not—is by noting the effects of climate change and glacier retreat on the tourism economy around Huascarán National Park.[24] But in many cases, the media reports misrepresent or oversimplify the human impacts. In one confusing example, the author mentions the loss of "picturesque glaciers" and then refers to local testimony from an elderly man who said, "We used to walk to those glaciers from my school. It would take six hours. Today I can walk there in two and a half hours. Some of the glaciers will be gone forever."[25] It is unclear how the glaciers got closer to this man's community—since everyone lives on slopes at elevations below glacier tongues—unless glaciers had advanced, not retreated. Such "evidence" to illustrate the effects of climate change impacts in and around Huascarán National Park actually undermines the point and clouds understanding of climate change and glacier shrinkage impacts. Another article argues that "the drastic melt forces people to farm at higher altitudes to grow their crops, adding to deforestation, which in turn undermines water sources and leads to soil erosion and putting the survival of Andean cultures at risk."[26] It's unclear why people would be *forced* to move higher because the glacier shrunk, especially because the water still runs downhill. But the insistence on linking park changes to human changes demonstrates the close connection between local residents and Huascarán. This also occurs when media accounts refer to so-called climate refugees, people around the park who will be displaced by climate change.[27] Articles about climate refugees near Huascarán do not definitively show how climate change affects the economic drivers of migration, which are usually identified as the most influential for triggering migration.[28] Nevertheless, the focus on human migration, the tourism economy, natural hazards, and human vulnerability demonstrates how depictions of the national park include people, which is a sharp contrast to the more common U.S. view of parks as isolated wilderness where nature and culture usually do not discursively connect.

Even though the climate discourse helps link people with parks in Peru, the media articles still present a view of nature and environmental processes that is devoid of human influence, behaviors, and decision making. This view is particularly notable in the portrayal of climate change impacts on freshwater supplies. Many articles note the potential negative effects of glacier shrinkage on regional water availability, which would also have a major effect on hydroelectricity generation and industrial-scale irrigation on Peru's Pacific coast.[29] There is little doubt that long-term continued glacier shrinkage will affect water supplies, but the media accounts often ignore the critical role that people play in the hydrologic cycle in and around the national park. They fail to ever mention the role of water rights, reservoirs and dams, and shifting land- and water-use practices that all affect watershed hydrology. For example, one article recounts the perspectives of a local resident who said that, as a child, he played in rivers that were too large to cross by foot but that are now easy to jump over "without ever touching the water."[30] He blames this river change on the retreating glaciers that are vanishing because of climate change. But it seems just as likely that this supposed river change could have resulted from fluctuations in upstream water-use practices (new water withdrawals), rather than glacier retreat. Another local resident living next to Huascarán National Park lamented, "We all get our water from there. But if the ice disappears, there won't be any more water."[31] This framing suggests that 100 percent of the region's water comes from glaciers, which is not the case. Other articles expand the impacts well beyond the national park by noting that glacier retreat in Huascarán is causing an "irreversible crisis because of water scarcity."[32]

But glacier runoff is a lot more complicated than these media accounts indicate, and the way journalists simplify hydrology unveils important insights about how they view human-environmental processes. Most scientific studies, in sharp contrast to the media accounts, suggest that glacier runoff has not yet begun to decline and that water flow from glaciers may, in fact, be in a period of increase, not decrease. These studies project a decrease of water flow of up to 23 percent by the year 2080 or 21 percent by 2050–59.[33] One new study, however, reports that seven of nine studied watersheds surrounding Huascarán National Park have seen reduced dry-season discharge that probably began around 1970. But the study maintains that even once these watersheds lose 100 percent of their glaciers—which will not occur for a long, long time—water flow for the western half of the Cordillera Blanca will decline between 2 and 30 percent, depending on the watershed.[34] This is a significant proportion of water, but a much smaller percentage than many news accounts imply when they suggest the imminent disappearance of all water in the region as glaciers shrink. The media and NGO reports also tend to overlook the important role that groundwater supplies play in the region's hydrology.[35] Moreover, they ignore how people affect those water supplies through subsistence and large-scale agriculture, human consumption, hydroelectricity generation, mining, reservoir management, and social action and protest.[36] News stories miss the ways in which these upstream water-use practices affect the rest of the watershed and interact with climate impacts.

These depictions of climate impacts and hydrologic fluctuations without much regard for the role of people reveals a broader trend that Mike Hulme refers to as environmental reductionism in climate change scenarios. As Hulme explains, this "new climate reductionism is driven by the hegemony exercised by the predictive natural sciences over contingent, imaginative, and humanistic accounts of social life and visions of the future."[37] In short, the increasing dominance of predictive quantitative models fails to account for social change and human ingenuity. This climatic reductionism is playing out in the media coverage of Huascarán National Park, and the rendering implies that Peruvians are simply passive victims waiting hopelessly for climate change to ruin their lives—whether through shrinking glaciers, glacier disasters, or evaporating water supplies. As one article exemplifying this view puts it: "The ice loss means less water, less food, and less hope for our future generation."[38] A more accurate interpretation of climate-glacier-water dynamics would suggest that Peruvians may be forced to change their water management strategies, and this will likely cause unequal impacts because some people are more (or less) vulnerable. And these impacts will be conditioned based on a variety of social, political, economic, cultural, and environmental factors that all intersect as the climate changes.

The portrayal of Peruvians as passive victims is also ahistorical. For one, it denies seventy years of successful Peruvian engineering and science to prevent GLOFs from within Huascarán National Park. Peruvians were enormously effective in adapting to the threat of outburst floods, although these accomplishments are rarely conveyed in media reports. Peruvian engineers have drained and dammed thirty-five glacial lakes in the park since 1951, and they developed flood prevention strategies that they are now increasingly sharing with the rest of the world, especially in the Himalayan region.[39] Categorizing Peruvians as waiting passively for their water to run out also overlooks how they have increased water use from Huascarán National Park rivers dramatically over the past half century. They expanded hydroelectricity generation, increased irrigated agriculture, and provided drinking water for a growing population. Human ingenuity, new technologies, economic investments, shifting management practices, and changing laws have all shaped historical water use in the region—and these factors no doubt will affect the future, even though most climate models and media accounts ignore them. This is not to say, of course, that climate change impacts in Huascarán National Park will not cause significant consequences. Rather, the point is that climate change does not occur in a social vacuum. Understanding the effects of climate in a national park (or anywhere) thus requires a much deeper analysis of human forces and the interconnected dynamics between coupled natural-human systems—precisely the kinds of insights that environmental historians have been offering for decades. Climate discourse does link parks and people in Peru, but only to a limited degree given these simplifications of the hydrosocial cycle and other environmental processes.

Static Scenery in Glacier

Glacier National Park has also attracted worldwide attention in the face of climate change. It was here, in 1997, where Vice President Al Gore hiked to the base of the shrinking Grinnell Glacier and pledged to fight against global warming.[40] Since then, the park's disappearing glaciers—among other climate impacts—have attracted consistent and increasing national media attention. While these news stories point to real climate-related problems in the park, they also reveal three trends in the conceptualization of national parks: the emphasis on lost scenery for tourists rather than the inclusion of local populations; the portrayal of static nature; and the misattribution of climate change impacts that perpetuate simplistic depictions of social-ecological systems.

The U.S. Congress established the 1-million-acre Glacier National Park in 1910 to protect its rugged and spectacular mountain scenery. In 1932 Glacier National Park combined with the adjacent Waterton Lakes National Park on the Canadian side of the border to form the world's first International Peace Park. Glacier became a UNESCO biosphere reserve in 1976. Since 1916, Glacier has been managed by the U.S. National Park Service, which was created "to conserve the scenery and the natural and historic objects and the wild life therein and to provide for the enjoyment of the same in such manner and by such means as will leave them unimpaired for the enjoyment of future generations."[41] This law and the 1910 enabling legislation for Glacier National Park set up contradictory objectives: to preserve nature but to manage wildlife and the natural and cultural scenery for tourists more than ecosystem health or scientific objectives. From the outset, the park founders also evicted Blackfeet Indians from Glacier, thereby establishing and perpetuating popular narratives (and policies) of national parks as devoid of people, even if that meant actively dispossessing Native peoples.[42] Such portrayals have spilled over onto other neighboring populations because, in the case of Glacier, the discourse rarely mentions how the park's natural resources help local populations as depictions in Peru do.

Recent climate change discourse about Glacier National Park reflects this historical legacy of the park in a key way: it portrays the park primarily as scenery for tourists. A typical news story lamenting climate change notes that glacier loss should motivate tourists "to take a road trip through Glacier National Park, Montana this summer before it is gone." After mentioning the likelihood of complete glacier loss in Glacier by 2020, the article concludes by noting that "with mountains not snow-capped as much or as long into the summer, the scenery that draws most visitors to Glacier—including stunning waterfalls and lakes—would be affected."[43] Glaciologists recognize that glaciers are indicators of long-term changes in the climate system because the ice responds to various climatic conditions.[44] But news articles do not always draw this connection. Instead, it is common for articles to focus on the mere loss of the ice as the main climate story, which then conveys the idea that visible scenery is more important than water supplies or habitats. Often these articles note how the number of park glaciers has decreased from 150 in 1850 to twenty-six in 2006, and so, as one article mentioned, the ice "simply faded away to expose bare mountainsides."[45] That article does not detail the consequences of this glacier retreat, nor does it suggest that glacier retreat is an indicator of climate change. Instead, the article primarily laments the loss of ice, implying that it is the replacement of ice with "bare mountainsides" that is tragic. This point of view, of course, is highly subjective, because a rock aficionado might appreciate the elimination of glaciers, just as ecologists were thrilled to establish Alaska's Glacier Bay National Park nearly a century ago precisely because the glaciers were retreating, which allowed them to study plant colonization and succession on newly opened landscapes.[46] Lamenting glacier retreat is a point of view that demonstrates how journalists value the tourist scenery in national parks. This glacier loss narrative is perhaps comparable to older narratives of wildfire, in which we were taught to see fires as pure destruction rather than regeneration.

This cultural construction of glaciers in recent years stands out when compared to past representations of glaciers in the national park. An analysis of a dozen guidebooks written about Glacier between 1910 and 1995 shows how past commentators did not apply such value judgments to glacier changes. For one thing, glaciers have only come to be the focus of Glacier National Park in the past decade or so. Previously, commentators remarked on the glaciers as one among many other remarkable features of the park. The geologist Marius R. Campbell pointed out in 1914 that glaciers "can hardly be considered [the park's] most striking feature. The traveler passing through it for the first time is generally impressed more by the ruggedness of the mountain tops, the great vertical walls which bound them, and the beauty of the forests, lakes, and streams, than by the glaciers."[47] Robert Sterling Yard noted in 1920 that "Glacier National Park is so named because in the hollow of its rugged mountain tops lie more than 60 *small* glaciers, the remainders of ancient monsters which once covered all but the highest mountain peaks."[48] Most guidebooks focused on the lakes, waterfalls, and majestic mountain vistas, not glaciers, in their description of tourist destinations in the park.[49] When authors did discuss glaciers, they usually portrayed them as dynamic, ever-changing bodies of ice that come and go, carving the magnificent landscape in the process.[50] As one wrote in 1963, "The growth and decay of the early glaciers was uneven and interrupted. There were numerous fluctuations and periods of little change. The ice may have completely disappeared from the area even within historic times."[51] The point to note in these depictions of Glacier's glaciers up until the late twentieth century is their recognition of glaciers that retreat and advance, even disappear—and they explain this without value judgment. In fact, these authors recognize that it is precisely because of significant glacier retreat that the park's scenery exists, thereby contrasting with today's lament for lost ice as if glaciers were static, unchanging living things.

When journalists underscore the rate of glacier retreat and show deep nostalgia or longing for the supposedly good old climate of the past, they convey a belief that national parks should remain static and unchanged. Commentators talk about "losing" parks to climate change, while others mention how climate change creates "an ecosystem out of balance" or how it upsets "intact ecosystems."[52] Some refer to past climatic conditions as allowing parks to be "healthy" or having "undisturbed ecosystems."[53] Global warming, on the other hand, causes national parks to lose their "natural condition" and become "an ecosystem out of balance."[54] This suggestion that past environmental conditions were static can occur with discussions of species migrations in national parks. Researchers note that global warming will drive species to higher areas to maintain their ecological and climatic niche. Few of the media accounts, however, indicate that climatic shifts have always occurred; they neglect to say that the real issue is the rapid rate of change in recent decades. Without discussing the various rates of change, readers are left to (mis)assume that static landscapes and climates existed until the last few decades.

Still others talk about seeing Glacier and other parks threatened by climate change and melting ice as destinations to see "before they [the parks] die."[55] Glaciers, pikas, and forest fires will likely undergo dramatic alterations from climate change.[56] But nature is also in constant flux, and the idea of static, unchanging nature is inaccurate.[57] The media accounts do little to clarify these differences, even if they are likely referring to the stunning rate of changes and the value of anything that is lost. Without such explanations, their messages convey the sentiment that national parks are unchanging—indeed, that they are places where the scenery and environments should not change. In many ways this view corresponds to the 1916 enabling legislation that created the National Park Service and set out to preserve parks "unimpaired for the enjoyment of future generations." Critics of wilderness have since pointed out that nature in national parks is as much a cultural construction as it is "wild," while they have also shown that nature is never static.[58] These traditional views of national parks as static wilderness, however, continue to exist, and the global warming narrative exemplifies how the static wilderness ideal both reflects this view and is perpetuated through such media accounts of climate change in Glacier.

Perceptions of the impact that glacier retreat in Glacier National Park will have on freshwater supplies helps illustrate the different views of parks in the United States and Peru. In Montana the impact of glacier shrinkage on human societies is largely absent, except for tourists, as news stories instead refer vaguely to environmental impacts. One news article about that process explained the rate of past glacier loss in Glacier, as well as the predicted outcome of having no glaciers by 2030. But the article never mentions any effects of these shrinking glaciers except to report that "climate change is eliminating glaciers and harming the park environment."[59] Another account explains that "there's more to glaciers than just beauty. They also play a crucial role in the ecosystem, and their disappearance may have widespread consequences."[60] It says glaciers provide water and help with the health of aquatic and riparian ecosystems, just as another article claims the disappearing glaciers are "endangering the region's plants and animals."[61] But how? Most articles do not explain precisely how, even though a few scientists have been studying climate-glacier-hydrology dynamics in Glacier.[62] Despite their studies, the U.S. Geological Survey still explains on its website that "few measurements of glacier volume or mass have been made. Measurements of area alone can be misleading; changes in mass and/or ice flux can result in significant changes to the glacier and to streamflow below the glacier even when glacier area remains stable. Though hydrologic changes such as these can have important ecologic effects downstream of the glaciers, the nature and extent of changes in runoff volume, and stream temperature have not been measured or analyzed."[63] In fact, it is not even clear what percent of the water supplied to Glacier National Park and surrounding areas comes from glaciers (versus snowpack, precipitation, or groundwater), or how much water supplies will decline if glaciers vanish altogether. This lack of evidence makes it difficult to determine if glacier shrinkage will, in fact, result in "losing an important source of fresh water."[64] Without much data available or analysis of the complex ways in which glacier volume affects downstream water supplies, it seems the media reports might be exaggerating the worries about glacier retreat for downstream hydrology.

But more relevant for what this climate discourse says about perceptions of national parks is the way the concerns about glacier-water fluctuations rarely mention local people—even though tens of thousands of people live outside Glacier National Park, including those on the Blackfeet Indian Reservation, and even though the Hungry Horse hydroelectric station outside Glacier generates more energy than the Cañón del Pato station outside Huascarán. Still, the discourse about Glacier largely overlooks the presence of local people, which contrasts markedly to portrayals of the national park in Peru. These views of climate change impacts thus illustrate how U.S. national parks can be viewed as an isolated landscape, where static nature is separate from people but preserved—preferably unchanged—for tourists. Despite a few decades of critical scholarship on national parks and wilderness, the traditional views of parks as untarnished by and disconnected from human beings remain prevalent, embedded in depictions of climate change impacts and continuing to drive NGO and environmental group agendas in parks.

Conclusions

Key similarities stand out in the analysis of popular media about Huascarán and Glacier National Parks. In both parks, climate discourse reveals embedded ideals about static nature and the way parks should preserve scenery primarily for tourists. Yet this discursive analysis also shows how perceptions of national parks in the two countries are also quite distinct. Unlike the discourse about Glacier that characterizes the park as an isolated, island wilderness area separate from human societies and culture, portrayals of Huascarán National Park tend to focus on its inextricable interconnections with surrounding societies. In particular, media and NGO accounts reveal that Huascarán is the source of many natural hazards that affect Peruvian populations; they also note how the park is central to the local tourism economy and a vital source of water and other ecosystem services. The severe human impacts of climate change that journalists discuss for Huascarán also reveal how much more acute climate effects are in the Andes than in northern Montana. What's more, nature and culture do not seem nearly so discursively divided in Peru as in the United States. Despite blurring nature-culture boundaries, however, journalists do nonetheless reveal a type of environmental reductionism for Huascarán that overlooks the role of cultural ingenuity, ignores the possibility of human adaptation to climate change in the future, and minimizes past Peruvian accomplishments. Overall, though, this utilitarian view of the socioeconomic and natural resource dimensions of Huascarán National Park contrasts markedly with the representation of Glacier National Park, which is portrayed primarily as tourist scenery and a place of biodiversity, with only limited mention of surrounding human populations.

Cautiously using these two parks as representations of national parks in Peru and the United States suggests key differences in the purpose and meaning of parks in the two countries, at least as uncovered in the popular media. The principal distinction lies at the heart of the U.S. Yellowstone model: evicting local people (both physically and discursively) from national parks to create supposed pristine wilderness landscapes.[65] In Latin America park developers have, to a degree, avoided this Yellowstone model through time. Instead, national parks with human residents have been common, and parks for utilitarian purposes—such as watershed conservation for urban populations—have been the norm.[66] Fortress conservation that blocks out everyone except tourists through policies and rhetoric has been the main practice in the United States. But in Peru as elsewhere in Latin America, national parks have never been as divorced from local populations as in the United States.

The analysis of climate discourse in national parks also has implications for responses to global warming. Many of the media accounts do not discuss broader implications of glacier retreat and species changes in these two parks. Nor do they recognize and convey the complexity of human societies or the role of both human and nonhuman variables that can cause environmental or societal changes. National park narratives thus penetrate the discussion of climate change impacts and affect the types of responses or solutions that might emerge. In Huascarán, a change in the representations could bring human variables into the climate change equation even more explicitly and meaningfully than they have been. More realistic identification of populations exposed to glacier hazards could target the placement of disaster prevention programs and direct attention to issues of socioeconomic and political inequality that exacerbate vulnerability and lead to disproportionate impacts of climate change. Recognition of past Peruvian achievements in the prevention of GLOFs could assist other regions. Finally, the acknowledgment of human water management practices alongside discussion of climate-glacier-hydrology frameworks would link societal and environmental forces while also projecting future scenarios that actually have people in them. In Glacier the discourse tends to focus on more cosmetic issues such as glacier (scenery) loss, rather than trying to discern the effects of glacier shrinkage for downstream water users, or the impacts of ice loss on stream ecology, or the potential impacts on the tourism economy. Media reports tend to recapitulate the same themes about lamentable glacier shrinkage without digging deeper into other human or ecosystem issues, such as species migration, forest fires, or precipitation and snowpack changes. A different narrative could shift the climate discourse toward actual impacts that require the implementation of adaptive measures. This different approach to climate change adaptation might be impossible, however, without first generating fresh narratives of nature and new perspectives on national parks.

Notes

1. Stephen Saunders, Tom Easley, and Suzanne Farver, *National Parks in Peril: The Threats of Climate Disruption* (Denver: Rocky Mountain Climate Organization and Natural Resources Defense Council, 2009), v.
2. Bill McKibben, *The End of Nature* (New York: Random House, 1989); Mike Hulme, *Why We Disagree about Climate Change: Understanding Controversy, Inaction, and Opportunity* (New York: Cambridge University Press, 2009).
3. For example, J. Baird Callicott and Michael P. Nelson, eds., *The Great New Wilderness Debate: An Expansive Collection of Writings Defining Wilderness from John Muir to Gary Snyder* (Athens: University of Georgia Press, 1998); Candace Slater, ed., *In Search of the Rain Forest* (Durham, N.C.: Duke University Press, 2003); William Cronon, "The Trouble with Wilderness; or, Getting Back to the Wrong Nature," in *Uncommon Ground: Rethinking the Human Place in Nature,* ed. William Cronon (New York: W. W. Norton, 1996), 69–90.
4. For more on coupled human-natural systems, hybrid landscapes, and social-ecological systems, see B. L. Turner and et al., "Illustrating the Coupled Human–Environment System for Vulnerability Analysis: Three Case Studies," *Proceedings of the National Academy of Sciences* 100, no. 14 (2003): 8080–85; Richard White, "From Wilderness to Hybrid Landscapes: The Cultural Turn in Environmental History," *Historian* 66, no. 3 (2004): 557–64; Oran R. Young et al., "The Globalization of Socio-Ecological Systems: An Agenda for Scientific Research," *Global Environmental Change* 16 (2006): 304–16.
5. For more on how parks became places for recreation more than species preservation or scientific research, see Susan R. Schrepfer, *Nature's Altars: Mountains, Gender, and American Environmentalism* (Lawrence: University Press of Kansas, 2005); Patrick Kupper, "Science and the National Parks: A Transatlantic Perspective on the Interwar Years," *Environmental History* 14, no. 1 (2009): 58–81.
6. Mark Carey, *In the Shadow of Melting Glaciers: Climate Change and Andean Society* (New York: Oxford University Press, 2010); Mark Carey, Adam French, and Elliott O'Brien, "Unintended Effects of Technology on Climate Change

Adaptation: An Historical Analysis of Water Conflicts below Andean Glaciers," *Journal of Historical Geography* 38, no. 2 (2012): 181–91.

7. For the work of two scholars who have critiqued both "camps," see Hulme, *Why We Disagree about Climate Change;* Roger Pielke, Jr., "Misdefining 'Climate Change': Consequences for Science and Action," *Environmental Science and Policy* 8 (2005): 548–61.

8. See, for example, Cronon, *Uncommon Ground*; Richard Drayton, *Nature's Government: Science, Imperial Britain, and the "Improvement" of the World* (New Haven, Conn.: Yale University Press, 2000); Nancy Leys Stepan, *Picturing Tropical Nature* (Ithaca, N.Y.: Cornell University Press, 2001).

9. For more park details, see United Nations Environment Programme and World Conservation Monitoring Centre, "World Heritage Sites: Huascarán National Park," http://whc.unesco.org/en/list/333 (accessed August 3, 2015).

10. Mark Carey, "Mountaineers and Engineers: An Environmental History of International Sport, Science, and Landscape Consumption in Twentieth-Century Peru," *Hispanic American Historical Review* 92, no. 1 (2012): 107–41; Thierry Lefebvre, "L'invention occidentale de la haute montagne andine," *Mappemonde* 79, no. 3 (2005), available at http://mappemonde.mgm.fr/num7/articles/art05307.html (accessed January 21, 2009); César Morales Arnao, *Andinismo en la Cordillera Blanca* (Callao, Peru: Imprenta Colegio Militar Leoncio Prado, Le Perla, 1968); Erwin Grötzbach, "Tourism in the Cordillera Blanca Region, Peru," *Revista Geográfica (Instituto Panamericano de Geografía e Historia, OEA)* 133 (2003).

11. For example, Carlos Ponce del Prado, *Resumen de los parques nacionales y reservas equivalentes del Perú* (Lima: Dirección General Forestal de Caza y Tierras, Ministerio de Agricultura, 1971). For the emphasis on forests, see assorted documents in Record Group 79 (National Park Service), Entry 11: Administrative Files, 1949–71, Box 2170, Folder: "[Foreign Parks] Land Planning Division, Peru, 1940," Location: RG 79/570/81/10/3, U.S. National Archives (NARA), College Park, Maryland.

12. Belén Delgado, "Excursion al cambio climático" (latam.msn.com: MSN Perú, 2010); Simeon Tegel, "La desglaciación de la cordillera andina," *El País*, September 29, 2011, http://sociedad.elpais.com/sociedad/2011/09/29/actualidad/1317247202_850215.html (accessed August 3, 2015); Rick Vecchio, "Global Warming Impact: Peru's Pastoruri Glacier Recedes into Two Patches of Ice," *Peruvian Times,* December 20, 2007, http://www.peruviantimes.com/20/global-warming-impact-perus-pastoruri-glacier-recedes-into-two-patches-of-ice/70 (accessed August 14, 2011); Norberto Ovando, "Impacts of Climate Change on Protected Areas and Glaciers: Focus Latin America," August 21, 2008 (accessed August 3, 2015); "El Parque Nacional Huascarán está en peligro: urge salvarlo," *La República,* November 21, 2004, http://archivo.larepublica.pe/21-11-2004/el-parque-nacional-huascaran-esta-en-peligro-urge-salvarlo (accessed August 3, 2015).

13. For example, Leslie Josephs, "Peru's Mountain Glaciers Are Melting Away," *MSNBC,* February 16, 2007, http://www.msnbc.msn.com/id/17113441/ns/us_news-environment/t/perus-mountain-glaciers-are-melting-away/ (accessed August 14, 2011).

14. Simeon Tegel, "The Changing Face of Andean Glaciers," *The Independent—Blogs,* September 2, 2011, http://blogs.independent.co.uk/2011/09/02/the-changing-face-of-andean-glaciers/ (accessed February 2, 2012).

15. "Goodbye Huascarán," July 10, 2006, http://enperublog.com/2006/07/10/goodbye-huascaran/ (accessed February 2, 2012). Also see "Andean Glaciers 'Could Disappear': World Bank," PhysOrg (February 18, 2009), http://www.physorg.com/news154187547.html (accessed August 15, 2011).

16. For the most recent scholarship, see Georg Kaser and Henry Osmaston, *Tropical Glaciers* (New York: Cambridge University Press, 2002); Adina E. Racoviteanu et al., "Decadal Changes in Glacier Parameters in the Cordillera Blanca, Peru, Derived from Remote Sensing," *Journal of Glaciology* 54, no. 186 (2008): 499–510; Mathias Vuille et al., "Climate Change and Tropical Andean Glaciers: Past, Present and Future," *Earth-Science Reviews* 89, no. 3–4 (2008): 79–96; Mathias Vuille, Georg Kaser, and Irmgard Juen, "Glacier Mass Balance Variability in the Cordillera Blanca, Peru and Its Relationship with Climate and the Large-Scale Circulation," *Global and Planetary Change* 62, no. 1–2 (2008): 14–28.

17. Aura Sabadus, "Call to Save Wonders of the World That Face Climate Catastrophe," *Scotsman (Edinburough),* July 10, 2006, http://news.scotsman.com/everest/Call-to-save-wonders-of.2790961.jp (accessed February 2, 2012). Also see Ovando, "Impacts of Climate Change.

18. Nicole E. Heller and Erika S. Zavaleta, "Biodiversity Management in the Face of Climate Change: A Review of 22 Years of Recommendations," *Biological Conservation* 142, no. 1 (2009): 18; Lee Hannah et al., "Protected Area Needs in a Changing Climate," *Frontiers in Ecology and the Environment* 5, no. 3 (2007): 131–38; Craig Moritz et al., "Impact of a Century of Climate Change on Small-Mammal Communities in Yosemite National Park, USA," *Science* 322, no. 5899 (2008): 261–64.

19. Alcides Ames, "A Documentation of Glacier Tongue Variations and Lake Development in the Cordillera Blanca, Peru," *Zeitschrift für Gletscherkunde und Glazialgeologie* 34, no. 1 (1998): 1–36; Mark Carey, *In the Shadow of Melting Glaciers;* Marco Zapata Luyo, "La dinámica glaciar en lagunas de la Cordillera Blanca," *Acta Montana* (Czech Republic) 19, no. 123 (2002): 37–60.

20. Augustin Colette et al., *Case Studies in Climate Change and World Heritage* (Paris: UNESCO World Heritage Centre, 2007).

21. Sabadus, "Call to Save Wonders of the World."

22. Jeffrey Bury et al., "Glacier Recession and Human Vulnerability in the Yanamarey Watershed of the Cordillera Blanca, Peru," *Climatic Change* (2010), doi:10.1007/s10584-010-9870-1; Bryan G. Mark et al., "Climate Change and Tropical Andean Glacier Recession: Evaluating Hydrologic Changes and Livelihood Vulnerability in the Cordillera Blanca, Peru," *Annals of the Association of American Geographers* 100, no. 4 (2010): 794–805; INEI, *The 2007 National Census: XI of Population and VI of Houses* (Lima: Institute of National Statistics and Information, 2007).

23. Sabadus, "Call to Save Wonders of the World"; also see Gregory J. Rummo, "Climate Change in the Peruvian Andes," *New Jersey News,* December 8, 2010, http://www.newjerseynewsroom.com/international/climate-change-in-the-peruvian-andes (accessed August 14, 2011); George E. Ericksen, George Plafker,

and Jaime Fernández Concha, "Preliminary Report on the Geological Events Associated with the May 31, 1970, Peru Earthquake," *United States Geological Survey Circular* 639 (1970): 1–25; S. G. Evans et al., "A Re-Examination of the Mechanism and Human Impact of Catastrophic Mass Flos Originating on Nevado Huascarán, Cordillera Blanca, Peru in 1962 and 1970," *Engineering Geology* 108, no. 102 (2009): 96–118.

24. "'La ruta del cambio climático,' un proyecto que busca recuperar el turismo en Pastoruri," *El Comercio,* July 5, 2010, http://elcomercio.pe/planeta/505913/noticia-ruta-cambio-climatico-proyecto-que-busca-recuperar-turismo-pastoruri (accessed February 17, 2012); Vanessa Roma Espinoza, "Soldados de hielo: encargados de cuidar el Parque Nacional Huascarán," *El Comercio,* August 1, 2011, http://elcomercio.pe/planeta/958366/noticia-soldados-hielo-encargados-proteger-parque-nacional-huascaran (accessed February 17, 2012); Martin Riepl, "El guardián del hielo," *Etiqueta Negra*, February 1, 2012, http://etiquetanegra.com.pe/articulos/el-guardian-del-hielo (accessed August 3, 2015).

25. "Current Threats to Traditional Medicine in the Andes (and Elsewhere)," *Exploring Traditional Andean Medicine in Peru*, http://traditionalmedicineinperuandes.weebly.com/current-threats.html (accessed February 3, 2012).

26. John Vidal, "Cities in Peril as Andean Glaciers Melt," *The Guardian,* August 28, 2006, http://www.theguardian.com/environment/2006/aug/29/glaciers.climatechange (accessed February 3, 2012).

27. Dava Castillo, "Glaciers Are Melting: Causes, Consequences and Innovation," *Climate Himalaya,* June 10, 2011, http://chimalaya.org/2011/06/10/glaciers-are-melting-causes-consequences-and-innovation (accessed August 3, 2015).

28. Richard Black et al., "The Effect of Environmental Change on Human Migration," *Global Environmental Change* 21, Supplement 1 (2011): 3–11.

29. For example, Vecchio, "Global Warming Impact"; Josephs, "Peru's Mountain Glaciers Are Melting Away"; "El Parque Nacional Huascarán está en peligro."

30. Rummo, "Climate Change in the Peruvian Andes."

31. Heather Somerville, "Glacier Melt in Peru Becomes More than a Climate Issue," *Washington Post*, January 16, 2011, http://www.washingtonpost.com/wp-dyn/content/article/2011/01/16/AR2011011604900.html (accessed July 27, 2011).

32. Jorge Zavaleta, "Cordillera Blanca: Negro futuro," *La Primera,* September 17, 2008, http://www.diariolaprimeraperu.com/online/informe-especial/cordillera-blanca-negro-futuro_23668.html (accessed August 3, 2015). Also see Vidal, "Cities in Peril as Andean Glaciers Melt"; Josephs, "Peru's Mountain Glaciers Are Melting Away"; Somerville, "Glacier Melt in Peru."

33. Irmgard Juen, Georg Kaser, and Christian Georges, "Modelling Observed and Future Runoff from a Glacierized Tropical Catchment (Cordillera Blanca, Perú)," *Global and Planetary Change* 59, no. 1–4 (2007): 37–48; Walter Vergara et al., *Assessment of the Impacts of Climate Change on Mountain Hydrology: Development of a Methodology through a Case Study in Peru* (Washington, D.C.:The World Bank, 2010); Pierre Chevallier et al., "Climate Change Threats to Environment in the Tropical Andes: Glaciers and Water Resources," *Regional Environmental Change* 11, no. Supplement 1 (2011): S179–S187.

34. Michel Baraer et al., "Glacier Recession and Water Resources in Peru's Cordillera Blanca," *Journal of Glaciology* 58, no. 207 (2012): 134–50.

35. M. Baraer et al., "Characterizing Contributions of Glacier Melt and Groundwater during the Dry Season in a Poorly Gauged Catchment of the Cordillera Blanca (Peru)," *Advances in Geosciences* 22 (2009): 41–49.

36. Jeffrey Bury et al., "New Geographies of Water and Climate Change in Peru: Coupled Natural and Social Transformations in the Rio Santa Watershed," *Annals of the Association of American Geographers* 103 (2013): 363–74; Mark Carey et al., "Toward Hydro-Social Modeling: Merging Human Variables and the Social Sciences with Climate-Glacier Runoff Models (Santa River, Peru)," *Journal of Hydrology* 518, Part A (2013): 60–70.

37. Mike Hulme, "Reducing the Future to Climate: A Story of Climate Determinism and Reductionism," *Osiris* 26 (2011): 245.

38. Sarah Alyssa, "Peru's Vanishing Glaciers," Student News Action Network, April 8, 2011, http://newsaction.tigweb.org/article/peru-s-vanishing-glaciers (accessed August 15, 2011). Also see Somerville, "Glacier Melt in Peru."

39. Carey, *In the Shadow of Melting Glaciers.*

40. The full text of Gore's speech is at http://clinton3.nara.gov/WH/EOP/OVP/speeches/glacier.html (accessed January 27, 2012). Gore reiterates, refines, and broadens these comments in his film, *An Inconvenient Truth* (2006).

41. U.S. National Park Service Organic Act (16 U.S.C. 12 3, and 4), August 25, 1916, http://www.nps.gov/parkhistory/online_books/anps/anps_1i.htm (accessed August 3, 2015).

42. Robert H. Keller and Michael F. Turek, *American Indians and National Parks* (Tucson: University of Arizona Press, 1998); Mark David Spence, *Dispossessing the Wilderness: Indian Removal and the Making of the National Parks* (New York: Oxford University Press, 2000).

43. Lindsay Godfree, "Climate Change in Glacier National Park Threatens to Melt Glaciers, Harm Tourism in Near Future," *Cross Country Travel Examiner,* April 23, 2010, http://www.examiner.com/cross-country-travel-in-national/climate-change-glacier-national-park-threatens-to-melt-glaciers-harm-tourism-near-future (accessed February 15, 2012).

44. Gregory T. Pederson et al., "Decadal-Scale Climate Drivers for Glacial Dynamics in Glacier National Park, Montana, USA," *Geophysical Research Letters* 31, no. 12 (2004): L12203; Joel Harper, Joel Brown, and Neil Humphrey, "Cirque Glacier Sensitivity to 21st Century Warming: Sperry Glacier, Rocky Mountains, USA," *Global and Planetary Change* 74, no. 2 (2010): 8; Joseph M. Shea, Shawn J. Marshall, and Joanne M. Livingston, "Glacier Distributions and Climate in the Canadian Rockies," *Arctic, Antarctic, and Alpine Research* 36, no. 2 (2004): 7; Myrna H. P. Hall and Daniel B. Fagre, "Modeled Climate-induced Glacier Change in Glacier National Park, 1850–2100," *BioScience* 53, no. 2 (2003): 9.

45. Bjorn Carey, "Glaciers Disappear in Before and After Photos," LiveScience, March 24, 2006, http://www.livescience.com/674-glaciers-disappear-photos.html. For other examples of articles that stress glaciers as scenery, see Chris Peterson, "Global Warming Rally in Glacier," *Hungry Horse News,* April 11, 2007, http://www.flatheadnewsgroup.com/hungryhorsenews/news/global-warming-rally-in-glacier/article_cd5ae19f-ed25-5b6c-a4cb-09775bd0f7c4.html (accessed August 3, 2015); Dan Shapley, "Endangered Vacations," *Daily Green,* 2009, http://preview.www.thedailygreen.com/environmental-news/latest/endangered-vacations-video

(accessed August 3, 2015); Kate Sheppard, "What Global Warming Could Do to National Parks," *Grist,* July 17, 2007, http://grist.org/article/it-aint-natural1 (accessed August 3, 2015); Daniel B. Wood, "How Climate Change May Be Threatening National Parks," *Christian Science Monitor,* April 9, 2009, http://www.csmonitor.com/Environment/Global-Warming/2009/0409/how-climate-change-may-be-threatening-national-parks (accessed August 3, 2015).

46. Theodore Catton, *Inhabited Wilderness: Indians, Eskimos, and National Parks in Alaska* (Albuquerque: University of New Mexico Press, 1997).

47. Marius R. Campbell, *Origin of the Scenic Features of the Glacier National Park* (Washington, D.C.: Department of the Interior/Government Printing Office, 1914), 3.

48. Robert Sterling Yard, *Glimpses of Our National Parks* (Washington, D.C.: Government Printing Office, 1920), 40 (emphasis added).

49. Mathilde Edith Holtz and Katharine Isabel Bemis, *Glacier National Park: Its Trails and Treasures* (New York: George H. Doran, 1917); Mitchell Mannering, "The New Glacier National Park," *National Magazine* 37 (1913): 69–76; George Ruhle, *Guide to Glacier National Park,* rev. ed. (Minneapolis: John W. Forney, 1963).

50. Yard, *Glimpses of Our National Parks,* 42.

51. Ruhle, *Guide to Glacier National Park,* 156. Also see David Rockwell, *Glacier National Park: A Natural History Guide:* (Boston: Houghton Mifflin, 1995), vii–viii; Holtz and Bemis, *Glacier National Park*; Nancy Trejos, "Glacier National Park on Ice," *Washington Post,* August 5, 2011.

52. Mary Bruno, "Our Parks in Peril," *Grist,* October 2, 2009; Leslie Burliant, "America's National Parks: Canaries in the Climate Change Coal Mine," Inside-Climate News, August 7, 2009, http://solveclimatenews.com/news/20090807/americas-national-parks-canaries-climate-change-coal-mine (accessed August 3, 2015); Jessie Lussier, "Climate Change vs. National Parks," Backpacker.com, 2009 (no longer available).

53. Wood, "How Climate Change May Be Threatening National Parks."

54. Burliant, "America's National Parks."

55. Shapley, "Endangered Vacations." Also see Daniel Terdiman, "Climate Change Taking a Toll on Glacier National Park" CNET, July 16, 2009, http://www.cnet.com/news/climate-change-taking-toll-on-glacier-national-park/ (accessed August 3, 2015); "Climate Change: Making the Nation's Bears Hungry?" National Public Radio, July 16, 2011, http://www.npr.org/2011/04/16/135468901/climate-change-making-the-nations-bears-hungry (accessed August 3, 2015); Mark Wenzler, "Helping America's Naitonal Parks Survive Climate Change," *Grist,* May 8, 2009, http://grist.org/article/2009-05-08-national-parks-climate-change/ (accessed August 3, 2015); "Warmer Climate Could Spark More Severe Yellowstone Fires," Livescience.com, July 25, 2011, http://www.livescience.com/15219-yellowstone-wildfires-increased-climate-change.html (accessed August 3, 2015).

56. Moritz et al., "Impact of a Century of Climate Change"; Anthony L. Westerling et al., "Continued Warming Could Transform Greater Yellowstone Fire Regimes by Mid-21st Century," *Proceedings of the National Academy of Sciences* 108, no. 32 (2011): 13165–70.

57. See, for example, P. C. D. Milly et al., "Stationarity Is Dead: Whither Water Management?" *Science* 319, no. 5863 (2008): 573–74.

58. See, for example, Cronon, "Trouble with Wilderness."

59. Associated Press, "Endangered Status for Glacier Park?" NBC News, February 16, 2006, http://www.nbcnews.com/id/11389665/ns/us_news-environment/t/endangered-status-glacier-national-park/#.Vb_m67VK8sI (accessed August 3, 2015). Also see Stephen P. Nash, "Twilight of the Glaciers," *New York Times* (July 29, 2011), http://travel.nytimes.com/2011/07/31/travel/glacier-national-park-montana-fading-glaciers.html?pagewanted=all (accessed February 12, 2012).

60. Terdiman, "Climate Change Taking a Toll on Glacier National Park."

61. Joseph Romm, "Glacier National Park to Go Glacier-Free a Decade Early," March 3, 2009, http://grist.org/article/welcome-to-_____-national-park/ (accessed August 3, 2015).

62. Scientists have examined some of these issues: Daniel B. Fagre et al., "Watershed Responses to Climate Change at Glacier National Park," *JAWRA Journal of the American Water Resources Association* 33, no. 4 (1997): 755–65; Frederick I. Klasner and Daniel B. Fagre, "A Half Century of Change in Alpine Treeline Patterns at Glacier National Park, Montana, USA," *Arctic, Antarctic, and Alpine Research* 34, no. 1 (2002): 7; Gregory T. Pederson et al., "Long-Duration Drought Variability and Impacts on Ecosystem Services: A Case Study from Glacier National Park, Montana," *Earth Interactions* 10, no. 4 (2006): 28.

63. U.S. Geological Survey, Northern Rocky Mountain Science Center, "Glacier Monitoring Studies: Monitoring and Assessing Glacier Changes and Their Associated Hydrologic and Ecologic Effects in Glacier National Park," http://www.nrmsc.usgs.gov/research/glaciers.htm (accessed February 12, 2012).

64. Terdiman, "Climate Change Taking a Toll on Glacier National Park."

65. Stan Stevens, "The Legacy of Yellowstone," in *Conservation through Cultural Survival: Indigenous Peoples and Protected Areas,* ed. Stan Stevens (Washington, D.C.: Island Press, 1997), 13–32.

66. Stephan Amend and Thora Amend, eds., *National Parks without People? The South American Experience* (Quito, Ecuador: IUCN/Parques Nacionales y Conservación Ambiental, 1995); Emily Wakild, *Revolutionary Parks: Conservation, Social Justice, and Mexico's National Parks, 1910–1940* (Tucson: University of Arizona Press, 2011).

CONCLUSION | # Geographies of Hope
Lessons from a World of National Parks

PAUL S. SUTTER

BEFORE DIGGING INTO THE ACCOMPLISHMENTS of the preceding essays in pushing our historical understanding of the national parks beyond the nation, I want to dwell briefly on the obvious foil for this volume, and the conference project from which it originated: the six-part, twelve-hour documentary by Ken Burns, with script by Dayton Duncan, titled *The National Parks: America's Best Idea.* I must confess that I had a hard time bringing myself to sit in front of the television when the series initially aired on PBS, mostly because I knew what I would be getting and I knew that I would not be satisfied. But when a prominent history journal asked me to write a review of the series and sent me the DVDs, I sequestered myself in my office and watched the whole thing through a couple of times. The series was beautiful and inspiring, but also problematic in the ways that I expected it would be; I have a larger critique of it that I will not rehash here.[1] For the purposes of this conclusion, though, let me sum up the major problems with a quick riff on the notion of "America's Best Idea." This particular phrase is borrowed from Wallace Stegner, who called the national parks "the best idea we ever had," but what is less well known is that Stegner himself credited the phrase to Lord James Bryce, one-time British ambassador to the United States, who in the early 1910s apparently called our then-fledgling system of national parks the "best idea America ever had." Alan MacEachern has, in this volume and elsewhere, raised some intriguing questions about whether Bryce ever wrote or uttered these words, suggesting that this is a myth whose origins are worth further historical exploration. Nonetheless, it's telling, though perhaps not surprising, that the Burns team chose to credit Stegner without mentioning Bryce, and thus to highlight national congratulation rather than international compliment. Indeed, the whole series is an exercise in national congratulation, one that is walled off from the larger world and its histories of park thinking and park preservation. Tearing down that wall and using the global history of national parks to interrogate congratulatory park nationalism in the United States is one of the great strengths of this volume.

As it turns out, the phrase "America's Best Idea" is packed with assumptions that need to be spelled out if we are to move beyond congratulatory park nationalism. Let's examine the phrase and its implications, taking each word in turn. First, there's the possessive "America's": the claim here is that the national parks were and are uniquely American—or, more specifically, to borrow a phrase discussed extensively at the symposium that produced this book, "U.S. American," which separates this nation from the more expansive geography of the Americas. The assumption here is that the national park idea is U.S. intellectual property, though generously unpatented, and that wherever we find national parks in the world, we are witness to the sincerest form of flattery. More than that, the use of a possessive that is coterminous with the entire national community suggests that the national park idea was an organic product of the national mind, a result of national consensus. U.S. historians have long since abandoned such consensus thinking as incommensurable with our contested history, but it clearly remains popular outside of the academy. Second, there is the word "best," which suggests not only that national parks have been an unmitigated good but that the idea was the most superior discrete result of cogitating that this imagined community has ever produced. For U.S. historians, the combination of these two words raises the immediate specter of American exceptionalism: the claim, to which U.S. Americans are ever-clingy, that the United States has a uniquely enlightened history and mission in the world, and that we are not only different from but better than other nations—more committed to freedom, democracy, and equality; incapable of impure motives; and apparently uniquely capable of protecting nature. What has concerned U.S. historians in the last couple of decades about this exceptionalist streak is not only its constant presence in political rhetoric describing our place in the world, but also that it has too often produced histories of the United States that have lacked the many transnational contact points that are absolutely necessary to a fuller rendering of our history.[2] In the national parks and their history we find a particularly rich vein of this exceptionalist intellectual tradition, one that the Burns series was intent on mining. Finally, there is the word "idea," which suggests that the national parks are a singular and unitary intellectual thing, born fully formed at an identifiable moment in time. As I will show in more detail below, this claim too is difficult to sustain.

My exegesis of this brief phrase raises three questions that seem to me to be central to the essays in this volume, and that will guide much of my concluding discussion. First, were the national parks a uniquely American invention, and how might we characterize and trace U.S. influence in the creation of national

The World's Protected Areas

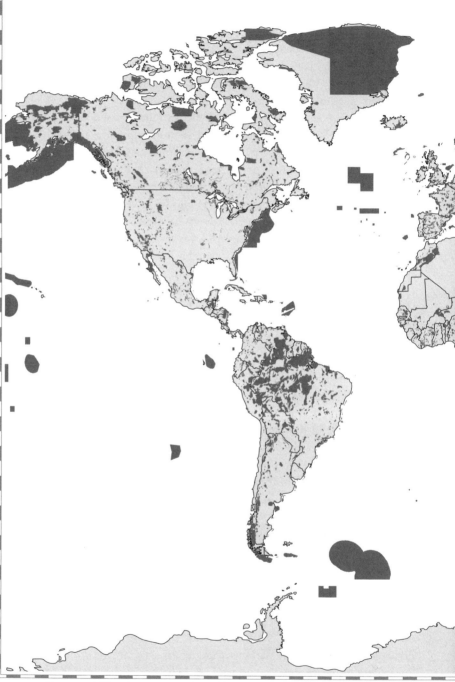

Data source: IUCN and UNEP-WCMC, The World Database on Protected Areas (WDPA)
(Cambridge, U.K.: UNEP-WCMC, March 2015) www.protectedplanet.net.

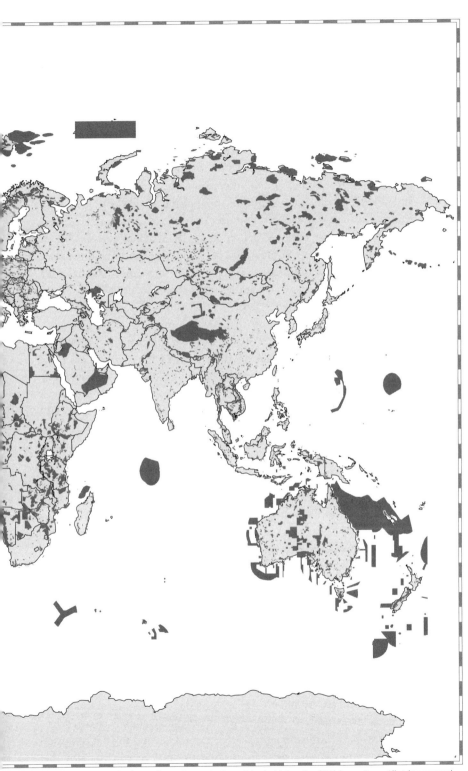

parks elsewhere in the world? Second, have the creation and management of national parks in the United States, and in other parts of the world where U.S. influence has been strong, always been indicative of a national consensus or of our political culture at its best? In other words, have national parks, wherever they might be found, always been a great idea? Third, is the "national park" a singular and coherent idea that has held steady across time and space? The quick answers to these questions are, I think, "no, no, and no." As the essays in this volume have demonstrated in myriad ways, all of the assumptions packed into this "America's Best Idea" sloganeering need to be thoroughly rethought.

The genius of this volume, and of the stimulating conference that gave birth to it, has been to take the histories of some of the world's national parks, put them in a transnational context, and see what we see once we have shed our national exceptionalist blinders. The first part of that project—in fact, a surprisingly big part of it—is a comparative "U.S. and the world" inquiry, as a number of the volume's chapters either compare U.S. and other national park preservation efforts or examine transnational connections between U.S. actors and other national and nonnational actors in the realms of park creation and management. In that sense, the nation that this volume seeks to go beyond is the United States—even as the United States remains a comparative touchstone. That is necessary work, and the volume does it well, but it is also worth pointing out that it is but one step on a path toward more fully global and transnational histories of national parks.[3]

A second project of this volume—one that is omnipresent but a bit less well developed—is to examine how park preservation has itself transcended the national. As many of the essays suggest, the last century has not only seen many other nations creating their own national parks, but also the birth of an international community dedicated to park protection—and to global conservation more broadly. We have NGOs such as National Geographic and the World Wildlife Fund (known internationally as the World Wide Fund for Nature) working for various forms of nature protection, often in the developing world. We have new institutions of international governance (IGOs), such as the International Union for the Conservation of Nature and various United Nations agencies, which work both with and beyond nations to conserve nature. We have parks that are themselves transnational, in that they cross borders that separate nations, and, as the introduction to this volume notes, we also have national parks along borders whose very territoriality is constantly challenged by border crossers of all sorts. We have fairly recent designations such as biosphere reserves and World Heritage Sites, both of which define areas as globally significant. And we have a few places, such as Antarctica, which are complex global territories anchored by preservationist agendas. We also have a new set of environmental concerns, from biodiversity loss to global climate change, that demand preservation or mitigation efforts that extend beyond the nation, or at least take into account dynamic natural forces that cannot be contained by the nation. This internationalization of the national park mission is built on the notion that nature protection is an imperative not only of national citizenship but increasingly of global citizenship as well.

Finally, the third aspect of moving "beyond the nation" that this volume embraces is an impulse in a direction opposite from the global; it is to break down the nation and to examine how national parks have, and have not, worked as national projects in different locales. In the United States, the impulse has been to see how the imagined consensus of "America's Best Idea" has masked some important disagreements about and contests over national park preservation—contests shaped by factors such as class, race, ethnicity, and gender. In this context we also need to recognize, as Ann McGrath reminds us in her essay, that there are nations within nations, and that an important part of the transnational history of the national parks has been how indigenous peoples have fought their exclusion from and then negotiated their inclusion within the nations that surround their own. Moreover, as the essays by Jane Carruthers and Steve Rodriguez suggest, the nationalisms that have guided national park creation in other parts of the world have sometimes been authoritarian, exclusionary, and deeply contested. In all these ways, efforts to understand national parks beyond the nation must also involve carefully looking *within* nations, as units of governance and sovereignty as well as inclusive communities, to see what national and subnational identities lurk therein.

Before getting to a fuller discussion of how the scholars in this volume have challenged the "America's Best Idea" claim by moving beyond the nation in these various ways, it is worth discussing why the nation has been such a persistent and powerful frame for park protection. At its most basic level, a national park requires a nation, even as it helps constitute that nation. In that sense, it is not surprising that the creation of national parks, in both the United States and throughout the world, has been coincident with the history of nation-states and their rise. We also ought to understand that national parks usually are designed to be landscapes of *national* importance, and that they thus need to be seen as existing on a continuum of park types and jurisdictions, from town commons and city parks to state parks and early open space reserves to national parks—that is a U.S. typology, but as many of the essays in this volume have shown, national parks in most national contexts exist as part of a larger conservation spectrum. To argue that a *nation* needs parks, rather than a town or a city or even a state or a province, raises an interesting series of historical questions about what nations are, what they need from the natural world, and why national parks arise when they do in particular contexts. It is also to ask what it means to attach the protection of nature to the celebration of nation. In the United States, it is not surprising that the first national parks originated during and immediately after the Civil War, when the nation was ascendant and when the West promised regional landscapes of iconographic unity and healing. It is also perhaps not surprising that the national parks gained appeal and political support in the midst of westward expansion and the Second Industrial Revolution, when Americans were transforming landscapes and resources in unprecedented and dramatic ways, or that the recreational use of these lands took hold in the United States at a time when many feared that Americans were becoming overcivilized and feminized. It is also worth noting that, in the U.S. context, most national parks were carved from the public lands as a kind of national territory. The national parks and the nation grew up together in the United States, then, and as they did, the parks provided Americans—or at least many Americans—with all sorts of cultural touchstones for imagining their national citizenship.

These are all familiar arguments to historians of America's national parks—we realize that aside from protecting wild nature, the national parks are cultural institutions in all sorts of wonderful and problematic ways. But what's equally important to recognize about the relationship between national parks and the rise of the nation in the United States is that American economic and political development came to be seen by many Americans, and by some outside of the United States as well, as a normative developmental path that other nations ought to follow, and that one of the things that a modern liberal democracy did was to protect certain portions of its national landscape as parks. Seeing the national park as "America's Best Idea" has thus served to link national park preservation, and conservation more broadly, to the ideal of development as Americanization, thus conflating American influence with modernization itself. It is this hubris—a hubris that can be as present in criticisms of Americanization as it is in celebrations of it—that has led some to assume that national parks in other places must be the intellectual offspring of a uniquely American idea, in the same way that we assume that consumer culture is American culture. As a result of this conflation, the national park often has been coded as American in ways that have thwarted inquiries into whether such was and is actually the case. Given the power of that narrative, the simple act of decentering, even provincializing, the United States in the global history of nature protection can be revelatory. Again, the essays in this volume provide a sampling of the larger possibilities here.

To argue that the national park idea is "America's Best Idea" is also to suggest that we can find a discrete point at which the park "idea" emerged in our nation's history, but that is not as easy as it might seem. Let's take five points of origin in the U.S. context and see what they add up to before we assume a more global frame. Scholars of the national park idea in the United States like to point to a big idea that the artist and western explorer George Catlin had in 1832 to make the Great Plains—then often referred to as the Great American Desert—into a vast "nation's park" where native peoples and bison could coexist as they presumably had since time immemorial. Ignoring for a moment that the bison-hunting cultures of the Great Plains were a new thing under the sun, made possible by horses, diseases, and expansive commercial markets of Euro-American origin, what's notable about Catlin's idea is that, in a highly romanticized way, he mixed the preservation of nature and indigenous culture, a mix that would be decidedly missing among later national park thinking in the United States—though, as we have learned from several essays in this volume, much more present in other national park traditions. Moreover, Catlin's was not a scenic ideal. Nonetheless, the phrase "nation's park" was important in its connection of park preservation to the nation, and so it has become an important genealogical moment in U.S. national park history.

A second point of origin can be found in an unlikely place at the same moment. It is a little-known fact that the first area within the current national park system to be withdrawn from the public domain and closed to settlement for the purpose of permanent public protection was Arkansas Hot Springs, now Hot Springs National Park, which was set aside as a federal reservation of national significance in 1832, largely for its alleged medicinal values. Although it did not become a national park until 1921, Arkansas Hot Springs has been publically preserved longer than any other unit in the system, and thus it has a kind of claim to the mantle of the first national park in the United States. The year 1832 was an important one, then, in the history of the national park "idea," but the two points of origin could not have been more different from each other in terms of intellectual content. Moreover, neither embodied or foreshadowed what would come next.

Let's move into more recognizable territory. The third point of origin lies in the Yosemite Valley, which was set aside in 1864, though not as a national park but as a state park. Nonetheless, we often cite this as a critical point of origin because Yosemite would not only become a national park—in 1890 Congress created a large national park surrounding the state reserve, and then, in 1905, the state reserve was folded into the national park, giving us Yosemite National Park as we know it today—but it is also generally recognized as one of the crown jewels of the system and thus one of the true embodiments of the national park idea. But in 1864 the nationalization of this spectacularly scenic landscape was still a ways off.

My fourth point of origin comes eight years later, in 1872, when Congress created Yellowstone National Park. This is the moment when many would argue that the national park idea was unequivocally born, but even here Yellowstone's creation appears paradigmatic only in retrospect. Yellowstone was initially established as a "public park," and as even the Burns series mentioned, Yellowstone became a *national* park by default, because it was within U.S. territorial holdings rather than an established state; had it been within a state, it likely would have been, like Yosemite, protected as a state park. (As an aside, it is worth mentioning that the next large-scale effort at park preservation in the United States after Yellowstone—the Adirondack Park—also occurred back at the state level.) It is certainly true that the phrase "national park" originated with Yellowstone, but does that phrase an idea make? One wonders if primacy here is more semantic than substantive.

Importantly, Yosemite and Yellowstone shared several characteristics that have, in retrospect, made them seem the germ of a larger idea. First, they were landscapes of monumental scenery protected for public use and from commercialization. Many park advocates and supporters would come to argue that scenic "monumentalism"—sublime scenery of national importance—defined landscapes eligible for inclusion in the national park system.[4] Second, these scenic jewels glistened most marvelously in a wilderness setting, and wilderness thus became a necessary—though not a sufficient—national park quality.[5] Moreover, and less flatteringly, the histories of both Yellowstone and Yosemite involved removing native peoples and their ecological legacies from national park landscapes, thereby burnishing their "wilderness" qualities.[6] These parks, in other words, decidedly did not follow a key part of Catlin's vision. Finally, it's worth noting here that even after Yellowstone apparently pioneered the idea of the national park, the nation's appetite for further national park creation was decidedly suppressed. There would be few other additions to the national park system before the end of the century, and one of the most important exceptions to that rule would only confuse matters. Few people today know that America's second national park, created by an act of Congress in 1875, was Mackinac National Park, which protected a portion of Michigan's Mackinac Island, then fast becoming a popular summer resort and which included the U.S. Army's Fort Mackinac. Almost nothing about Mackinac National Park hewed to the Yellowstone model. Thus, if we are to see Yellowstone as having pioneered "America's Best Idea," the Mackinac legislation seems to have ignored most of its particulars. Mackinac National Park lasted twenty years, until 1895, when it was transferred to Michigan and became a state park. Its curious creation, coming on the heels of Yellowstone's preservation, suggests that in the late nineteenth century the national park "idea" was still largely undefined if not downright incoherent.[7]

Only in the early 1900s would the U.S. national parks start to take on some definitional coherence as a system. And here the key moment—and the fifth point of origin in my scheme—was the creation of the National Park Service in 1916, when the U.S. national parks finally gained federal institutional support and a clear policy voice. It is also worth noting that the creation of forest reserves and then their redefinition as utilitarian "national forests" in the 1890s and early 1900s—and to a lesser extent the creation of other national-level landscape protection schemes such as national monuments and national wildlife refuges—were critical developments in creating the necessary contrast that helped give definitional clarity to what a U.S. national park was and was not. Even then, such definitional clarity required the sustained intellectual and promotional work by a handful of important advocates, from John Muir, whose *Our National Parks* (1901) helped define and publicize the system (although in 1901 Muir still lumped together the national parks and "forest reservations"), to lesser-known figures such as Robert Sterling Yard, who fought to define and defend national park standards in the late 1910s and 1920s.[8]

I present these five moments of plausible origin to make a simple but important point: the national park "idea" in the United States did not emerge as a coherent template for U.S. land preservation, let alone as a policy model neatly packaged for export, until the years just before World War I. And even then it is debatable just what the United States achieved in terms of coherence and international influence. Moreover, not long after the end of World War I, all sorts of other additions to the system—battlefields, historic sites, civic monuments—muddied the coherence of the national park idea, while wilderness advocates, ecologists and other scientists, wildlife preservationists, and outdoor recreationists continually remade the U.S. national park idea and the national park system across the twentieth century. It is very difficult, in other words, to find the national park idea and fix its origins in a particular moment in time.

As it turns out, it is also difficult to fix the origins of the national park idea in space. Even if we do, for the sake of argument, assume a coherent national park idea had emerged in the United States by the early 1900s, examining that development in an international context raises some important questions about U.S. primacy. In a recent essay in the *Journal of American Studies,* Ian Tyrrell, an Australian who is a leading scholar of transnational U.S. history, argues that the Progressive Era creation of national parks was a decidedly transnational achievement, and that a unique American origin of the idea can only be sustained with a selective rendering of the evidence. It may be true that Yellowstone was the first preserve to be called a national park, but there were lots of other proximate moments of creation that are not so tidily chalked up to U.S. influence. Australians created Audley National Park in 1879 in New South Wales; Canadians created what would become Banff National Park soon thereafter, first protecting it as a hot springs reserve in 1885 and then as a national park in 1887; and in New Zealand that same year, as Ted Catton details in his essay, the Māori donated tribal land that would become Tongariro National Park. Tyrrell not only argues, quite bluntly, that "the United States provided no model for global diffusion of the idea" that could explain these other park creations, but he also points to the diversity of models and influences traversing these transnational circuits during the late nineteenth and early twentieth centuries. Tyrrell's most important and instructive point is that arguments for primacy ought to be beside the point, for they do violence to the transnational richness and diversity of the late nineteenth and early twentieth centuries as a period when all sorts of concerns about environmental change and ideas about how best to protect natural landscapes and wildlife manifest themselves over much of the globe—and particularly in settler societies and in other imperial settings. Or, to put it another way, to isolate the national park as a singular achievement rooted in the American model is to miss how these diverse traditions have flowed in and out of national parks and their management, both in the United States and throughout the rest of the world, over the last century. The essays in this volume build on Tyrrell's argument to demonstrate that the transnational richness and diversity of national park history has been the rule, from the nineteenth century into the twenty-first century.[9]

A number of the chapters in this volume expand on and deepen Tyrrell's basic point. Jane Carruthers, Emily Wakild, and José Drummond suggest that while the American example did exert some influence in the creation of South African and Latin American national parks, that influence was selective and meshed with local conditions and influences. In the Latin American context, Wakild suggests that while the United States was an important model of national park protection, national parks in Latin America were also products of their own regional traditions of field science and even policies aimed at protecting indigenous peoples. Indeed, Wakild's recent book on the history of national parks in Mexico, *Revolutionary Parks,* shows how different Mexican national parks have been from their U.S. analogues, particularly in their attention to issues of social justice.[10] But difference is not Wakild's only compelling point; in her essay in this volume, she also intimates that there was a Pan-American dimension to be found in the history of national park creation in the Americas, one influenced, for instance, by Humboldtian science. Meanwhile, Carruthers shows how the national park ideal operated in South Africa in the service of apartheid, creating distinctively racialized spaces of nature conservation—and particularly "fortress" wildlife conservation—that few Americans would agree embodied the best of American intellectual influence. That the United States similarly used park creation to dispossess native peoples suggests a germ of insidious influence, but Carruthers's point is a more exciting one than merely showing how an unflattering part of the U.S. national park legacy found its way to South Africa. Rather, she insists, as Wakild does for Latin America, that we need to understand national park creation as a result of specific national contexts and cultures, not merely spreading U.S. influence. Indeed, Carruthers suggests that we see national parks, in South Africa and elsewhere, less as an American export than as an emergent multinational brand that was, and continues to be, deployed flexibly in different national settings. This seems an idea worth thinking with. To that we might add Chris Conte's analysis of the role that *National Geographic* has played in presenting a historically and ecologically inaccurate rendering of the Great Rift Valley and the need to protect its landscapes from African peoples. Despite its name, the National Geographic Society is an international actor in the world of conservation; as importantly, it is an international media operation that trucks in the national park brand. Conte reminds us that the national park idea is hopelessly entangled in media representations and goals, and that we ought to pay attention to the relationship between the media and the message. If we do not, as Conte evocatively suggests, we get national parks that are, to the people that surround them, "dead zones," places hostile to their history and contemporary presence.

Ted Catton and Ann McGrath explore some similar issues in their essays on national parks in New Zealand and Australia. Catton does note that the

U.S. National Park Service provided considerable technical guidance to New Zealanders; indeed, his is one of a number of essays in this volume that make me wonder if the greatest influence that the United States has had on the national parks of the rest of the world has been as a bureaucratic model rather than an intellectual inspiration. But, more importantly, Catton argues that the early creation of national parks in New Zealand is more accurately chalked up to a complex and layered series of influences: what he calls a "world culture" of romanticism and the global growth of tourism; a heritage of wilderness thinking that seems common across settler societies; a transnational culture of science; and the particulars of New Zealand's natural and cultural history. In this last sense, two points stand out. First, there is the centrality of endemism and what Catton calls the "weird and unsettled ecology" of the islands. Much more so than in the United States, New Zealand's national parks have been about protecting a unique legacy of native species particularly susceptible to recent waves of human colonization, Polynesian as well as European. Second, Catton's essay wonderfully illustrates the central role that the Māori played, and continue to play, in national park creation and management in New Zealand. Catton settles on the nation and a distinctive New Zealand nationalism, though one that is increasingly bicultural (perhaps even binational), as the major force for explaining the origins and development of New Zealand's national park system.

McGrath provides a thoughtful "transnational juxtaposition" of debates over two national landmarks that have been contested by internal nations—Devils Tower, the first national monument created in the United States, and Ayers Rock in Australia. Both of these rockscapes have attached to them competing claims to sovereignty by indigenous peoples, and even competing names—Mato Tipila and Uluru—and both have been the sites of conflict over what it means for white settler nationalists to use these rock formations as climbing challenges. Like Catton and several of the other authors, McGrath argues that we ought to see settler societies as particularly keen national park creators, and that national parks and other natural landmarks work in complicated ways to celebrate the conquests of lands and peoples as well as working to help forge new national cultures distinctive from home cultures in Europe. McGrath really brings out an idea that I think is critical to the present and future of national parks: that, despite their frequent origins in colonial contexts in which the preservation of nature often resulted in the dispossession of indigenous peoples, national parks have also been flexible spaces—she evocatively refers to them as "elastic signifiers of the sacred." Or, to put it another way, while national parks in many places, the United States included, have not always lived up to their billing as national spaces in the most inclusive possible sense, the very claim that they represent national public spaces means that they are open to contest and policy revisions that can rectify past exclusions. In his previous work, Catton has shown how such a revisionist approach played out in Alaska, resulting in an ideal of "inhabited wilderness" in which native peoples came to enjoy rights within preserved parks.[11] More recently, in a portion of Badlands National Park, we have the first national park area in the United States comanaged by native peoples—in this case the Lakota. What is perhaps most interesting about the evolution of these policies in the transnational context is how relatively late the United States was to rethink the national park ideal in relation to native claims—Australia, New Zealand, and Latin America recognized these claims much earlier and incorporated them to varying degrees in their preservation efforts. And while, in South Africa, national parks functioned as apartheid institutions into the 1990s, Carruthers shows that even there they are being redefined in some fascinating ways now that South Africa is a democratic nation. Indeed, as these elastic spaces evolve to be more inclusive, we may even come to see the dispossessive tendencies of early U.S. national park history as the exception rather than the rule.

Several other contributors show that the United States was actually slow in creating a national park system and administration, both of which were essential to bringing definition and coherence to the "idea." Alan MacEachern, for instance, shows that the Canadians beat the United States to the punch in creating a park service, and that advocates for a U.S. national park service in fact often invoked the Canadian model, if sometimes more condescendingly than admiringly. While MacEachern notes that the Dominion Parks Branch in Canada, created in 1911, was "the first agency in the world devoted to national parks," his essay is less about claiming this place of primacy than it is about a subtler but more intriguing point. The creation of both the Dominion Parks Branch and the National Park Service, he points out, occurred in the midst of a strong "See America First" (SAF) campaign, one designed to keep American tourists at home rather than seeing them head off for a European grand tour, and a chief part of the SAF claim on tourists was that American nature was a formidable competitor with European culture. Scholars such as Peggy Shaffer have argued convincingly that the SAF campaigns were a critical moment in the attachment of strong U.S. patriotism and nationalist identity to park nature.[12] But MacEachern found something surprising in this movement during the 1910s: SAF campaigns often included Canadian as well as American national parks, and they sometimes even included Mexico as well. For a brief moment at least, See America First was a truly continental imperative, and many U.S. Americans seemed as eager to see the Canadian Rockies as they were to see the American Rockies. MacEachern ultimately does suggest that the United States had a strong influence on the Canadian administration of its parks, despite Canadian primacy in creating a parks bureau, but he also demonstrates how much Americans were paying attention to Canada and other park examples as they moved toward the creation of the National Park Service. Finally, MacEachern's point about the surprising number of Americans who visited Canadian national parks in the early twentieth century raises a larger point, one raised briefly in chapter 1 but otherwise left unexamined: as I have visited the national parks of the United States over the past several decades, I have always been amazed at how international they are—they are polyglot places filled with tourists from other nations. I have even been struck with how often the concessionaire employees are themselves foreign nationals, often with countries of origin identified on their nametags. It would be curious to know what has motivated international visitors and workers to come to American national parks, to get some of their reactions to their visits, and to see how their perspectives would intersect with the "America's Best Idea" narrative. Clearly our national parks are important American destinations for visitors from other nations; it would be interesting to know what they learn about our nature and our nation on their visits, and what they carry home with them from their experiences here.

While most national parks have had tourism as a major motivating force, science has also had an important place in national park creation and management. In his wide-ranging essay, Patrick Kupper uses park science to destabilize the "America's Best Idea" narrative in several productive ways. First, Kupper shows that national parks were networked beyond national borders through science and that, as such, we ought to see conservation scientists as an international community of interest, at least aspirationally if not always in fact.[13] Second, he shows how the goals and discoveries of scientists altered the national park idea across the twentieth century, though to varying degrees in different places. Third, he shows that while science was an important justification for park creation in some parts of the world, the United States was not one of them. Indeed, he begins his analysis with an attention-getting claim: "When the United States built its national park system in the late nineteenth and early twentieth centuries," science "played only a peripheral role." Echoing the work of Richard Sellars, who argued that the early professional culture of the NPS was dominated by landscape architects rather than scientists, Kupper suggests that facilitating outdoor recreation was the much more important purpose of U.S. national parks during most of their history.[14] Moreover, Kupper convincingly shows that a scientific rationale for national park protection was perhaps the dominant one in the early creation of certain European national parks, and that its influence spread to other parts of the world—particularly Europe's colonies. As Sterling Evans has shown, we could add the example of Costa Rica to this story as well.[15] In distinguishing between the national parks for science in places such as Switzerland and the national parks for tourism in the United States, Kupper suggests that different parts of the world developed different models for national park preservation—America's idea, in other words, was one idea among many.

To the extent that national parks have moved in increasingly scientific directions, we have to wonder whether Europe has a claim to primacy, at least for a key strand of the park "idea." But beyond that, Kupper suggests that we need to see scientists as a transnational community of practitioners whose research in and on national parks not only created international networks but also was an "interfering force" in national park management. In many places scientists emerged as important critics of national park priorities, claiming for the scientist a more dispassionate and objective perspective on defining the best interests of nature, and in other instances they were one among several user groups competing for priority in parks management. It is also the case that in recent years—and this is something Kupper only hints at—conservation biologists (practitioners of a branch of science birthed by nature conservation efforts) have fundamentally challenged the ecological efficacy of the bounded national park. Instead, some of the most ambitious conservation schemes today are about linking national parks and buffering them with other sorts of conservation lands, often across national borders. Scientists, then, have taught us that national parks do not always work ecologically; they have pushed us to think about the need for preservation beyond the parks. Even more than that, scientists have been a driving force in the globalization of landscape and species conservation, sometimes with controversial results. The South Asian scholar Ramachandra Guha warns about what he calls "authoritarian biologists"—scientists, often American ones, who have moved into developing nations and imperiously gained control over landscapes for conservation.[16] As a transnational community, then, conservation scientists deserve the careful attention of national park historians not only because they tend to be transnational in professional orientation but also because they function as discrete and often disruptive epistemic communities.

While scientists have their own interests—interests that sometimes jibe with nationalist conservation agendas and sometimes do not—it is also important to recognize that they pay attention to the natural world and evince a loyalty to rendering its processes as objectively as possible. As such, they help give voice to environmental dynamics that challenge how we have constructed the natural world intellectually. Increasingly those environmental dynamics are extranational. Karen Routledge provides a nice example of how grizzly bears—a form of what Mark Fiege, in another context, has called "mobile nature"—have pushed park managers on either side of the U.S.–Canadian border to interact and cooperate with each other.[17] Her setting is the Waterton-Glacier International Peace Park, which, beginning in 1932, linked Glacier National Park in the United States (created in 1910) with Waterton Lakes National Park in Canada (created in 1895, a full decade and a half before Glacier), and her key players are nonhuman border crossers. Indeed, if one of the central contributions of environmental history has been to give causal power to the nonhuman world, Routledge's essay is but one of many examples of how mobile nature constantly defies the boundedness of park preservation. Beyond pointing out that grizzly bears (and other forces such as fire) created managerial challenges that refused to respect international borders, Routledge's chapter is notable for an intriguing archival point—she was able to discover a hidden history of transnational managerial cooperation that existed quite intentionally outside of the official park service records. Park rangers, in other words, are not so different from those wandering grizzlies (which, perhaps, suggests the need to radio-collar them and follow their movements).

Mark Carey tackles perhaps the most important impending environmental challenge to national park preservation all over the world—global climate change. Carey's essay, which examines popular and environmentalist responses to disappearing glaciers in places such as Glacier National Park in the United States and Huascarán National Park in Peru, locates important disconnects between landscape iconography, ecological processes, and the social dimensions of climate change. Receding glaciers in national parks, he argues, may well be poignant indicators of a warming global climate, but our undue focus on them reveals an attachment to what he calls, provocatively, a "wilderness climate." Glaciers, Carey suggests, may be iconic scenery, but they have never been static, and, from the standpoint of conservation science, their recession may not be that meaningful. Lamenting the ice itself, a media and environmentalist strategy that works the scenic at the expense of the scientific, has functioned to distract us from the broader ecological and social costs of climate change. In this sense, Carey provocatively argues that the national park idea, insofar as it is defined by iconic and unchanging scenery, has become an obstacle to more fully understanding the environmental and social challenges posed by global climate change.

Where Carey uses global warming science to critique the cultural commitments of environmentalists and the media, Steve Rodriguez's discussion of national parks in Indonesia does something of the opposite. Rodriguez's is a story of failure—how the national park idea did *not* work in Indonesia. But while the traditional narrative in such settings has been about the arrogance of first-world conservationists in imposing their conservation regimes on the developing world, Rodriguez's story is more focused on what happens when national parks come to the developing world under authoritarian home rule and fail to meet the cultural and aesthetic needs of the people. In this case, Suharto, who ruled Indonesia for more than thirty years, presided over what seemed at the time to be one of the most remarkable cases of a developing-world nation adopting the national park agenda. In consultation with a number of NGOs and IGOs (and the articulation of authoritarian home rule and international conservation organizations is one of the most fascinating aspects of his case study), Indonesia set aside millions of acres as national parks, prioritizing biodiversity protection and ecotourism development. Indonesia achieved such a high profile in the international conservation community that the Third World Parks Congress was held in Bali in 1982—the first time it was held outside of the United States (and as an aside, these world congresses appear in a number of the chapters and seem to be one of the most important venues for witnessing the globalization of national park ideals in all their complexity). But Suharto's national park program paid almost no attention to what Indonesians themselves wanted from their parks, and, as a result, Indonesians became hostile to the national parks as symbols of Suharto's repressive regime. In a different sense of the phrase, then, Suharto created "national parks beyond the nation."

A few important lessons emerge from Rodriguez's case study. First, while national park programs in democratic nations have been imperfect, and those imperfections have generally mirrored inequalities in access to full national citizenship, these programs have been more open to contestation and renovation in democracies than they have been in authoritarian regimes. Indeed, Rodriguez shows that Suharto's efforts in Indonesia created "paper parks" that largely served to increase the authoritarian regime's control over national space while ingratiating itself with the international conservation community. Second, while much of the literature critical of the national parks in the developing world has contrasted parks as spaces of nature conservation for the recreational desires of wealthy urbanites with the resource needs of people who live in and around parks, Rodriguez's essay reminds us that we risk essentializing developing-world peoples by assuming they are all resource users hostile to recreational preservation. In the case of Indonesia, Rodriguez argues that one reason for the failure of national parks had to do with their inability to meet the *recreational* preferences of Indonesians and that, as such, they "failed to . . . provide meaningful locations for national identity." The upshot of such an analysis, it seems to me, is that if national parks are to function in developing-world contexts while also serving global conservation agendas, they must be not only democratic spaces but also spaces that speak to and reflect national culture. Or, to put it another way, maybe aesthetic and cultural constructions of nature, which Mark Carey dismisses as an opiate in the case of global warming and its "real" impacts, function in important ways to give nature preservation democratic legitimacy. This is to say that one of the most important tensions that emerges in these essays is that between discrete national cultures of nature preservation—cultures the authors describe in ways that vary from admiring to deeply critical—and a universalizing culture of scientific concern.

Let me move toward a conclusion by summarizing what I have learned from all of these excellent essays. The first lesson is that America's place in the history of the national parks of the world has not been as large as many have assumed, though exactly how we should characterize that influence remains to be worked out and may differ widely from place to place and across time. More than that, as national parks are increasingly spaces shaped by international forces and influences, we need to be careful to not assume too heavy a U.S. hand. As we have watched, for instance, the World Parks Congresses shift from Seattle and Yellowstone to Bali, Caracas, Durban, and, most recently, Sydney in 2014, we have seen a real shift in who controls the international discourse on national parks. This shift has paralleled how national parks have themselves evolved so as to respect and better account for marginal peoples who had often experienced park preservation as a kind of violence against them. Second, while what we might call second-generation conservation scholarship on national parks and conservation has been deeply critical, challenging the "best" part of the "America's Best Idea" narrative as well as fundamentally questioning the social costs of conservation,

I have been struck by what I can only identify as a third generation of national parks and conservation scholarship that seems to be emerging in this volume. One flaw of second-generation studies is that, in dismantling ideas such as the national park or wilderness, they too often themselves assumed the fixity and ideological power of those ideas, when, in fact, careful attention to their history often reveals flexibility and pragmatism. Most of the scholars gathered here seem to believe in national parks as spaces that are fundamentally improvable, if not quite perfectible. Indeed, I am struck by the pervasive optimism of this volume, even as its authors have discussed cases in which national park preservation has been a troubled and troubling enterprise. That optimism, it seems to me, is the final lesson, and perhaps the great gift, of really putting the national park idea into a global historical context. Are the national parks "America's Best Idea?" I do not think that this simplistic and jingoistic phrase holds up well when we absorb the international histories of national park creation offered here. Instead, we ought to be focused on the ideas of the world, on transnational exchanges, and on what we learn from each other when we move beyond the nation. Doing so will only strengthen the national parks as natural and civic spaces. Indeed, we might do well to apply to the national parks of the world another of Wallace Stegner's famous phrases, for, in the wonderful essays in this volume, they emerge, warts and all, as "geographies of hope."[18]

Notes

1. Paul S. Sutter, review of *The National Parks: America's Best Idea,* directed by Ken Burns, *Journal of American History* 97, no. 3 (December 2010): 892–96.
2. For an introduction to transnational U.S. scholarship as a challenge to American exceptionalism, see Thomas Bender, ed., *Rethinking American History in a Global Age* (Berkeley: University of California Press, 2002).
3. Since I first wrote this essay, another important collection on the global and transnational history of national parks has appeared. See Bernhard Gissibl, Sabine Höhler, and Patrick Kupper, eds., *Civilizing Nature: National Parks in Global Historical Perspective* (New York: Berghahn, 2012).
4. I borrow the term "monumentalism" from Alfred Runte, *National Parks: The American Experience* (Lincoln: University of Nebraska Press, 1979).
5. I discuss the relationship between national parks and wilderness in Paul S. Sutter, *Driven Wild: How the Fight against Automobiles Launched the Modern Wilderness Movement* (Seattle: University of Washington Press, 2002).
6. See Mark David Spence, *Dispossessing the Wilderness: Indian Removal and the Making of National Parks* (New York: Oxford University Press, 1999).
7. My point about the Yellowstone model not being quickly emulated is influenced by Ian Tyrrell, "America's National Parks: The Transnational Creation of National Space in the Progressive Era," *Journal of American Studies* 46, no. 1 (February 2012): 1–21. I want to thank Patrick Kupper for bringing the Mackinac example to my attention. On Mackinac's brief history, see Barry Mackintosh,

The National Parks: Shaping the System (Washington, D.C.: Department of the Interior, 2005), 13–14.

8. On Muir, see Donald Worster, *A Passion for Nature: The Life of John Muir* (New York: Oxford University Press, 2008). On Yard's career, see Sutter, *Driven Wild,* 100–141; Marguerite Shaffer, *See America First: Tourism and National Identity, 1880–1940* (Washington, D.C.: Smithsonian Institution Press, 2001); John Miles, *Guardians of the Parks: A History of the National Parks and Conservation Association* (Washington, D.C.: Taylor & Francis, 1995).

9. Ian Tyrrell, "America's National Parks: The Transnational Creation of National Space in the Progressive Era," *Journal of American Studies* 46, no. 1 (February 2012): 1–21.

10. Emily Wakild, *Revolutionary Parks: Conservation, Social Justice, and Mexico's National Parks, 1910–1940* (Tucson: University of Arizona Press, 2011).

11. Theodore Catton, *Inhabited Wilderness: Indians, Eskimos, and National Parks in Alaska* (Albuquerque: University of New Mexico Press, 1997).

12. Shaffer, *See America First.*

13. For a counterexample of nationalism getting in the way of transnational science, see Michael Lewis, *Inventing Global Ecology: Tracking the Biodiversity Ideal in India, 1947–1997* (Athens: Ohio University Press, 2004).

14. Richard West Sellars, *Preserving Nature in the National Parks: A History* (New Haven, Conn.: Yale University Press, 1997).

15. Sterling Evans, *The Green Republic: A Conservation History of Costa Rica* (Austin: University of Texas Press, 1999).

16. This notion has appeared several places in Guha's work. See, for instance, his chapter, "Authoritarianism in the Wild," in Guha, *How Much Should a Person Consume: Environmentalism in India and the United States* (Berkeley: University of California Press, 2006).

17. Mark Fiege, "The Weedy West: Mobile Nature, Boundaries, and Common Space in the Montana Landscape," *Western Historical Quarterly* 36 (Spring 2005): 22–47.

18. Stegner used this phrase in his famous "Wilderness Letter," written in 1960 to the Outdoor Recreation Resources Review Commission and later published as "Coda: Wilderness Letter," in Stegner, *The Sound of Mountain Water* (Garden City, N.Y.: Doubleday, 1969).

Selected Bibliography
Recommended Sources on the International History of National Parks

The following is a selective list of readings that have helped the authors and editors think about national parks beyond the nation. We make no attempt to be comprehensive. Rather, our goal is to identify those sources that have been useful to us and that might be useful to others who are working on the international history of national parks and other protected areas.

Adams, William M., and Mulligan, Martin, eds. *Decolonizing Nature: Strategies for Conservation in a Post-colonial Era.* London: Earthscan, 2003.

Ali, Saleem H., ed. *Peace Parks: Conservation and Conflict Resolution.* Cambridge, Mass.: MIT Press, 2007.

Anderson, Benedict. *Imagined Communities: Reflections on the Origin and Spread of Nationalism.* Rev. ed. London: Verso, 2006.

Banivanua Mar, Tracey, and Penelope Edmonds, eds. *Making Settler Colonial Space: Perspectives on Race, Place, and Identity.* New York: Palgrave Macmillan, 2010.

Banner, Stuart. *Possessing the Pacific: Land, Settlers, and Indigenous People from Australia to Alaska.* Cambridge, Mass.: Harvard University Press, 2007.

Barton, Gregory. *Empire Forestry and the Origins of Environmentalism.* New York: Cambridge University Press, 2002.

Beinart, William, and Lotte Hughes. *Environment and Empire.* Oxford: Oxford University Press, 2007.

Belich, James. *Replenishing the Earth: The Settler Revolution and the Rise of the Anglo-World, 1783–1939.* Oxford: Oxford University Press, 2009.

Bender, Thomas. *A Nation among Nations: America's Place in World History.* New York: Hill & Wang, 2006.

———, ed. *Rethinking American History in a Global Age.* Berkeley: University of California Press, 2002.

Benson, Etienne. *Wired Wilderness: Technologies of Tracking and the Making of Modern Wildlife.* Baltimore: Johns Hopkins University Press, 2010.

Borchers, Henning. *Jurassic Wilderness: Ecotourism as a Conservation Strategy in Komodo National Park, Indonesia.* Stuttgart: Ibidem, 2004.

Brown, Kate. "Gridded Lives: Why Kazakhstan and Montana are Nearly the Same Place." *American Historical Review* 106 (February 2001): 17–48.

———. *Plutopia: Nuclear Families, Atomic Cities, and the Great Soviet and American Plutonium Disasters.* New York: Oxford University Press, 2013.

Bsumek, Erika Marie, David Kinkela, and Mark Atwood Lawrence, eds. *Nation-States and the Global Environment: New Approaches to International Environmental History.* New York: Oxford University Press, 2013.

Burke, Edmund, III, and Kenneth Pomeranz, eds. *The Environment and World History.* Berkeley: University of California Press, 2009.

Campbell, Claire Elizabeth, ed. *A Century of Parks Canada, 1911–2011.* Calgary, Alberta: University of Calgary Press, 2011.

Campbell, SueEllen. *The Face of the Earth: Natural Landscapes, Science, and Culture.* Berkeley: University of California Press, 2011.

Carey, Mark. *In the Shadow of Melting Glaciers: Climate Change and Andean Society.* New York: Oxford University Press, 2010.

Carr, Ethan, Shaun Eyring, and Richard Guy Wilson, eds. *Public Nature: Scenery, History, and Park Design.* Charlottesville: University of Virginia Press, 2013.

Carruthers, Jane. *The Kruger National Park: A Social and Political History.* Pietermaritzburg, South Africa: University of Natal Press, 1995.

Catton, Theodore. *Inhabited Wilderness: Indians, Eskimos, and National Parks in Alaska.* Albuquerque: University of New Mexico Press, 1997.

Cioc, Marc. *The Game of Conservation: International Treaties to Protect the World's Migratory Animals.* Athens: Ohio University Press, 2009.

Conte, Christopher. "Creating Wild Landscapes from Domesticated Landscapes: The Internationalization of the American Wilderness Concept." In *American Wilderness: A New History,* ed. Michael Lewis, 223–41. New York: Oxford University Press, 2007.

Cullather, Nick. *Feeding a Hungry World: America's Cold War Battle against Poverty in Asia.* Cambridge, Mass.: Harvard University Press, 2010.

Diefendorf, Jeffry, and Kurk Dorsey, eds. *City, Country, Empire: Landscapes in Environmental History.* Pittsburgh: University of Pittsburgh Press, 2005.

Dorsey, Kurkpatrick. *Whales and Nations: Environmental Diplomacy on the High Seas.* Seattle: University of Washington Press, 2013.

Dunlap, Thomas R. "Beyond the Parks, Beyond the Borders: Some of the Places to Take Ian Tyrrell's Perspective." *Journal of American Studies* 46 (February 2012): 31–36.

———. *Nature and the English Diaspora: Environment and History in the United States, Canada, Australia, and New Zealand.* New York: Cambridge University Press, 1999.

Earhart, H. Byron. *Mount Fuji: Icon of Japan.* Columbia: University of South Carolina Press, 2011.

Edwards, Paul. *A Vast Machine: Computer Models, Climate Data, and the Politics of Global Warming.* Cambridge, Mass.: MIT Press, 2010.

Ekbladh, David. *The Great American Mission: Modernization and the Construction of an American World Order.* Princeton, N.J.: Princeton University Press, 2010.

Fiege, Mark. "The Nature of the West and the World." *Western Historical Quarterly* 42 (Autumn 2011): 305–12.

Fink, Leon, ed. *Workers across the Americas: The Transnational Turn in Labor History.* New York: Oxford University Press, 2011.

Foresta, Ronald A. *Amazon Conservation in the Age of Development: The Limits of Providence.* Gainesville: University of Florida Press, 1991.

Frost, Warwick, and C. Michael Hall, eds. *Tourism and National Parks: International Perspectives on Development, Histories, and Change.* New York: Routledge, 2009.

Gissibl, Bernhard, Sabine Höhler, and Patrick Kupper, eds. *Civilizing Nature: National Parks in Global Historical Perspective.* New York: Berghahn, 2012.

Grove, Richard. *Ecology, Climate, and Empire: Colonialism and Global Environmental History, 1400–1940.* Cambridge: White Horse Press, 1993.

————. *Green Imperialism: Colonial Expansion, Tropical Island Edens, and the Origins of Environmentalism, 1600–1860.* Cambridge: Cambridge University Press, 1995.

Hall, Marcus. *Earth Repair: A Transatlantic History of Environmental Restoration.* Charlottesville: University of Virginia Press, 2005.

Hall, Melanie. *Towards World Heritage: International Origins of the Preservation Movement, 1870–1930.* Burlington, Vt.: Ashgate, 2011.

Holdgate, Martin W. *The Green Web: A Union for World Conservation.* London: Earthscan, 1999.

Hornborg, Alf, J. R. McNeil, and Joan Martinez-Alier. *Rethinking Environmental History: World-System History and Global Environmental Change.* Lanham, Md.: AltaMira Press, 2007.

Howkins, Adrian. "The Significance of the Frontier in Antarctic History: How the U.S. West Has Shaped the Geopolitics of the Far South." *Polar Journal* 3, no. 1 (2013): 9–30.

Hughes, J. Donald. *What Is Environmental History?* Cambridge, Mass.: Polity Press, 2006.

Igoe, Jim. *Conservation and Globalization: A Study of National Parks and Indigenous Communities from East Africa to South Dakota.* Toronto: Wadsworth/Thompson Learning, 2004.

Isserman, Maurice, and Stewart Weaver. *Fallen Giants: A History of Himalayan Mountaineering from the Age of Empire to the Age of Extremes.* New Haven, Conn.: Yale University Press, 2008.

Jacobs, Margaret D. *White Mother to a Dark Race: Settler Colonialism, Maternalism, and the Removal of Indigenous Children in the American West and Australia, 1880–1940.* Lincoln: University of Nebraska Press, 2009.

Jepson, Paul, and Robert J. Whittaker. "Histories of Protected Areas: Internationalism of Conservationist Values and Their Adoption in the Netherlands Antilles (Indonesia)." *Environment and History* 8, no. 2 (2002): 129–72.

Jones, Karen R. *Wolf Mountains: A History of Wolves along the Great Divide.* Calgary: University of Calgary Press, 2003.

Jones, Karen R., and John Wills. *The Invention of the Park: From the Garden of Eden to Disney's Magic Kingdom.* Cambridge, Mass.: Polity Press, 2005.

Jørgensen, Dolly, Finn Arne Jørgensen, and Sara B. Pritchard, eds. *New Natures: Joining Environmental History with Science and Technology Studies.* Pittsburgh: University of Pittsburgh Press, 2013.

Kupper, Patrick. *Creating Wilderness: A Transnational History of the Swiss National Park.* New York: Berghahn, 2014.

Lake, Marilyn, and Henry Reynolds. *Drawing the Global Colour Line: White Men's Countries and the International Challenge of Racial Equality.* New York: Cambridge University Press, 2008.

Lewis, Michael L. *Inventing Global Ecology: Tracking the Biodiversity Ideal in India, 1947–1997.* Athens: Ohio University Press, 2004.

MacEachern, Alan. *Natural Selections: National Parks in Atlantic Canada, 1935–1970.* Montreal: McGill-Queen's University Press, 2001.

Mauch, Christoph, Nathan Stoltzfus, and Doug Weiner, eds. *Shades of Green: Environmental Activism around the Globe.* Oxford: Rowman & Littlefield, 2006.

McNeil, John Robert, and Erin Stewart Mauldin, eds. *A Companion to Global Environmental History.* Hoboken, N.J.: Wiley, 2012.

Maier, Charles S. "Consigning the Twentieth Century to History: Alternative Narratives for the Modern Era." *American Historical Review* 105 (June 2000): 807–31.

Mels, Tom. *Wild Landscapes: The Cultural Nature of Swedish National Parks.* Lund, Sweden: Lund University Press, 1999.

Monahan, William J., and Nicholas Fisichelli. "Climate Exposure of U.S. National Parks in a New Era of Change." *PLoS ONE* 9, no. 7 (2014): e101302. doi:10.1371/journal.pone.0101302.

Morris, Stephen, and Jonathan Putnam, eds. "Fulfilling the International Mission of the U.S. National Park Service, a Special Issue." *George Wright Forum* 28, no. 3 (2011): 257–306.

"The Nation and Beyond: Transnational Perspectives on United States History, a Special Issue." *Journal of American History* 86 (December 1999): 965–1307.

Netz, Reviel. *Barbed Wire: An Ecology of Modernity.* Middletown, Conn.: Wesleyan University Press, 2004.

Orsi, Jared. "Construction and Contestation: Toward a Unifying Methodology for Borderlands History." *History Compass* 12 (May 2014): 433–34.

Radding, Cynthia. *Landscapes of Power and Identity: Comparative Histories in the Sonoran Desert and the Forests of Amazonia from Colony to Republic.* Durham, N.C.: Duke University Press, 2005.

Radkau, Joachim. *Nature and Power: A Global History of the Environment.* New York: Cambridge University Press, 2008.

"Rethinking History and the Nation-State: Mexico and the United States as a Case Study, a Special Issue." *Journal of American History* 86 (September 1999): 438–697.

Robin, Libby, and Tom Griffiths, eds. *Ecology and Empire: Environmental History of Settler Societies.* Seattle: University of Washington Press, 1998.

Robin, Libby, Sverker Sörlin, and Paul Warde, eds. *The Future of Nature: Documents of Global Change.* New Haven, Conn.: Yale University Press, 2013.

Scott, James C. *Seeing Like a State: How Certain Schemes to Improve the Human Condition Have Failed.* New Haven, Conn.: Yale University Press, 1998.

Sheaill, John. *Nature's Spectacle: The World's First National Parks and Protected Places.* London: Earthscan, 2010.

Sörlin, Sverker, and Paul Warde, eds. *Nature's End: History and the Environment.* London: Palgrave Macmillan, 2009.

Sutter, Paul. "Review of *The National Parks: America's Best Idea.*" *Journal of American History* 97 (December 2010): 892–96.

———. "The Trouble with 'America's National Parks'"; or, Going Back to the Wrong Historiography: A Response to Ian Tyrrell." *Journal of American Studies* 46 (February 2012): 23–29.

Swenson, Astrid. "Response to Ian Tyrrell, 'America's National Parks: The Transnational Creation of National Space in the Progressive Era.'" *Journal of American Studies* 46 (February 2012): 37–43.

———. *The Rise of Heritage: Preserving the Past in France, Germany, and England, 1789–1914.* Cambridge: Cambridge University Press, 2014.

Swenson, Astrid, and Peter Mandler. *From Plunder to Preservation: Britain and the Heritage of Empire, 1800–1950.* Oxford: Oxford University Press, 2013.

Teisch, Jessica. *Engineering Nature: Water, Development, and the Global Spread of American Environmental Expertise.* Chapel Hill: University of North Carolina Press, 2011.

Thom, David. *Heritage: The Parks of the People.* Auckland: Landsdowne Press, 1987.

Tolba, Mostafa K. *Global Environmental Diplomacy: Negotiating Environmental Agreements for the World, 1973–1992.* Cambridge, Mass.: MIT Press, 2008.

Trigger, David, and Gareth Griffiths, eds. *Disputed Territories: Land, Culture, and Identity in Settler Societies.* Hong Kong: Hong Kong University Press, 2003.

Tucker, Richard. *Insatiable Appetite: The United States and the Ecological Degradation of the Tropical World.* Berkeley: University of California Press, 2000.

Turner, James Morton. "Rethinking American Exceptionalism: Toward a Transnational History of National Parks, Wilderness, and Protected Areas." In Andrew C. Isenberg, ed., *The Oxford Handbook of Environmental History.* Oxford: Oxford University Press, 2014.

Tyrrell, Ian. "America's National Parks: The Transnational Creation of National Space in the Progressive Era." *Journal of American Studies* 46 (February 2012): 1–21.

———. *Crisis of a Wasteful Nation: Empire and Conservation in Theodore Roosevelt's America.* Chicago: University of Chicago Press, 2015.

———. "Ian Tyrrell Replies." *Journal of American Studies* 46 (February 2012): 45–49.

———. *True Gardens of the Gods: Californian-Australian Environmental Reform, 1860–1930.* Berkeley: University of California Press, 1999.

Uekötter, Frank, ed. *The Turning Points of Environmental History.* Pittsburgh: University of Pittsburgh Press, 2010.

Vitalis, Robert. *America's Kingdom: Mythmaking on the Saudi Oil Frontier.* Stanford, Calif.: Stanford University Press, 2007.

Wakild, Emily. "Border Chasm: International Boundary Parks and Mexican Conservation, 1935–1945." *Environmental History* 14 (July 2009): 453–75.

———. *Revolutionary Parks: Conservation, Social Justice, and Mexico's National Parks, 1910–1940.* Tucson: University of Arizona Press, 2011.

Walker, Brett L. *The Lost Wolves of Japan.* Seattle: University of Washington Press, 2005.

Walls, Laura Dassow. *Passage to Cosmos: Alexander Humboldt and the Shaping of America.* Chicago: University of Chicago Press, 2009.

Weaver, John C. *The Great Land Rush and the Making of the Modern World, 1650–1900.* Montreal: McGill-Queen's University Press, 2003.

White, Richard. "The Nationalization of Nature." *Journal of American History* 86 (December 1999): 976–86.

Worster, Donald, ed. *The Ends of the Earth: Perspectives on Modern Environmental History.* Cambridge: Cambridge University Press, 1989.

Young, Terence, and Lary M. Dilsaver. "Collecting and Diffusing 'the World's Best Thought': International Cooperation by the National Park Service," *George Wright Forum* 28, no. 3 (2011): 269–78.

CONTRIBUTORS

MARK CAREY is an associate professor of history and associate dean in the Robert D. Clark Honors College, University of Oregon, Eugene. A specialist in environmental history and the history of science, he is the author of *In the Shadow of Melting Glaciers: Climate Change and Andean Society* (2010). He thanks Kerry Snodgrass, Kelsey Ward, and Brandon Luedtke for research assistance. He is also grateful for excellent feedback from this book's editors and contributors, especially Ted Catton and Alan MacEachern. His chapter is based on work supported by the National Science Foundation grants No. 1010550 and No. 1253779.

JANE CARRUTHERS is professor emeritus in the Department of History, University of South Africa (Unisa), Pretoria. She was one of South Africa's first environmental historians and has authored and edited numerous books, most recently *Thomas Baines: Exploring Tropical Australia, 1855 to 1857* (2012). She thanks the organizers of the symposium, including Maren Bzdek; Mark Carey, Patrick Kupper, and the other participants for their comments and suggestions; Brandon Luedtke, the Unisa Library (Mary-Lynn Suttie and Marié Coetzee), the National Archives of South Africa, and SANParks (Glenn Phillips); and, for financial support, Colorado State University, the University of South Africa, and the National Research Foundation, South Africa.

THEODORE CATTON is an associate research professor in the Department of History at the University of Montana, Missoula. A public historian, he specializes in the history of the U.S. National Park Service and has written seven administrative histories of specific national park units, most recently *To Make a Better Nation: An Administrative History of the Timbisha Shoshone Homeland Act* (2009). In addition, he is the author of *Inhabited Wilderness: Indians, Eskimos,* *and National Parks in Alaska* (1997), and *National Park, City Playground: Mount Rainier in the Twentieth Century* (2006).

CHRIS CONTE is an associate professor in the Department of History at Utah State University, Logan. A specialist in environmental history and African history, Conte holds a joint appointment in the Department of Environment and Society in the College of Natural Resources at Utah State. He is the author of *Highland Sanctuary: Environmental History in Tanzania's Usambara Mountains* (2004).

JOSÉ DRUMMOND is an associate professor at the Center for Sustainable Development, Universidade de Brasília, Brazil. He holds a PhD in land resources from the University of Wisconsin, Madison. Since 2009 he has been a research fellow of Brazil's National Research Council. In 2011 he was visiting scholar in the Department of History and School of Global Environmental Sustainability, Colorado State University, Fort Collins. He has published numerous articles, most recently in *Environmental History, Environment and History, Historia Ambiental Latinoamericana y Caribeña (HALAC), Ambiente e Sociedade,* and *Novos Cadernos NAEA.* Most of his publications are available in PDF format at https://brasilia.academia.edu/JoseDrummond.

MARK FIEGE is a professor of history and a council member in the Public Lands History Center at Colorado State University, Fort Collins. He is the author of *Irrigated Eden: The Making of an Agricultural Landscape in the American West* (1999) and *The Republic of Nature: An Environmental History of the United States* (2012). From 2008 to 2013 he held the William E. Morgan Chair of Liberal Arts at Colorado State University. A specialist in the environmental history of the U.S. and North American West, he is working on the history of the U.S. national parks and changing conceptions of conservation. He thanks Adrian Howkins for the ideas that guided this project, and he thanks Jared Orsi for joining him and Adrian on the journey.

ADRIAN HOWKINS is an associate professor of history and a council member in the Public Lands History Center at Colorado State University, Fort Collins, and is a specialist in international environmental history and the environmental history of Antarctica. He is author of *The Polar Regions: An Environmental History* (2015), and he is completing an environmental history of the Antarctic Peninsula.

PATRICK KUPPER is a lecturer in modern history, with special emphasis on environmental and technology history, at Swiss Federal Institute of Technology (ETH Technigeschichte), Zurich. He has published numerous works on the history of nature and the environment, including a transnational history of

Swiss National Park: *Wildnis schaffen: Eine transnationale Geschichte des Schweizerischen Nationalparks* (2012).

ALAN MACEACHERN teaches history at the University of Western Ontario in London. He has written extensively on the history of Canada's national parks system, including *Natural Selections: National Parks in Atlantic Canada*. He is also the founding director of NiCHE: Network in Canadian History and Environment, and editor of the Canadian History and Environment series at University of Calgary Press. He thanks Mark Carey, Karen Routledge, Paul Sutter, the volume's editors, and all of its contributors for their insightful comments on his essay; Terry Young and Lary Dilsaver for sharing research material; and Curt Buchholtz and James Pickering for information on the naming of Rocky Mountain National Park.

ANN MCGRATH is a professor of history and director of the Australian Centre for Indigenous History, College of Arts and Social Sciences, Australian National University, Canberra. A fellow of the Academy of Social Sciences, she was awarded an Order of Australia Medal for services to history, especially indigenous history. She has been leading a digital history project that explores Indigenous landscape histories. With Mary Anne Jebb, she edited *Long History, Deep Time* (ANU Press, 2015). She recently codirected and coproduced a film with Andrew Pike, *Message from Mungo* (Ronin Films, 2014), which explores the conflict between scientists and Indigenous custodians over ancient remains. She was the Louise and John Steffens Founding Circle Member at the Institute for Advanced Study Princeton 2013–14 and was honored with a scholarly residency at the Rockefeller Foundation Center at Bellagio, Italy, in June–July 2014.

JARED ORSI is a professor of history and a council member in the Public Lands History Center at Colorado State University, Fort Collins. A specialist in environmental history and the history of the U.S.-Mexico borderlands, he is the author of *Hazardous Metropolis: Flooding and Urban Ecology in Los Angeles* (2004) and *Citizen Explorer: The Life of Zebulon Pike* (2014). He thanks Eric Bittner, Caitlyn Carrillo, Connie Gibson, Brandon Luedtke, and Clarissa Trap for research assistance, and, for inspiration, Mark Fiege, Mike Wilson, Margaret Regan, and the students in History 492 Capstone Seminar, spring semester 2012.

STEVEN RODRIGUEZ completed his PhD at UCLA under the direction of Lynn Hunt. His recent publications have examined the history of wildlife extermination programs in British India. Currently his research focuses on the development of national parks and environmental movements in Indonesia and Vietnam.

KAREN ROUTLEDGE is a historian for Parks Canada, Calgary, Alberta. An environmental and cultural historian, her research includes the experiences of Inuit and Americans who traveled to each other's homelands. She thanks Brandon Luedtke for research assistance; Ted Catton, for comments and for sources on Glacier National Park; Dennis Madsen and Rob Watt at Waterton Lakes National Park; and Ann McGrath, Emily Wakild, the editors, and other symposium participants for considerate and helpful comments and suggestions.

PAUL S. SUTTER is an associate professor of history at the University of Colorado, Boulder. In addition to many notable essays and critiques, he is the author of *Driven Wild: How the Fight against Automobiles Launched the Modern Wilderness Movement* (2002), *The Art of Managing Longleaf: A Personal History of the Stoddard-Neel Approach* (with Leon Neel and Albert Way, 2010), and *Let Us Now Praise Famous Gullies: Providence Canyon and the Soils of the South* (2015). His current research examines the environmental and public health history of the Panama Canal's construction, and he is series editor for Weyerhaeuser Environmental Books, published by the University of Washington Press.

EMILY WAKILD is an associate professor in the Department of History at Boise State University, Boise, Idaho. Her research and teaching specialties include environmental history and Latin American history. She is the author of *Revolutionary Parks: Conservation, Social Justice, and Mexico's National Parks, 1910–1940* (2011), which won book awards from the Conference on Latin American History, the Forest History Society, and the Southeastern Conference on Latin American Studies. She would like to thank Brandon Luedtke for research assistance and conference organizers and participants for excellent suggestions. Support for her essay came from National Science Foundation Grant No. 1230911 and a National Endowment for the Humanities Fellowship. She is now completing a comparative history of transnational conservation and scientific research in Amazonia and Patagonia.

INDEX

Abel Tasman National Park (New Zealand), 85

aboriginal Australians. *See* native peoples: Australia

accidents, fatal. *See* fatalities, human

acclimatization societies, 71, 72

Adams, Bill, 123

Adirondack Park, 285

advertising, 57, 58, 59, 60, 66n29, 142, 246, 250

Advisory Commission for the International Protection of Nature, 117–18

Africa, 70, 115, 116, 118, 123–24, 126, 170–87, 288. *See also* South Africa

agriculture, 115; Brazil, 215, 217, 229–30n11; East Africa, 175, 176, 177, 181, 182, 184; New Zealand, 79, 81; South Africa, 142; Switzerland, 122

Alaska, 267, 289–90

Albert National Park (Belgian Congo), 118, 120–21

Albright, Horace, 62, 67n50, 77, 121

alien species. *See* invasive species

Alto da Serra Biological Station, 214

Amazon Conservation in the Age of Development (Foresta), 233n42

Amazonia, 92, 101–5, 213, 218, 225–27, 233n42

American Civic Association, 60, 61

American Committee for International Wild Life Protection, 114

American exceptionalism, 3, 5, 8, 11, 279, 282

American Indians. *See* Native Americans

American Mount Everest Expedition (AMEE) (1963), 23, 24–26

American Museum of National History, 91

"America's Best Idea" (phrase), 3, 29, 37–38, 51–52, 63n3, 278–79, 284, 295

Anangu. *See* native peoples: Australia

Anglo-Boer War (1899–1902), 137, 141

Antarctica, 16, 26–33, 282

Antarctic Treaty, 10, 16, 26, 27, 29–32

anticommunism, 27, 76–77

Antiquities Act of 1906, 34, 96, 237

Aparados da Serra National Park, 216, 217

apartheid, 136, 137, 147, 148, 150, 151, 288, 290; Antarctic Treaty and, 27; U.S. and, 27, 138, 146, 148

Araguaia National Park (Brazil), 216, 217, 218

Archer, Manuel Gomes, 213

Argentina, 91, 92, 93, 96–100, 105, 109n12, 217, 220–21

Arkansas Hot Springs, 285

Arthur's Pass National Park (New Zealand), 73, 83

Atlantic Forest, 217, 231n21

attendance. *See* park attendance

Audley National Park (Australia), 287

Australia, 32, 70, 105, 116, 217, 235–57, 287

307

Maasai Mara National Park (Kenya), 173, 176

Maasai people, 172, 173, 174, 176, 185, 185n5

Maasai Reserve, 173

Mabo judgment (1992), 239, 241, 255n29

MacEachern, Alan, 278, 290

Mackinac National Park, 286

Mackinnon, John, 159–60

MacKinnon, Kathy, 168n17

Madagascar, 118, 123–24

Madrid Environmental Protocol, 32

Magnanini, Alceo, 220, 221, 225, 226

Magome, Hector, 135

Maier, Charles, 36, 41

Malaysia, 31

Man and the Biosphere (MAB) program. See UNESCO: Man and the Biosphere (MAB) program

Mandela, Nelson, 150

Māori, 70–71, 72, 73, 83–85, 86, 240, 287, 289

maps and mapping, 6–7, 171, 223–24, 226, 244, 280–81

Mar, Tracey Banivanua. See Banivanua Mar, Tracey

Marshall, Bob, 106

Martinka, Cliff, 200

Mather, Stephen, 57, 58, 59–60, 77, 121

Matsushita, Iwao, 21, 22–23

Mazey, John, 77

Mazuru, Mzee, 180

McCaskill, Lance, 78, 88n34

McFarland, J. Horace, 56, 61

McGrath, Ann, 288–89

Meany, Edmond, 17, 18, 19, 42n6

media, 269–71; global warming coverage, 258–67, 293; Indonesia, 168–69n25; U.S., 52–53, 58. See also documentary films; National Geographic

Merriam, C. Hart, 20

Merriam, John F., 120

Mexico, 217, 219, 288. See also U.S.-Mexico border region

Michigan, 286. See also Isle Royale National Park

middle class, 192; Indonesia, 161, 164, 166

migration, 16, 262, 268, 271; wildlife, 122, 126, 258, 262. See also human migration

Miles, Beryl, 247, 248

military dictatorship, 157–69, 233n33, 293–94

Millard, Candace, 106

minerals and mining, 30–31, 32, 79, 104, 142

Mitre, Bartolomé, 97

moa, 69, 71

Moi, Daniel Arap, 175

Montana, 94, 202, 268, 270. See also Glacier National Park (United States)

Monte Pascoal National Park (Brazil), 217

"monumentalism," 215–16, 217, 226, 230n14, 286

moose, 202

Moreno, Francisco P., 92, 93, 96–100, 101, 103, 105–6, 110n35, 112n91

mountaineering, 23, 24–26, 75, 97, 140, 248–49; Devils Tower, 235, 236, 238–39, 241, 243, 245–46, 250–51, 252; Indonesia, 163–65, 166; Peru, 261; Uluru, 235, 236, 238–39, 243, 244, 245, 247, 251, 252; Yosemite, 246

mountain lions. See cougars

Mount Bromo-Tengger-Semeru National Park (Indonesia), 165, 166

Mount Cook, 73, 75, 83

Mount Egmont, 73

Mount Everest, 23–26, 44n38, 249. See also Sagarmatha National Park (Nepal)

Mountford, Charles, 251

Mount Fuji, 21, 23

Mount Gorongosa, 176, 182, 184

Mount Longonot, 172, 175

Mount Rainier, 14–17, 19–23, 24, 40

Mount Rainier National Park, 14–17, 19–23, 24, 43n26

Mount Rushmore National Monument, 32

Mount Soekarno. See Puncak Jaya

Mount Suswa, 172, 173, 175

Mozambique, 170, 176, 182–83, 184

Muala, Domingos, 184

mudflows, 15

Muir, John, 31, 56, 106, 286

Muldoon, Robert, 80

Müller, Lauro, 102

Murie, Adolph, 201

Murie, Mardy, 63n3

Murie, Olaus, 88n34, 201

Murillo Aguila, Panfilo, 38

United Kingdom. *See* Great Britain
United Nations, 29, 30, 31, 148, 221,
    282. *See also* Food and Agriculture
    Organization of the United Nations
    (FAO)
United Nations Development Program
    (UNDP), 159, 160, 161, 225
University of California at Berkeley, 119
Upper Tugela Native Reserve (South Africa),
    140–41
Urrea, Luis Alberto, 39
U.S. Bureau of Biological Survey, 62, 67n52
U.S.-Canada border region, 10, 63, 191–
    209, 266, 292
U.S. Foreign Assistance Act of 1961, 77
U.S. foreign policy, 16, 26, 29, 76–77
U.S. Forest Service, 58, 72, 81, 193
U.S. Geographic Board, 19–20
U.S. Geological Survey, 202, 245, 268
U.S. immigrants. *See* immigrants, U.S.
U.S.-Mexico border region, 10, 16–17,
    33–40
U.S. National Park Service. *See* National
    Park Service (NPS)
U.S. State Department, 16, 26, 29

Vancouver, George, 17–18
Van Graan, H. S., 146
Vargas, Getulio, 104
Vicente Pérez Rosales National Park (Chile),
    107n3
Victor Emanuel III, king of Italy, 118
Victoria Forest Park (New Zealand), 80
Villas Bôas brothers, 104, 109n12
volcanoes, 15–21, 23; East Africa, 172, 173;
    Indonesia, 163, 165; New Zealand, 68,
    72–73

Wagner, Michele, 178
Waitangi Tribunal, 83, 84, 85
Wakild, Emily, 288
Washington, George, memorials, 19, 238,
    246
waterfalls, 140, 144, 184, 216, 266, 267
water supply and conservation, 160, 268–
    69, 270–71; Brazil, 213; Peru, 263–64
Waterton-Glacier International Peace Park,
    191–209, 266, 292
Waterton Lakes National Park (Canada), 57,
    60, 63, 191, 198, 200–203, 266, 292

Watrous, Richard W., 60
Waugh, Andrew, 23–24
Weaver, John, 244
Weems, Samuel P., 88n34
Westland National Park (New Zealand), 77,
    80, 85
West Papua, 163, 167n2
whaling and sealing, 29, 71
White, Richard, 238
Whittaker, Jim, 23, 24–26
Whittaker, Lou, 25
wildfire, 40, 121, 124, 157–58, 193, 258,
    268, 271
wildlife, 115, 119, 125; Belgian Congo,
    118; Brazil, 102, 218; East Africa, 173,
    175, 176, 183; global warming and,
    268; Indonesia, 157, 159, 162; Italy,
    118; Nepal, 122, 124; New Zealand,
    69, 71–72, 75, 79; Patagonia, 96–97;
    Peru, 261; Sonoran Desert, 35, 36;
    South Africa, 140, 141, 143, 144,
    147; surveys, 119, 123; Switzerland,
    122–23, 125; Waterton-Glacier, 191–
    92, 194–203, 205n18. *See also* bears;
    birds; hunting; predator control; radio
    tracking of animals; ungulates
wildlife preserves and refuges, 39, 96, 219,
    227, 286
Williams, Joseph, 207n46
Williams, Mabel, 55
Wilson, Mike, 36–37
World Bank, 160, 168n17
World Conference (Congress) on National
    Parks, 294; 1962 (Seattle), 77, 79, 114,
    156n62, 220; 1972 (Yellowstone/Grand
    Teton), 29, 30, 31, 32, 54, 122, 123,
    149, 220; 1982 (Bali), 161, 220, 293;
    1992 (Caracas), 139; 2003 (Durban),
    139, 150; 2014 (Sydney), 138–39
World Database on Protected Areas
    (WDPA), 122
World Health Organization (WHO), 221
World Heritage Convention, 159
World Heritage Sites, 68, 136, 150,
    153n20, 162, 236, 261
world parks (proposed), 29–32
World Parks Congress. *See* World
    Conference (Congress) on National
    Parks
World War I, 58, 60, 66n40, 117